AF327425

Cultivating Biodiversity to Transform Agriculture

Étienne Hainzelin

Editor

Cultivating Biodiversity to Transform Agriculture

 Springer

Editor
Étienne Hainzelin
Cirad, Direction générale
Montpellier
France

ISBN 978-94-007-7983-9   ISBN 978-94-007-7984-6   (eBook)
DOI 10.1007/978-94-007-7984-6
Springer Dordrecht Heidelberg New York London

Library of Congress Control Number: 2013955250

Éditions Quæ, R10, 78026 Versailles cedex, France http://www.quae.com

Printed on acid-free paper

Springer is part of Springer Science+Business Media (www.springer.com)

Foreword

Biodiversity? It is the living part of nature, one we absolutely cannot do without. And yet, as explained in this book's Introduction, it is vanishing, vanishing under humanity's constant assault at a frighteningly accelerated rate since decades and even centuries—since the beginning of the famous 'Anthropocene' era popularized by Paul Crutzen in 2000. And this is taking place in an atmosphere of general indifference! Our planet's biodiversity is built on its earlier geodiversity, which dates back to the Earth's formation some 4.6 billion years ago. Biodiversity is 'attached' to the planet Earth: it consists of the forms that Life has been able to differentiate since its origins, 3.9 billion years ago in the ancestral ocean, forms of life that are 'associated', in every sense of the term, in building ecosystems in close relationships with their environments. Today we can imagine, that over this period, the living world has been able to create over a billion species, many appearing then disappearing, some still with us, with their infinite shapes, sizes, colours, behaviours, adaptations, functions, etc. We estimate that 1–1.5 % of them are still with us today. The UN declared 2010 to be the 'International Year of Biodiversity' in order to encourage and manifest interest in the subject. But even after the recommendations of various international commissions over the past two decades, we have been unable to halt—or even slow—the loss of biodiversity by 2010, as previously targeted by, amongst others, the Earth Summit in Johannesburg in August 2002. Now, after the UNESCO International Year of Biodiversity conference in Paris in January 2010 and the one in Rio in June 2012, we have moved this goal to 2020. A realistic target or utopian dream? Indeed, given our reluctance to change our ways, should we even expect anything better?

It is quite clear that biodiversity cannot be represented in its entirety solely through an inventory of the living species inhabiting a particular ecosystem. This is specific diversity. Biodiversity has been variously explained, but its meaning always revolves around something like 'the genetic information contained in each elementary unit of diversity, whether it be an individual, a species or a population'. This determines its history: past, present and future. In addition, this history is influenced by processes that are themselves components of biodiversity. In fact, we now encompass various approaches under this term: the study of basic biological mechanisms to explain the diversity of species and their specificities and to force us to 'dissect' further the mechanisms of speciation and evolution; the recent promising approaches in functional ecology and in biocomplexity, including the study of matter and energy flows and of large biogeochemical cycles; work on nature that is 'useful' to humanity in its ability to provide food and substances with high value addition for products ranging from medicines to cosmetics to molecular probes, or to provide simpler and more original models for basic and applied research in order to solve agronomic or biomedical problems; and, finally, the implementation of conservation strategies to preserve and maintain a natural heritage, one that is for—and will be expected from us by—future generations. And yet, we are forced to still emphasize that inventories and catalogues are insufficient to specify what biodiversity is and how much more important are the relationships between living things and their environments! It is essential to know and understand paleobiodiversity (and paleo-associated habitats) in order to clarify the current situation and the dynamics of this diversity.

So where does agrobiodiversity figure in all this? Mankind developed gradually by hunting and gathering (and fishing more recently), but as soon as its population began to explode, in the Neolithic from 15,000 to 8,000 years ago, everything changed. But which was the cause and which was the effect? Humanity and agriculture are interrelated and inseparable and have always been so. How then to explain humanity's success today, on the one hand, and the serious agricultural situations we have ourselves caused,[1] on the other?

Étienne Hainzelin and Christine Nouaille write at the beginning of this book: '... The diversity of living organisms has long been the mainstay of agricultural activity and its innovations. However, since the late nineteenth century, particularly in industrialized countries, increases in yields have been based on radically new technologies which deny the biological reality of agriculture and end up artificializing environments. This greatly intensified agriculture is primarily based on fossil fuels (mainly petroleum). It now finds itself at an impasse because of its impacts on ecosystems and the dramatic increase in the prices of inputs and energy. Social inequalities and massive rural exoduses that it has caused are further reasons for concern....' And so the debate is launched! The book you are holding is crucial to this debate, originating, as it does, from a community of experienced scientists and 'on-the-field' practitioners aware of the ground realities. CIRAD, with its mission of applied research for development in the South, is a key organization for the topics discussed in this book. Its work, in partnership with scientists from these countries, makes it a repository of unique skills and exceptional experience for the themes addressed in the six chapters of this book.

Seven species of potatoes with 5,000 varieties grown in the Andes, 92 rice varieties listed in the Philippines... . The European agricultural landscape has been profoundly changed by the arrival of New World plants: tomatoes, strawberries from Chile, potatoes, maize, beans, and many more, as well as by the

---

[1] Toussaint J.-F., Swynghedauw B., Boeuf G. (coordinators), 2012. *L'homme peut-il s'adapter à lui-même? Versailles, Éditions Quæ, 176 p.*

transplantation of cocoa, banana, rubber, etc. Because it is the engine of cultivated ecosystems, agrodiversity is the main lever of agricultural change and a major factor for development. The dramatic increase in agricultural production after the 1950s was based on the idea that production could be advantageously reduced to chemical fluxes, and it was enough to make good nature's shortfall. From 1945 to 1985, the consumption of fertilizers and pesticides thus doubled every 10 years! The use of synthetic fertilizers, herbicides, insecticides and fungicides has artificialized agroecosystems by homogenizing the soil at the trophic level through the accumulation of these chemicals in the soil and water and by destroying many species necessary to maintain their balance. One chapter of this book is entitled 'Rethinking Plant Breeding' and discusses the improvement of plants. Should we not, in fact, come together to radically redefine all that is meant by the word 'improvement' and clarify with respect to what reference?

What does agrodiversity represent within biodiversity and what are the relationships between them? Humanity continues to grow in numbers and its needs are constantly increasing. How will we feed a population of nine billion tomorrow?[2] We cannot 'simply' advocate, as some have already done, the idea of making the Earth a huge agrosystem. In any case, I doubt that the oceans will be willing to participate! How can we develop agricultural land in harmony with the planet? How can we produce more without affecting human health? How can we reduce sharply—and perhaps even eliminate—the use of inputs? How can we stop wasting water and start sharing resources? How can we stop the almost limitless expansion of agricultural lands? The answer, as this book makes amply clear, is by using and respecting biodiversity. A highly biodiverse system could produce at least as much and would slow or prevent the intrusion of invasive species. Agrodiversity is essential and human intelligence and technology should be harnessed for developing it. The context today is of an infinite number of agricultural peoples, situations, customs, desires, practices, techniques, species, varieties and changes, all in a dangerous environment of globalization, arrogance, disharmony and selfishness. Is this context sustainable? The loss of biological diversity is synonymous with agricultural decline; the agricultural world remains essential for humanity!

This book makes some strong points, presents well-analysed thoughts, and introduces us to attractive and encouraging developments. We need to thank its authors who, without hiding the urgent necessity of taking action, show us the path to solutions. In any case, we do not want to 'save the planet'—in fact, it does not even consider us; what a disappointment to our ego!—but only want to ensure that everyone can find his or her place without too much 'discomfort'. To this end, we need to do intelligent work, use our abilities for something other than greed, clothe ourselves in humility, and engage in a constant struggle to maintain harmony and sharing and respect for others. Let us pay this noble 'price' for success, and we will

---

[2] Guillou M., Matheron G., 2011. *Neuf milliards d'hommes à nourrir. Un défi pour demain*, François Bourin éditeur, Paris, 421 p.

finally earn the name *sapiens* by which we call ourselves! Agriculture and the agricultural world are inseparable from such a commitment, both in the North as well as in the South, to the East and to the West, in a plurality of cultures and infinity of diversities.

Full Professor, Pierre and Marie Curie University, Paris   Gilles Boeuf
President, National Museum of Natural History, Paris

Acknowledgments

This book covers a vast topic, and while it is presented through 'six viewpoints', the discussions in it have relied on multiple interactions between many researchers, primarily from CIRAD but also from its partner organizations. This is what CIRAD likes to call its 'collective intelligence'. Therefore, in addition to the authors of each chapter, this book has drawn on the knowledge and expertise of many people—especially represented through the text boxes and figures—to present its arguments. Despite the long list of contributors, we would like to thank each of them for their unique contributions:

Raphaël Achard, Martine Antona, Jacques Avelino, Pierre Bonnet, Patrick Caron, Jacques Chantereau, Marie Chave, Danièle Clavel, Harouna Coulibaly, Michel Crétenet, Benoît Daviron, Peninna Deberdt, Jean-Philippe Deguine, Dominique Dessauw, Marc Dorel, Ghislaine Duqué, Sandrine Dury, Pierre-François Duyck, Hervé Etienne, Guy Faure, Philippe Feldmann, Paula Fernandes, Francis Ganry, Claude Garcia, Frédéric Goulet, Michel Griffon, Hervé Guibert, Henri Hocdé, Olivier Husson, Patrick Jagoret, Hélène Joly, Mamoutou Kouressy, Cheppudira Kushalappa, Marie-Dominique Lafond, Marie-Christine Lambert, Claire Lanaud, Jaques Lançon, Christian Lavigne, Fabrice Le Bellec, Christian Leclerc, Maya Leclercq, Delphine Luquet, Delphine Marie-Vivien, Pascale Moity-Maïzi, Béatrice Moppert, Krishna Naudin, Samuel Nibouche, Marc Piraux, Serge Quilici, Louis-Marie Raboin, Jean-François Rami, Bruno Rapidel, Vincent Requillart, Bernard Reynaud, Jean-Michel Risède, Manuel Ruiz, Éric Sabourin, Éric Scopel, Lucien Séguy, Luciano Silveira, Louis-Georges Soler, Mamy Soumaré, Ludovic Temple, Philippe Tixier, Emmanuel Torquebiau, Jean-Marc Touzard, Bernard Triomphe, Gilles Trouche, Gilles Trystram, Philippe Vaast, Michel Vaksmann, Maryon Vallaud, Isabelle Vagneron, Bernard Vercambre, Robin Villemaine, Christopher Viot and Kirsten Vom Brocke.

We would also like to extend our thanks to Kim Agrawal for translating this book from the original French.

CIRAD is a French research centre which addresses international issues of agriculture and development in association with countries of the South. Through partnerships with these countries, it produces and distributes new knowledge to support agricultural development and to make meaningful contributions to the debates on major global issues of agriculture, food and rural areas. CIRAD has a global network of partners and regional offices, from which it coordinates and undertakes joint operations in more than 90 countries.

Contents

Authors

Nourollah Ahmadi Department of Bios, CIRAD, Montpellier, France, e-mail: nourollah.ahmadi@cirad.fr

Didier Bazile Department of ES, CIRAD, Montpellier, France, e-mail: didier.bazile@cirad.fr

Benoît Bertrand Department of Bios, CIRAD, Montpellier, France, e-mail: benoit.bertrand@cirad.fr

Estelle Biénabe CIRAD, Montpellier, France, e-mail: estelle.bienabe@cirad.fr

Éric Blanchart IRD, Montpellier, France, e-mail: eric.blanchart@ird.fr

Gilles Boeuf Paris, France, e-mail: boeuf@mnhn.fr

Christian Gary Department Environment and Agronomy, INRA, Montpellier, France, e-mail: christian.gary@supagro.inra.fr

Jean-Christophe Glaszmann Department of Bios, CIRAD, Montpellier, France, e-mail: jean-christophe.glaszmann@cirad.fr

Étienne Hainzelin CIRAD, Montpellier, France, e-mail: etienne.hainzelin@cirad.fr

Philippe Lecomte Department of ES, CIRAD, Montpellier, France, e-mail: philippe.lecomte@cirad.fr

Sélim Louafi Department of Bios, CIRAD, Montpellier, France, e-mail: selim.louafi@cirad.fr

Éric Malézieux Department Persyst, CIRAD, Montpellier, France, e-mail: eric.malezieux@cirad.fr

Florent Maraux Department Persyst, CIRAD, Montpellier, France, e-mail: florent.maraux@cirad.fr

Christine Nouaille Montpellier, France, e-mail: christine.nouaille@cirad.fr

Jean-Louis Noyer Department of Bios, CIRAD, Montpellier, France, e-mail: jean-louis.noyer@cirad.fr

Alain Ratnadass Department Persyst, CIRAD, Montpellier, France, e-mail: alain.ratnadass@cirad.fr

Chapter 1
Introduction

Étienne Hainzelin

Even though biodiversity is a relatively new concept, it is now widely acknowledged to represent all the creativity of life on our planet. According to the Convention on Biological Diversity (CBD 1992), it encompasses all 'living organisms from all sources and the ecological complexes of which they are part: this includes diversity within species, between species and of ecosystems'. Over the last few decades, we have slowly come to realize how precious biodiversity is to humankind and how much it is currently under threat. Changing constantly, impossible to inventory exhaustively, branching into—and manifesting at—multiple levels, biodiversity is both a source of wonder and concern. Each one of us can try to better understand biodiversity by contemplating its extraordinary interweaving of production (food, fuel, material, etc.), essential services (water purification, air purification, climate regulation, renewal of soil fertility, etc.) and cultural and aesthetic resources. Source of wonder because, as the driving force of ecosystems, biodiversity—this 'thin layer of life'—is at the origin of a large number of goods and services and, indeed, of human existence and well-being. Source of concern because year after year alarm bells have sounded on the damage being caused to this biodiversity, mainly due to human activity. There is even talk of an inexorable rush towards mass extinction. Despite the limitless profusion and incredible generosity of the diversity of living organisms, we are beginning to see the emergence of biodiversity's finiteness in the form of impoverished landscapes and precarious and incomplete ecosystems. Are the numerous enormous challenges that humans have created for themselves by causing this massive destruction going to manifest multiplied, much larger than the sum of their parts? Have we learned all the lessons from the risks we are undertaking?

É. Hainzelin (✉)
Cirad, Montpellier, France
e-mail: etienne.hainzelin@cirad.fr

É. Hainzelin (ed.), *Cultivating Biodiversity to Transform Agriculture*,
DOI: 10.1007/978-94-007-7984-6_1, © Éditions Quæ 2013

1 Biodiversity has Always Been at the Heart of Agricultural Activity

Since the Neolithic Age, human beings have harnessed biodiversity by domesticating plants to feed themselves better, cure themselves and clothe themselves. They have relied on it to domesticate animals for their strength, their milk, their meat and their leather. They have taken advantage of the diversity of living organisms in a very wide range of activities, most notably for agricultural production at all latitudes, on all continents. As biodiversity has shaped the planet over billions of years, so farmers have shaped agricultural landscapes for the past 12,000 years by mobilizing and organizing 'agrobiodiversity'. The planet has thus gradually been 'anthropized', i.e., humanized, on nearly half of its landmass. The cultivated area—that is to say the area where humans directly plan and control the vegetation cover—today consists of over 1.5 billion hectares of annual and perennial crops. In addition, there are 1.4 billion hectares of improved pastures and nearly 300 million hectares of planted forests, i.e., more than 20 % of the total landmass.

From time immemorial, farmers have improved their techniques, sometimes with real technological breakthroughs, such as irrigation in 5000 BC or forage crops in the eighteenth-century. Most often they have used specific diversity in their innovations or new combinations of crops in space and time. Associations between crops and livestock, acclimatization of exotic species and the use of auxiliary species have all, at one time or another, been their paths to improved techniques. The last agricultural revolution, however, was based on selected varieties, synthetic fertilizers and pesticides, as well as on a massive mechanization in some countries. It has led to a veritable industrialization of agriculture. Heavily reliant on fossil fuels, it has resulted in a surge of artificialization of agricultural fields. There biodiversity is often reduced to a uniform and synchronous canopy, usually consisting of a single genotype of some major species, with the rest of the living organisms being systematically eliminated as 'limiting factors'. This transformation has affected not only most of the agricultural land in developed countries, but also a part of agriculture in countries of the South, given that the Green Revolution was based on the same principles of simplification and artificialization.

Nature and agriculture occupy two distinctly defined spaces which we have been keeping separate for a long time. There is one set of rules to govern the cultivated space and a separate set for 'natural' spaces. These two spaces are even categorized into different scientific disciplines (agronomy vs. ecology), and thus are considered to have distinct biodiversities. Until recently, we spoke of 'genetic resources' and selected varieties in the context of agricultural species and spaces, while biodiversity was relegated only to the natural space. We thus compartmentalized the world between a space for production—which was somehow condemned to impoverish its diversity in the pursuit of productivity—and a natural space to preserve. For some time, we even professed that by protecting the latter, we could somehow compensate for the damage and negative externalities of the former. But with growing awareness of the planet's limitations, of human pressure and of the widespread deterioration of cultivated and 'natural' ecosystems, we came to realize that this was not sufficient. The connectivity of all ecosystems and the intensity of threats biodiversity is under exclude solutions which envisage putting nature under glass, in discrete places, while the majority of ecosystems become impoverished. We know today that as useful as they are, protected natural areas will not be sufficient to restore ecosystems in their multiple functionalities and services. The only way forward seems to be to preserve biodiversity in all ecosystems, cultivated or not, and in some cases, to massively enrich it. This necessary integration between the two spaces is now manifesting, for example, in the convergence between agronomic and environmental disciplines or through the emergence of the concept of agrobiodiversity. Agrobiodiversity encompasses not only the cultivated plant communities but all living species—animal, vegetal or microbial; aggressive or useful; below- or above-ground; etc. —and their interactions, in and around the cultivated plot.

2 The Challenges of Agricultural Transformation

Agriculture is the primary human activity. If involves 1.3 billion people, nearly a quarter of the world's population and half of its labour force. By the sheer number of people who make their living through it, by the fact that it produces the only truly renewable biomass on the planet and because of its vast spatial extent, agriculture finds itself at the heart of many of the great challenges of our time. After decades of considering agriculture an archaic activity that we needed to 'get out of' in pursuit of development, we now realize its importance. We recognize that we have to mobilize agriculture to help meet humanity's core challenges: food security, energy sufficiency, human health, proper functioning of ecosystems, climate-change mitigation, poverty alleviation, rural development, etc.

But agricultural systems around the world also pertain to biodiversity and are, in fact, incredibly plural in character. They can be described according to climatic, historical or political criteria or they can be characterized by farm size, land tenure type, capital intensity, type of labour, the level of technification, etc. Consequently, when we take an interest in the working of fields and the role that farms attribute to biodiversity, the discriminant axis that dominates is the one that pits two divergent agricultural systems against one another. First is the so-called modern agriculture—productivist, industrialized, connected to global markets, capital intensive and not very labour-intensive. Then there is family-farm agriculture, habitually accused of being archaic, based on small farms, often very dynamic but suffering due to a lack of competitiveness and inadequate market access. There is a wide range between these two extremes which includes all kinds of intermediate situations. However, there still exists a real divide that largely transcends ecologies and North–South differences. This divide has to do mainly with the mode of intensification chosen, which is directly linked to biodiversity's role in the production processes. The agricultural systems of the first type rely on the path of specialization by focusing on intensification based on a very small number of species, mechanization and massive use of inputs. Those of the second type have not even begun the process of intensification and thus are still very 'biodiverse'.

What emerges as we delve deeper into the differences between these two types of agriculture is that both are concerned with the remobilization of agrobiodiversity. 'Modern' agriculture, despite its emphasis on production, is stymied by the limits of the very dynamics of intensification that led to its 'modernization': energy balance, finiteness of resources, considerable environmental and social externalities, stagnation of yields and decrease in ecosystem capital. It therefore has to radically transform itself to become more sustainable. Family-farm agriculture absolutely needs to increase its productivity and is forced to invent new ways of intensification that remain economically viable. Any new path this agriculture takes must meet the challenges of income and employment, preserve biological heritage as an insurance of future adaptation, and not increase dependence on exogenous technologies.

Furthermore, in addition to increasing production to ensure food security, all agricultural systems now also are expected to provide ecological services to society as a whole. There is now a clearer perception of these services, consequent to the analyses resulting from the Millennium Ecosystem Assessment (MEA 2005). Because agricultural systems occupy—along with livestock farming and forestry—an important and structuring place in the planet's territories, these services become one of their fully distinct missions, even if many of the services generate no income (Griffon 2007). Therefore agricultural systems must increase their production of goods and services while preparing to cope better with risks and uncertainties such as those caused by climate change.

3 Intensifying Ecological Processes to Transform Agricultural Performance

It is therefore no longer possible today to assess the performance of an agricultural production system through the sole measure of its yields or its economic performance. The requirement of sustainability compels us to include all products, services and externalities in any analysis and to gauge them through economic, environmental and social criteria over sufficiently large scales of time and space. However, it is not always easy to do so given that many of these elements are not easily quantifiable.

It is this widened appraisal of performance that has led to the development of multiple alternative paths to agricultural transformation, some of which have been implemented at large scales in various regions of the world. Most of these paths rely on the intensification of ecological processes in the cultivated space. They emphasize the supply of water and minerals to crops, maximization of photo-synthetic activity, control over pest populations, activation of nutrient-cycle loops by limiting the use of expensive inputs, avoidance as far as possible of losses at the plot level, etc. These systems are based specifically on the activation of soil biology, home to several drivers of production, hitherto largely neglected and unknown. Depending on the services being considered, this intensification is planned for and managed at the plot level or at a higher one, such as of the landscape. The contribution of synthetic fertilizers and pesticides is subsidiary to the ecosystem's own contributions and controls. The former no longer remain central to the producer's operations (IAASTD 2009).

One might think that the requirement to provide ecological services, over and above that of agricultural production, may hamstring the system and that a compromise may be necessary between harvested production and generation of these services. Surprisingly, it is far from always being the case. As shown by a collective expertise on biodiversity and agriculture (Le Roux et al. 2008), production processes benefit over the long term from the enrichment of biodiversity in and around the plot. This book includes several examples: reintroduction of trees in cereal fields, coffee- or cocoa-based agroforestry systems, diversified grasslands, and orchards with cover plants. They show that the yield per unit area of various crops through various mechanisms can be increased while, at the same time, improving the ecological services provided.

4 Agrobiodiversity, the Main Lever of this Ecological Intensification

The diversity of living organisms has served agriculture since its very inception. But ever since the last agricultural revolution, when it was primarily used for varietal improvement, we have tended to forget how much it is the driving force for processes of production and regulation in cultivated ecosystems. We will have to know and understand this biodiversity better, remobilize it, enrich it and plan it—in one word *cultivate* it—in order to intensify the agro-ecosystem's ecological functioning. The mobilization of agrobiodiversity and its management by the producer will lead to an in-depth transformation of agriculture and its stakeholders for several reasons, in both the North and the South.

To begin with, the local context, with all its strengths and limitations, once again will become fundamental for developing a strategy of intensification. It will no longer be possible to develop universal solutions, the so-called 'technological packages'. Instead, it will be necessary to customize solutions whenever required. These solutions will be 'knowledge-intensive' and take into account the environment, the biodiversity present and its functional vitality and mobilizable resources, as well as the producer's projects and strategies (Griffon 2007).

This new importance accorded to the local context will give producers once again the opportunity to decide what combination of technologies, technical interventions, species and varieties, etc. to use for each of their plots. This is thus a great incentive for producers to innovate and develop a patrimonial management of their farms, in terms of resources as well as of knowledge.

The scientists' role will also be transformed by these changes as they will have to face a diversity of situations and unprecedented complexity. Agronomists will have to gain a better understanding of the processes at work in complex biodiversities and will have to be able to extract generic knowledge from local contextual interactions. Plant breeders, having worked for decades on monoculture prototypes in pursuit of production at all costs, will have to take advantage of the enormous advances in the in-depth knowledge of living organisms to 'open up the playing field' and imagine entirely new ideotypes. Biologists too will have to grasp the complexity of living communities, and contribute to widening the range of relevant technologies. Agronomists, plant breeders and biologists will have to rely more on local knowledge and build new innovation systems in partnership with local actors in order to be really useful to innovative producers.

This new measure of agricultural performance will have to be taken into account at regional and national levels by producer associations and by market and exchange organizations, and be considered when standards are being drawn up and when sectoral public policies are being formulated. Suitable regulatory tools will be required as well as those for encouraging sustainability. This is particularly true of African countries, which have yet to address large increases in population growth and which will soon need to generate agricultural jobs in large numbers.

This new consideration of agrobiodiversity therefore calls upon first the producers, i.e., the farmers, to define their evolutionary path. Biologists (genetics, plant breeding, pathology, physiology, ecology, etc.), agronomists (ecological intensification, cropping systems, agricultural practices, performance analysis, etc.) and researchers in the human and social sciences (public policy, collective action, etc.) then follow. We need to revise our perception of relationships between human societies and nature and its resources by placing agriculture at the forefront of this changed perception. This book shows how this transformation is not limited to just plots and their crops, but rather how the transformation affects the strong links between farmer communities and their living heritage. It depends on their ways of preserving this agrobiodiversity and of innovating to draw benefit from it. This is far from being a purely technical issue since choices and production models are imposed by the models of ownership of living organisms, innovation systems, market organizaion, distribution channels and consumption patterns.

5 Ecological Intensification, a Strategic Priority for CIRAD

We know it—the signs are all around us. Yes, it is urgent. If we do not act soon, we will be unable to reverse the trend of agrobiodiversity degradation. We must choose to act quickly to transform agriculture and make biodiversity a true component of development. We must acknowledge that, so far, producer organizations and technical associations—and not researchers—have been at the forefront of exploring ways of transforming agriculture. CIRAD is a research institution which has been present in the tropical world for several decades. Its goal is rural development in countries of the South. Based on its diagnosis of agriculture in these countries, CIRAD has chosen to devote a large part of its resources to ecological intensification over the past several years. The remobilization of biodiversity in agriculture in the South is fundamental to this intensification and requires significant amounts of research. This book takes stock of the discussions and work in progress in this area, presenting a shared vision, mainly developed through participatory processes, with producer partners in the South.

6 A Book with Six Viewpoints

We address a vast issue in this book. Our goal is probably unattainable and the viewpoints of the limited number of experts, agronomists, geneticists, pathologists, entomologists, ecologists, economists and innovation specialists that are presented here do not purport to cover all the complexities of such an immense topic. These viewpoints provide rather a summary of attempts at finding parts of the solution. They also describe the new knowledge that we must mobilize in this area and particularly the enormous challenge of integrating all its aspects together. The authors have focused on family-farm agriculture in the South and their cropping systems. Plant cropping is therefore in the foreground in this book, but a similar approach for aquaculture, forestry or livestock farming also could have been proposed. The authors of each chapter explore, according to their particular point of view, the role of biodiversity in the transformation of all types of agriculture. These range from types of agriculture which have been artificialized so much that they can no longer return to sustainability to those which have no other choice but to increase their productivity and ability to generate development. In between these two extremes can be found all other types of agriculture.

In the Chap. 2, Étienne Hainzelin and Christine Nouaille delve into the recent emergence of the concept of biodiversity and show that the diversity of living organisms has always been intertwined with the history of humankind and its agriculture in particular. They remind us of the difficulty of inventorying—and thus to actually quantifying the erosion of—this diversity of living organisms, especially of biodiversity relating to agriculture, i.e., agrobiodiversity. They show through which mechanisms this erosion occurs and how violent and fast it is. Because it is the engine of cultivated ecosystems, agrobiodiversity is the main lever of agricultural evolution and a major component of development.

In Chap. 3, Florent Maraux, Éric Malézieux and Christian Gary look at the functioning of plots and farms. They first analyze the major known impasses of conventional methods of agricultural intensification which are based on the simplification of agricultural systems and massive recourse to chemical inputs. They then inventory production systems existing today that rely on a broader agrobiodiversity, whether within plots or across landscapes. These systems, which provide a multiplicity of functions, usually generate benefits both in terms of material production as well as of ecosystem services. However, despite their ability to adapt and their improved overall performance, these systems also exhibit limitations and constraints. In explaining the various phenomena which generate them, the authors show the rational basis on which agrobiodiversity can be mobilized to 'complexify' cultivated ecosystems and better manage the tradeoffs between the different services provided, without losing sight of performance exigencies. They thus revisit the agronomists' role in the process of transformation and evolution of an agricultural system. This system has to be based on a rational enrichment of biodiversity in close relationship with producers. The authors discuss the innovations that need to be implemented in this regard at the territorial level.

To meet the technical challenges of this transformation and mobilize cultivated plant material in an intelligent manner, we most probably will have to rethink plant breeding in some detail. In the Chap. 4, Nour Ahmadi, Benoît Bertrand and Jean-Christophe Glaszmann place plant breeding in the context of the unprecedented accumulation of knowledge of the inner workings of living organisms and revisit the role of genetic improvement in the evolution of agroecosystems. To implement ecological intensification, it is necessary to develop new types of varieties that optimize biological interactions, leverage resources more efficiently, including in constrained environments, resist bio-aggressors better, etc. However, these new varieties will have to be designed to also fit into more biodiverse communities, taking advantage of each local context, thereby enhancing the range and complexity of possible varietal solutions (mixture of genotypes, complex populations, new species, etc.). In all cases, the definition of these new ideotypes requires a considerable effort of integration between various perspectives on plot productivity and fundamentally new systems of innovation and varietal dissemination. To be successful, this effort has to include the participation of producers.

However, the functioning of an agricultural plot is not limited to the functioning of its plant cover. In Chap. 5, Alain Ratnadass, Éric Blanchart and Philippe Lecomte look at how the community of cultivated plants continuously interacts with other complex living communities present in the plot or around it. They first look at communities of all kinds of bio-aggressors (diseases, parasites, predators, weeds, etc.) and show that the ecological interactions within agrobiodiversity can be finely controlled and indeed be used to protect agricultural production. Several illustrative examples are included. They then explore interactions with living communities in the soil, which probably represents a 'new knowledge frontier' since so little is known about these communities despite the key role they play. Living soil organisms are too often considered the soil-based enemies of crops, but their incredible diversity can be mobilized to improve the performance of the plant cover through their mechanical and trophic functioning. Finally, interactions between agriculture and livestock animals, almost or completely eliminated in modern agriculture, can represent a real gold mine for improving performances at the plot, farm and landscape levels. Appropriate management of these interactions can lead to marked improvements in production and resilience, but the bottom line is that this is only possible through an improved understanding of the mechanisms at work.

Issues of intellectual property and the ownership of living organisms today go hand in hand with the use of cultivated species and their ancestors and of all useful species and genes derived from agrobiodiversity. These critical issues have to be resolved otherwise they can become unsurmountable blocks to innovation. In Chap. 6, Selim Louafi, Didier Bazile and Jean-Louis Noyer analyze the evolution of the status of genetic resources of agricultural species as both private and public properties, within the framework of the Convention on Biological Diversity. They evoke the clear inadequacies of these various regimes and outline possible solutions. Furthermore, agrobiodiversity conservation, closely linked to the issue of ownership, is no longer approached today in terms of number of samples, sizes of cold rooms, or collections. The authors show how farmers are key actors in this indispensable in situ dynamic conservation, allowing diversity to continue to evolve and forming an essential complement to ex situ conservation.

Finally, in Chap. 7, Estelle Biénabe describes the roles of the various actors and forms of governance to accompany the transformation of agriculture. Transformation needs to take place not only at the technical level but also at the social and institutional levels. She shows that the promotion of more 'biodiverse' agricultural systems implicitly refers to very different innovation systems. The agronomy of 'ecologization' is the opposite of the agronomy of 'artificialization'. The former requires a deconcentration of innovation, a smaller emphasis on technology, a different link with industry, a more active role for producers and a different role for research. In these transformations, the divide between a productivist model—intensive, agro-industrial, criticized for its environmental and social impacts—and alternative models—highly biodiverse and implementing a wide range of local practices—remains deep. These models confront each other and the author describes the power relationships and regulatory processes involved both via public standards and through markets and consumers. With the challenge of transforming agriculture extending far beyond the issue of marketable goods, she analyzes the links necessary between public action and market regulation.

References

CBD (Convention on Biological Diversity) (1992). United Nations, p. 3. Retrieved April 1, 2013 from www.cbd.int/doc/legal/cbd-en.pdf.

GRIFFON, M. (2007). Pour des agricultures écologiquement intensives. *In Les défis de l'agriculture au xxie siècle*, Leçons inaugurales du Groupe ESA, Angers.

IAASTD (International Assessment of Agricultural Knowledge, Science and Technology for Development), (2009). , In B.D. MacIntyre, H.R. Herren, J. Wakhungu, & R. T. Watson, (Eds.), *Agriculture at a Crossroads, Global Report*, Island Press, Washington DC, 606 p.

Le Roux, X., Barbault, R., Baudry, J., Burel, F., Doussan, I., Garnier, E., Herzog, F., Lavorel, S., Lifran, R., Roger-Estrade, J., Sarthou, J.-P., & Trommetter, M. (2008). Agriculture et biodiversité: valoriser les synergies. In Expertise scientifique collective INRA, Editions Quae, 178 p.
MEA (Millenium Ecosystem Assessment) (2005). Ecosystems and Human Well-Being: Synthesis. Retrieved April 1, 2013 from http://www.maweb.org/documents/document.356.aspx.pdf.

Chapter 2
The Diversity of Living Organisms:
The Engine for Ecological Functioning

Étienne Hainzelin and Christine Nouaille

The diversity of living organisms has long been the mainstay of agricultural activity and its innovations. However, since the late nineteenth century, particularly in industrialized countries, increases in yields have been based on radically new technologies which deny the biological reality of agriculture and end up artificializing environments. This greatly intensified agriculture is primarily based on fossil fuels (mainly petroleum). It now finds itself at an impasse because of its impacts on ecosystems and the dramatic increase in the prices of inputs and energy. Social inequalities and massive rural exoduses that it has caused are further reasons for concern. Scientists, politicians and NGOs have striven, mainly over the last 20 years, to come up with alternative approaches for developing countries to overcome these energy, economic and environmental crises, and in order to ensure food security for the most vulnerable populations. There is now a widespread conviction that these countries must develop the capacity to ensure sustainable food security. The intensification of their production is therefore essential but has to be based on new approaches. Often grouped under the all-encompassing term 'agroecology', these new approaches rely on both the most modern advances in agricultural sciences and the traditional know-how of rural populations.

Paths to ecological intensification today mainly depend on biological diversity. They appear promising not only in terms of yields and economic efficiency but also in terms of sustainability, especially in vulnerable areas (Pretty et al. 2011). We find ourselves in a context of ecosystems with radically altered functioning. Large biological cycles are no longer able to provide services and sufficient renewable resources to meet our needs. And, most worryingly, biodiversity is

É. Hainzelin (✉)
Cirad, Montpellier, France
e-mail: etienne.hainzelin@cirad.fr

C. Nouaille
Cirad, Communication Service/Office of the Director General
in Charge of Research and Strategy, Montpellier, France
e-mail: christine.nouaille@cirad.fr

É. Hainzelin (ed.), *Cultivating Biodiversity to Transform Agriculture*,
DOI: 10.1007/978-94-007-7984-6_2, © Éditions Quæ 2013

eroding away at an alarming rate. Given this situation, we propose in this book ways of producing more biodiversity, by doing more than just preserving our resources, in fact by explicitly cultivating them. Ecological intensification must produce biodiversity, in all environments ranging from industrial production to small family farms.

This chapter aims to show that the evolution of cultivated biodiversity, i.e., agrobiodiversity, is inseparable from the history of agriculture and of the life sciences.

1 Diversity and Unity of Living Organisms: The Successive Revolutions of the Biological Sciences

If interest in the diversity of living organisms—and ecological thinking—goes back to antiquity, the term 'biodiversity' itself is of very recent origin. Scientific ecology has really developed along with other biological disciplines in the second half of the twentieth century. The idea that almost all of the resources we use are directly dependent on the activity of living organisms from the very origins of life, and that human activities have a major impact on their renewal, received widespread international exposure at the Rio Summit in 1992.

1.1 The Concept of Diversity Explored Over the History of Science

Compared to other sciences, biology is in its infancy. Chemistry was already present in prehistory: thanks to fire, discovered 800,000 years ago, metals and metal-working were mastered early on (gold and silver in 7000 BC, bronze in 5000 BC, iron in 2500 BC). The first physicist-astronomers, on the banks of the Tigris and Euphrates, described the rules of cyclical astronomical phenomena (diurnal, lunar, annual) 5,000 years ago. Greek mathematicians mastered abstraction to establish theorems we still cannot prove today. But it was not until 1854 that G.A. Thuret described gamete fertilization for the first time even though farmers had been using sexual reproduction by practicing empirical selection for 12,000 years.

Aristotle (384–322 BC) is considered the founder of ecology and botany, and to his pupil Theophrastus we owe the first *History of Plants* (320 BC). But it was not until the seventeenth century that naturalists began to identify and classify species, and it is Linnaeus (1707–1778) who devised the binomial system of nomenclature that designates each plant by a generic and specific name.

In the nineteenth century, species were fixed once and for all for naturalists. Charles Darwin (1809–1890) published his groundbreaking *The Descent of Man, and Selection in Relation to Sex* in 1871, in which he laid the foundations of the theory of evolution. Meanwhile, Gregor Mendel (1822–1884) had described the laws of heredity. Published after his death, his work was rediscovered in the early twentieth century, revealing through Mendelian genetics, a law common to all living organisms. In 1918, Ronald Fisher used the laws of heredity to establish the theoretical basis of evolutionary biology which has led to numerous applications in plant breeding.

The revolution in molecular biology, which began with the discovery of DNA—and thus of the profound unity of all living things—by James D. Watson and Francis Crick in 1953, contributed to the considerable development of genetics of today (Box 1). Paradoxically, this revolution has led us to revise and expand the concept of the diversity of living organisms beyond the strict specific diversity.

Box 1. The genome, a computer program of living organisms or a toolbox?

The genome has often been compared to a computer program.

The development of amplification techniques for cell proliferation and gene analysis[*] opened up significant opportunities for genome analysis. With genetic engineering and clonal propagation, it became possible to build and multiply customized genotypes. These techniques have led to the development of new medical treatments (diabetes, vaccines, antibiotics, etc.). In agriculture, the seed sector and the agrochemical industry have transferred specific characteristics into varieties using gene transfer: resistance to herbicides, various forms of the Bt toxin, etc. But the applications of genetic engineering in plants mainly use molecular markers for genetic selection. Indeed, transgenic varieties—even setting aside the controversies surrounding them—do not always express the genes that are transferred to them. Why does then transgenesis not lead to more applications, and to what extent the comparison of the cell with a computer has done a disservice to biology? This is the question posed in 2011 by the philosopher and physician Henri Atlan in his book *Le vivant post-génomique. Ou qu'est-ce que l'auto-organisation ? (The Post-Genomic Living Organism. Where is the Self-Organization?)* He poses the question asked by a growing number of biologists and addressed by Thomas Heams, molecular biologist, in a *Monde* op-ed on 22 September 2012: '[comparing DNA to a computer program] has been the topic of a vast research programme since the 1950s, that of molecular biology and its thousands of genes, first studied one by one, culminating in large sequencing programs, including that of the human genome, at the turn of the millennium', an issue that extends into the current vogue of synthetic biology. Today, 'genetics is drowning in data'. We take recourse (yet again) to computers to try to bring some coherence in systems biology that is struggling to emerge: 'Apart from a few pioneering studies, multiscale syntheses (from the molecule to the organism) have not been forthcoming.' (Heams 2012).

Thus, are not scientists going down the wrong path in according a major programming role to DNA?

Recent studies militate towards a paradigm shift: the genome, far from being a program 'written in advance', is rather a toolbox that each cell could use with more or less degrees of freedom (Cohen et al. 2009; Ruault et al. 2008). This new 'post-genomic' vision, which emphasizes the 'self-organization of living organisms' (Atlan 1999, 2011) and is consistent with the Darwinian theory of evolution, offers an alternative to the technological approach to decryption: 'By restoring cellular disorder to its proper place in biological explanation, one need not seek non-existent programs [...]. Would we expect to understand the climate using an atlas of all the clouds and all the raindrops on Earth?' (Heams 2012).

[*]In 1993, K. Mullis was awarded the Nobel Prize in chemistry for the invention of polymerase chain reaction (PCR).

Thanks to new scientific and technological methods, the rate of new scientific discoveries has been accelerating since the 1970s. Our knowledge of evolution has grown by leaps and bounds and we now realize the key role of living organisms in the history of our planet and of mankind.

On a geological time scale, it was the life of cyanobacteria that created a breathable atmosphere, patiently accumulating oxygen for several billions of years. This led to an evolutionary revolution by opening the door to aerobic organisms. It is the diversity of living organisms that has shaped our planet by allowing the accumulation of metal ores through oxidation. They are also responsible for the vast sedimentary formations of limestone and shale, and for coral reefs. Finally, the very same diversity of living organisms is at the origin of deposits of coal, oil, natural gas and phosphates—veritable storehouses of energy or chemicals.

Nowadays, biodiversity is considered in its temporal dimension, as a dynamic process. It is an ever-evolving system, from the point of view of the species as well as of the individual. The mean half-life of a species is estimated to be 1 million years and a full 99 % of species that have lived upon Earth are now extinct.

Biodiversity is also considered through its spatial component: it is not distributed evenly on Earth (Box 2). Flora and fauna differ depending on various criteria such as climate, altitude, soil or intervention by man or by other species. At the local or regional level, it evolves within ecosystems, which are associations between a given biophysical environment, the habitat and populations of living organisms—the biotic community—in perpetual co-evolution.

As our knowledge grows of the dynamics of interactions between species or populations of species within ecosystems and of the multiple functions of production, regulation and services they provide shows that the diversity of living organisms is indeed the engine of ecological functioning.

> **Box 2. Areas of megabiodiversity (hotspots)**
>
> Today, there are estimated to be over 8.7 million living species on Earth. They are not evenly distributed: thus, in 1988, the American primatologist Mittermeier discovered that four 'megadiversity' countries—Brazil, Indonesia, Madagascar and the Democratic Republic of Congo—accounted for, just by themselves, two-thirds of primate species on the planet (Mittermeier and Goettsch Mittermeier 2005). With his colleagues at Conservation International, Mittermeier expanded his research to other species. In 1997, 17 countries were recognized as megadiverse because they each host at least three thousand species of endemic vascular plants.
>
> 'Hotspots' of biodiversity are locations where the largest concentrations of plant and animal species, often endemic, are found. They are mainly concentrated in the tropics: the Amazon Basin, the Congo Basin, Madagascar, islands of Melanesia and of the East Indies, Amazonian foothills of the Andes, coral reefs, and forests of Borneo and New Guinea.
>
> The concept of megabiodiversity applies to these hotspots at the national level and are usually demarcated by political boundaries. Thus the Amazon region extends over six countries: Bolivia, Brazil, Colombia, Ecuador, Venezuela, and Peru (French Guiana makes France their neighbour!). Costa Rica and Mexico in America; China, India, Indonesia, Malaysia, Nepal and the Philippines in Asia; Kenya, the Democratic Republic of Congo, Madagascar and South Africa in Africa joined these six countries in Cancun in 2002 to form the 'Group of Like-Minded Megadiverse Countries'. This group serves as a mechanism for consultation and cooperation so that these countries' interests and priorities related to the preservation and sustainable use of biological diversity can be promoted. France, through its overseas *departments*, is also part of the movement.
>
> It is worth noting that the biologically megadiverse countries are also those where cultural forms (language, arts, etc.) abound in greatest numbers.

1.2 Biodiversity on the International Stage

The term 'biodiversity' was born in 1986 due to heightened concerns about the extinction of species and due to a growing perception of its role as an engine in the planet's functioning.

In June 1992, the Rio de Janeiro Summit brought these concerns onto the international stage. Biological diversity was viewed as 'the variability among living organisms from all sources including, inter alia, terrestrial, marine and other aquatic ecosystems and the ecological complexes of which they are part: this includes diversity within species, between species and of ecosystems' (Convention on Biological Diversity, CBD 1992). The Convention on Biological Diversity (CBD) became available for adoption at the Rio Summit. Today, it has been signed by 193 countries committed to the conservation of biological diversity, its sustainable use, and fair and equitable sharing of benefits arising from the use of genetic resources. A comprehensive work was undertaken to highlight three levels of understanding the diversity of living organisms in terms of their social, economic, cultural or scientific importance:

- ecosystems and habitats which (i) host high diversity or large numbers of endemic or threatened species, or encompass wilderness areas, (ii) are necessary for migratory species and (iii) are representative, unique or associated with key biological processes;
- threatened species and communities; wild species related to domesticated or cultivated species which are of medicinal, agricultural or economic interest; species of interest to research, such as indicator species;
- genomes and genes of social, scientific or economic importance.

In (1996), the Conference of the Parties of the CBD held in Buenos Aires recognized 'the special nature of agricultural biodiversity, its features, and problems needing distinctive solutions'. Man and his agricultural activities are indeed at the heart of these ecosystems.

The concept of agrobiodiversity was thus born, defined as 'the set of components of biological diversity in relation to the production of goods in agricultural systems'. It encompasses the genetic resources of plants, animals and microorganisms (programmed biodiversity), genes, species and ecosystems necessary to maintain the functions, structures and key processes of agricultural systems (associated biodiversity), in particular the pollinators, parasites, symbionts, pests and competitors. Agrobiodiversity has a critical socio-economic and cultural dimension because it is heavily influenced by human activities and management practices. Local and traditional knowledge, cultural factors, participation processes and tourism are also taken into account (Conference of the Parties 2000).

Our understanding of agrobiodiversity has changed: we now put the interactions between organisms and their functional relationships at the heart of the agronomic approach. We consider the cultivated plot as a dynamic system and integrate cultivated areas into the landscape. Agrobiodiversity is not only the genetic resources of domesticated plants and animal species or gene stocks to keep available for their improvement and future adaptation. It is also the diversity of all species 'auxiliary' to crops, aerial or soil-based, in and around the plot. In the agroecosystem, man too is very much present through his interventions, production activities, and his cropping practices. He programmes a part of the diversity, one that he cultivates or raises (plants and animals). But in an integrated approach, he must henceforth take into account, the 'other' side too, i.e., species that share this space with crops, irrespective of whether they are 'useful' or 'harmful'. In addition, since the space is not compartmentalized, there exist flows between agrobiodiversity and 'natural' biodiversity through numerous interfaces and interactions.

How does one manage such a diversity? How to enrich, optimize, guide—in one word 'cultivate'—this diversity? A further step forward was taken with the Millennium Ecosystem Assessment (MEA 2005), a monumental undertaking by over 1,300 experts from around the world. In arriving at an assessment of the planet's resources and spaces, they have helped us understand the importance of local customs and cultures in the maintenance of biodiversity and the many services it provides.

2 A History Closely Linked to Man's

It is now established that agriculture arose simultaneously in several centres, on all continents, about 12,000 years ago. It marked the transition from man's use of natural plant and animal resources—hunting and gathering over vast spaces—to the domestication of useful species, i.e., their multiplication and cultivation on cleared and demarcated plots. The descendants of plants that had different favourable usage characteristics (food, medicines, fibres, transmission, production cycles, harvesting, storage, etc.) were grown through seeds or cuttings. The dawn of agriculture also corresponded with the settlement of a portion of human populations and a new division of labour between farmers and non-farmers. Agriculture also made them probably aware of the necessity of managing resources in a limited space, and thus encouraged them to innovate in production techniques in these very first stages of 'intensification'.

2.1 The Origins of Agrobiodiversity

Anthropologists have drawn attention to the role of traditional communities in the conservation of agricultural diversity. Women in particular play a very important role through their detailed knowledge of plants and the myriad criteria by which they choose and exchange them: behaviour depending on soil; altitude; sunshine and rainfall; yield and ease of being processed; pest and disease resistance; organoleptic and aromatic qualities; emotional, social and aesthetic values; etc.

This is why in regions where some species have originated or diversified, there is a staggering variety of cultivars: seven recognized species and 5,000 varieties of potatoes still cultivated in the Andes (FAO 2008a), 92 names of rice varieties listed with the Hanunóo in the Philippines (Conklin 1957), etc.

Selecting plants like this, according to very varied criteria, enables continuous genetic hybridization, including with wild relatives. Indeed, this is the basis of the adaptation of crops and their renewal.

Peasant communities have drawn from the genetic diversity of species for over six hundred generations. They first domesticated them, then organized production into 'cropping and livestock systems' and, finally, improved their performance through increasingly focused work on the selection of varieties or races. By specializing the production, first the farmers, then agronomists and scientists, and finally the seed industry, have ended up maintaining a 'gene pool' of crops and livestock.

2.2 The Great Voyages of Exploration and Redistribution of Cultivated Species from the 15th to the 19th Century

Even though species and varieties have travelled at all times, their movements accelerated in an unprecedented way in the fifteenth century. The history of the cultivated plants commonly found today begins in 1453 with the great navigational explorations, when the fall of Constantinople led Europeans to leave Turkey and undertake sea journeys to obtain food commodities that they needed (Volper 2011).

With the discovery of America, the range of useful plants in Europe became considerably enlarged. Tomatoes, maize, beans and potatoes were introduced to Europe in the sixteenth century. The English, French, Dutch, Spaniards, Portuguese and Danes all stepped up their efforts to successfully 'acclimatize' these new species, which soon changed profoundly the landscape of European agriculture. But many species collected were not always cultivable in European climates, and the territories conquered in the tropics were instead asked to grow them. Thus sugar cane, native to New Guinea and grown in the Pacific, was spread across the New World. This species is typical of the colonial era. In fact, to satisfy their passion for sugar, the colonial powers disrupted international trade, imposing appalling terms on their new possessions: the so-called 'triangular trade' (Europe, Africa and the New World). The same pattern was repeated for other large strategic tropical productions—cotton, coffee, cocoa, rubber, oil palm, etc.—which led to frenzied commercial competitions around these commodities.

The European industrial revolution in the nineteenth century accelerated the pressure on crops, increasing the need for raw materials and new markets. Europe turned to Africa and Southeast Asia and, at the beginning of the twentieth century, embarked on major undertakings for the exploitation of agricultural potential of their colonies.

Botanical gardens, in particular the National Museum of Natural History in Paris (and the colonial garden it set up in nearby Nogent-sur-Marne) and the Royal Botanic Gardens at Kew in the United Kingdom, played a role in the acclimatization of species. Colonial enterprises headquartered in the home countries started expanding in size and reach from 1925 onwards. Colonial exhibitions attracted investors who in turn relied on the professionals: the tropical agronomists. And so, with the intensification of production, was born tropical agronomy, based on the scientific knowledge of the time, i.e., tested in temperate industrialized countries. Agricultural systems in the former colonies, even after they gained independence, remained profoundly and lastingly marked by these developments.

During this time, the seeds travelled and were exchanged (Box 3). Genetics arose as a major component of agricultural intensification, especially for improving productivity and for combating the many crop diseases and pests.

Box 3. Origins of a few iconic tropical crops

Cocoa, discovered by Christopher Columbus in Nicaragua, was introduced to Europe by Cortez in 1528 to his king, Charles V of Spain. Initially, the preserve of the elite, its consumption became popular only in the twentieth century. The primary producers of cocoa today are Côte d'Ivoire, Ghana and Malaysia.

Coffee, which is of Ethiopian origin, was introduced to Europe in 1615, 5 years after tea. It soon gained a following there and spread widely. The history of its cultivation is long and convoluted. It is now produced by nine major countries including Brazil, Colombia, Indonesia, and Vietnam (Ethiopia is in sixth place).

The **rubber tree** and its latex became known to Europeans with the discovery of America. Yet it was only in the eighteenth century that it attracted any real interest with the first industrial applications and the first processes, developed in 1790. It is the automotive industry which, in the late nineteenth century, caused global production to explode, mainly due to the development of the vulcanization process. Native to the Amazon, it is now produced in Asia, with Thailand, Malaysia and Indonesia accounting for three quarters of global production. Rubber cultivation provides 44 % of the world's production of elastomer and is grown on 10 million hectares maintained by millions of smallholders. Eighty percent of the 10 million tonnes of natural rubber they produce is used in the automotive industry.

The **banana**, discovered in India by Alexander the Great, also existed in China at that time. In 650, Islamic conquerors imported it into Palestine and to the island of Madagascar. From there, traders carried it all over Africa. In 1402, Portuguese sailors planted it on the Canary Islands. A Portuguese monk brought it to the island of Santo Domingo in 1516. It did not take long before it became popular throughout the Caribbean and Central America: in the eighteenth century, more than three million banana trees were growing on Martinique. The banana is the third biggest tropical fruit crop, with about 15 % of production being exported and 85 % being consumed locally, especially in the poorest countries of Africa, Latin America and Asia. Over a thousand banana varieties currently exist. Staple food in many tropical countries, locally consumed banana plays a major role in food security.

The **oil palm** originated in West Africa, where its consumption in food probably dates back over 5,000 years. In 1959–1960, the government of Côte d'Ivoire launched a major development programme for industrial and village plantations of selected oil palm varieties. But African production, while flourishing at the beginning, has been overtaken by the explosion of production in Asia in a market dominated today by Malaysia, Indonesia, Nigeria and Thailand. The primary oilseed crop in the world, the oil palm has become a strategic crop for many tropical countries. With a production of 42 million tonnes in 2008, it represents more than a third of global production of vegetable oils. Its rapid expansion, like that of rubber, often at the expense of the primary forest, has raised new research questions.

2.3 Agricultural Revolutions and Genetic Resources in Europe in the 20th Century

Since the Neolithic age, agriculture has experienced some major technical revolutions (Mazoyer and Roudart 2002). These have included use of flood rivers in antiquity, the advent of irrigation systems, the development of complex cropping systems such as triennial crop rotations with fallow in the Middle Ages and animal draught cultivation. More recently, in the seventeenth century, animal feed, mechanization and organic manure arrived on the agricultural scene. This dominant model persisted in France until the early twentieth century, even in the period leading up to the Second World War, whereas the United States began the modern agricultural revolution as far back as the 1930s.

It should however be noted that in France, the Vilmorin-Andrieux family of seed merchants published a catalogue describing and comparing wheat varieties as far back as 1850. By cross-breeding wheat and *Aegilops*, they played a pioneering role in plant breeding. In 1873, Henry de Vilmorin (1843–1899) started improving wheat varieties through systematic and reasoned cross-breeding and, 10 years later, marketed the first wheat variety originating from a genealogical selection.

The last phase of modernization and industrialization of agriculture has greatly increased yields and lowered production costs in industrialized countries. Between the two World Wars, the goal was to better integrate agricultural production into market economics and improve farmers' living conditions. Increasingly powerful mechanization, the development of agrochemicals and agrifood industries and genetic advances have enabled this transformation of agriculture. High yields have been accompanied with extraordinary increases in labour productivity but also, unfortunately, in energy consumption. By modernizing itself, agriculture has become dependent on fossil fuel resources.

The pesticides industry has developed, thanks to research in organic chemistry during the two wars. Military research had already perfected combat poison gases which, after the war, were used against insects. In the 1950s, insecticides such as DDD and DDT were used in large quantities in preventive medicine (mosquito disinfestation against malaria) and agriculture (elimination of the Colorado potato beetle). The use of these products then experienced very strong growth, making them virtually indispensable to most agricultural practices. This dramatic intensification was based on the idea that agricultural production could be beneficially reduced to chemical flows, and that all that was necessary was to complement what nature had difficulty providing. From 1945 to 1985, the consumption of fertilizers and pesticides doubled every 10 years. Pesticides proved to be a powerful means for increasing agricultural yields and for helping ensure an abundance of food while limiting deforestation. But the use of synthetic fertilizers, herbicides, insecticides and fungicides has artificialized agroecosystems by homogenizing the soil at the trophic level. These synthetic products have accumulated in soils and waters and have destroyed, along with pests, many useful species necessary to maintain agroecosystem balance.

Genetics has contributed greatly to the development of this intensive agriculture. Geneticists have relied on the richness and diversity of cultivated varieties from around the world to find characteristics that can provide ever increasing yields in a non-limiting nutritive context and to fight against various agricultural pests. In 1933, the first hybrid maize from cross-breeding came to market in the United States. Today these high-yielding varieties allow compounded year-on-year gains of more than 1 % over long periods.

To improve crop species, geneticists study their history, their origins and the role of farmers in their selection. Surveys are organized, private and public collections of genetic resources are created, and seed companies arrange to exchange samples.

The intensification of agriculture has had the rapid effect of reducing considerably the number of cultivated species and varieties. This has led a concerned scientific community to scramble to preserve their genetic resources. The first collection of material obtained from surveys was created by Russian geneticist Vavilov (1887–1943), following botanical and agronomic expeditions to help support the theory on the origin of cultivated plants. It had 250,000 accessions in 1940, of which 30,000 for wheat. Today, it has 400,000. There are now many gene banks around the world, with the Svalbard Global Seed Vault in Norway being probably the best known. In addition to these major banks, a few small treasures of genetic material are being maintained and enriched by enthusiastic and concerned producers (Box 4).

Vavilov developed the concepts of 'centres of origin' and 'zones of domestication' of a given species, where genetic intermixing is the richest. These concepts, still in vogue today, were further advanced and refined by numerous studies on the phylogeny of different cultivated species with the help of the tools of molecular biology. The history of 'complex species' and the role mankind and populations have played in their selection and dissemination have been studied by geneticists with the help of ethnobotanists, joined nowadays by linguists and anthropologists.

The 'programmed biodiversity' of agroecosystems is thus stored in gene banks since the advent of genetics, even though the ideas and modes of preservation and conservation of 'useful' genetic diversity have changed over time.

2.4 The Green Revolution

After the Second World War, the colonies took the path of independence and development. But even though by then industrialized countries had adopted intensive agriculture, its transfer to developing countries did not prove so simple, mainly because of its dependence on inputs.

The Green Revolution was thus born out of a political will to transform agriculture in developing countries (FAO 1996; Griffon 2006, 2011). Modelled on the systems in industrialized countries, its main goal was intensification and the use of high-yield cereal varieties. It was orchestrated by international agricultural research centres and large foundations associated with major American universities.

The Green Revolution was based on three factors: use of high-yield varieties, use of inputs—fertilizers and pesticides—and irrigation in areas at risk of water stress. This profound transformation of agriculture has led to increased energy costs in developing nations, but not in the same measure as in industrialized countries because mechanization has remained ultimately limited. This revolution was also based on a massive support from public policies, both for infrastructure investments as well as for price guarantees and technical training. Its beginnings can be traced to Mexico in 1943, where the government, with support from the Rockefeller Foundation, achieved a dramatic increase in wheat production. Self-sufficient in 1951, the country became an exporter the following year, even though during the same period its population grew significantly.

Norman Borlaug, who developed the high-yield wheat varieties in the Office of Special Studies (OSS), Mexico City, and then disseminated them in Asia, is considered the father of the Green Revolution. His work earned him the Nobel Prize for Peace in 1970. The Rockefeller Foundation endeavoured to spread the idea of the Green Revolution by helping set up new international research centres in the world: CIMMYT (Centro Internacional de Mejoramiento de Maiz y Trigo), succeeded the OSS in Mexico in 1963, and IRRI (International Rice Research Institute) was established in the Philippines in 1960. The latter was instrumental in helping spread the use of high-yield rice varieties in Asia.

The Green Revolution, undeniably successesful in Asia—though less obviously so on other continents—has long seemed to be the most effective model of development in developing countries. India is the most cited example: it has increased wheat production ten-fold and that of rice three-fold. Areas of chronic hunger have turned exporters, but under conditions in which sustainability remains challenged due to the requirements the varieties grown have for water, fertilizers and pesticides.

The externalities of the Green Revolution have also gradually emerged: social, with a massive rural exodus; and environmental, with widespread soil degradation, misuse of pesticides and subsequent pollution, and the homogenization of cultivated varieties. This homogenization has led to a large loss of traditional knowledge and agricultural biodiversity, most notably in local cultivars. Concerns have also emerged about the resilience of these varieties to the emergence of new pathogens.

The spectre of genetic erosion is now haunting the varieties of the countries of the South. To address these concerns, seed banks have been established, modelled on the International Plant Genetic Resources Institute (IPGRI), now Bioversity International.

In fact, many developing countries have drawn little or no benefits from the advantages expected from modern agriculture and have seen no wealth flow from it. The reasons cited most often are unfavourable soils and climate, and lack of water, financial capital and appropriate training. To these can be added the unfavourable political, economic and legal environments in a number of countries. Imbalances caused by the protection of certain markets, most notably the massive subsidies given to industrial agriculture in rich countries, also share the blame. In the case of Africa, the dynamics of intensification were abruptly halted by the period of structural adjustment in the 1980s.

3 Documented Risks of Erosion of Agrobiodiversity

Many studies have attempted to assess the general effects of pressures on biodiversity. But biodiversity, as we have mentioned, is a dynamic process, wherein agriculture plays a special role. In this book, we are particularly interested in the biodiversity of agroecosystems.

The difficulty of quantifying diversity depends on the level we approach it at: allelic, specific or ecosystemic (Le Roux et al. 2008). Documenting changes in diversity is just as difficult and controversial as quantifying it in the first place. The figures quoted below are mainly drawn from the Millennium Ecosystem Assessment (MEA 2005) and FAO. The latter was made responsible in 1999 to assess periodically the state of Earth's plant genetic resources for food and agriculture in the world. Its brief is to propose action plans to the international community, guided by the international Commission on Genetic Resources for Food and Agriculture. In 1996, a global action plan was drawn up and to date 170 countries have adopted it. According to the Second Report on the State of Plant Genetic Resources for Food and Agriculture in the World (Commission on Genetic Resources for Food and Agriculture 2010), the main reason for the erosion of genetic diversity is the replacement of local varieties by modern varieties, a trend that has accelerated in the last 50 years. Other causes, such as environmental degradation, urbanization, and land clearing through deforestation and bush fires are also highlighted.

The need for relevant indicators that can be reliably defined and acquired has made them the topic of research programmes and international negotiations.

3.1 Agriculture, the Planet's Most Important Landscape

Today, of Earth's entire surface of about 51 billion hectares (including the oceans) an estimated 12 billion hectares (land and water) are bioproductive in the sense that they create a certain amount of organic matter each year through photosynthesis. The landmass covers 14.9 billion hectares, of which 70 % is directly subject to human activity. Photosynthesis also exists in deserts and most parts of the ocean, but is spread out too thinly for its products to be exploited by man (Ecological Footprint Atlas 2009).

Ten percent of the landmass area, i.e., around 1.5 billion hectares, is cultivated land, of which one–third is dedicated to livestock feed. Added to this area are 3.4 billion hectares of pastures (consisting of 1.4 billion hectares of improved grasslands and 2 billion hectares of natural pastures and rangelands) and 4 billion hectares of forest cover (including 1.4 billion hectares of primary forests). Farmers have therefore shaped a very large part of our planet over the centuries.

The number of farmers in the world is estimated at 2.3 billion,[1] i.e., almost one quarter of the world's population and half its workforce. A large majority of these farmers is in developing countries; in developed countries, farmers represent only 2–3 % of the population. The World Bank (2008) considers that in the least developed countries (LDCs), two-thirds of jobs remain directly linked to agricultural activity. Most of these countries are in sub-Saharan Africa, where agriculture employs 65 % of the workforce and accounts for nearly a third of GDP growth.

3.2 A World Heritage Under Threat

According to the FAO (2010), about 7,000 *plant species* have been grown for consumption over the span of human history, with other sources putting this number much higher (it must be noted that for plants, we speak of complex of species rather than of individual species). For nearly 12,000 years, the large diversity of varieties maintained or domesticated by man have provided him food, fibres, material and energy. It has ensured the survival and development of human populations inspite of pests, diseases, climatic fluctuations, droughts or other hazards.

---

[1] Most of the figures mentioner here are from Millennium Ecosystem Assessment. See also: Convention on biological diversity, 2008. Biodiversity and agriculture—Protecting biodiversity and ensuring food security, www.cbd.int (retrieved: 29 November 2012).

Yet, today, only about thirty species meet 95 % of human and animal food needs. Four of them—rice, wheat, maize and potato—satisfy more than 60 % of food energy needs. The loss of diversity within cultivated species is also widespread, as illustrated by the case of the apple (Box 4).

Box 4. From industrialization and erosion because of standardization to diversification by farmers: the case of the apple in France

Known and appreciated since ancient times, the apple experienced spectacular growth in the nineteenth century, with France playing a major role[*]. André Leroy, a famous French nurseryman of the early part of that century, described 527 well-differentiated varieties in his catalogue. And plant breeders did the rest: there are today nearly six thousand varieties worldwide.

Between the two World Wars, urbanization led to the disappearance of many town orchards and small producers. More distant producers were then tasked with producing more fruit with increased tolerance to transport and which could remain fresher for longer periods. In this way, a large number of varieties fell by the wayside.

After the Second World War, grubbing-up premiums were offered to farmers for replacing their apple orchards with intensive agriculture. In 1960, the official catalogue of species and varieties was created by the Permanent Technical Committee for Selection. It drew up the list of varieties that could be marketed. Only one French apple was listed as Class I and could become part of modern distribution supply chains. The other varieties (11,000 worldwide) were no longer cultivated except only occasionally by a few small producers who slowly succumbed to economic pressure.

Faced with such a significant loss of heritage, pomological associations were created by enthusiasts in the late 1970s. One of them, the Pomological Association of Upper Normandy[**], conducted a census in France: there are 37 conservatory orchards growing 987 apple varieties. Two hundred of these varieties are used, half of them as 'table fruit'.

[*]See http://www.lapomme.org (retrieved: 4 April 2013).

[**]*Association pomologique de Haute Normandie*, http://www.aphn.net/ (retrieved: 4 April 2013).

About 35 *animal species* have been domesticated for agriculture and food production. Their intraspecific diversity is reflected in the many indigenous breeds. These chosen breeds are well-suited to local conditions because of their resistance to climatic stress, diseases and parasites, or because of their adaptation to specific agroecosystems. However, according to *The State of the World's Animal Genetic Resources for Food and Agriculture* (FAO 2008b), 20 % of livestock breeds, i.e., around 1,500 of the 7,600 races in the world, could disappear forever in the near future due to inability to adapt, inbreeding, unsustainably small populations, etc.

Even though *aquatic biodiversity* plays a vital role in human livelihood, it is currently under threat from overfishing, resource depletion, destructive practices, the introduction of exotic species, and habitat destruction and degradation. In 2008, an estimated 1,731 species or groups of aquatic species (finfish, shellfish, molluscs, etc.) were commercially fished, many of which are destined to disappear before the middle of this century. With the trawlers going further out into the oceans and becoming better equipped, the FAO estimates that about one fish species in three is threatened with extinction. Only pisciculture can compensate for the expected drop of fished quantities. *Aquaculture* is one of the food production sectors experiencing the fastest growth. More than 360 species of fish, invertebrates and plants are grown in the world, most only since the last 100 years or so.

Two agroecosystem compartments are particularly vulnerable to the effects of intensive agriculture: (i) the soil and (ii) the habitats of auxiliary species of crops. It is this biodiversity—whose importance we are only now realizing—which is emphasized in the concept of agrobiodiversity. Because it is poorly understood and because we have an incomplete grasp of its functioning, it is particularly difficult to estimate its erosion.

Soil biodiversity reflects the variability among living organisms. It ranges from micro-organisms (e.g., bacteria, fungi, protozoa and nematodes) to the larger meso-fauna (e.g., acari and springtails) to the more familiar macro-fauna (e.g., earthworms and termites). Plant roots can also be considered soil organisms in view of their symbiotic relationships and interactions with other soil components. These diverse organisms interact with one another and with the various plants and animals to ensure the ecological functions of the soil through trophic exchanges, information flows, etc. In this way, they contribute to the provision of ecosystem services essential for life.

Microbes and invertebrates form the group of species which are the most numerous on the planet (World Conservation Monitoring Centre, WCMC 1992). It seems impossible to inventory them, given the difficulty in even quantifying the number of species. It is now estimated that there are about 10–50 million undescribed species of microbes and invertebrates. Food and agricultural production depends on multiple interactions with this 'hidden' biodiversity, whose functional role has been completely ignored by intensive agriculture. Bees, butterflies and other insects pollinate fruits and vegetables. Microorganisms form symbioses with the roots of cultivated plants and some fungi, or with animal organisms whose intestines they inhabit and whose assimilation functions and health they help regulate. They allow livestock ruminants—bovines, ovines and caprines—to assimilate cellulose. They help conserve and enhance protein in foods, especially through fermentation. Microorganisms and invertebrates are essential for breaking down dead matter and for the recycling of organic matter in soils. They can even be used as biological control agents. They are thus indeed at the heart of the ecosystems' basic operating mechanisms.

3.3 Ecosystems and Habitats Under Pressure

Natural forests are a source of income for many of the poorest countries, representing more than 10 % of GDP for some of them. One billion people make their living directly from them. They are also the most important reserves of terrestrial biodiversity. Despite this crucial economic role, the loss and degradation of tropical forests continues at an alarming rate of more than 10 million hectares per year. The loss of forest diversity imperils its future valorisation in terms of medicines, foodstuffs and raw materials, and it jeopardizes the well-being of many populations since it impacts the very basis of their livelihood. The use of forest plantations can meet some of these needs (timber, fuel wood) by sparing natural forests, but it cannot recreate the complex biodiversity of natural forest ecosystems.

At the *agroecosystem* level, the industrialization of agriculture generally results in the dissociation of crops and livestock through the specialization of farms and a homogenization of landscapes. This very important aspect is discussed further below (see 'Effects of the evolution of landscapes').

Finally, at the landscape scale, the phenomena of *biological invasion* (Box 5) also comes into play. More widespread today than ever before because of the globalization of trade, they are now considered by the UN as one of the major causes of the loss of biodiversity, along with pollution, the ecological fragmentation of ecosystems, hunting, fishing and the overexploitation of certain species. Reductions in the number of individuals of endangered species in particular has a very significant impact on their intraspecific diversity.

Box 5. Invasive species and biodiversity in island environments: example of the French Antilles

The French Antilles (also known as the French West Indies) have a climate conducive to extreme weather events such as cyclones. In addition they run the risk of volcanic activity. During the colonial period, massive clearing and overexploitation of forests took place and this was followed by a period of intensive agricultural production with heavy pesticide use. Its biodiversity thus underwent great stress (Sastre et al. 2007). Like other tropical islands, the islands of the French Antilles too host a unique biodiversity. But because of their insularity and pressures of rapid population growth and development, these ecosystems have become particularly vulnerable.

The destruction of natural habitats has led to the disappearance of most of the dry forests for purposes of urbanization and agriculture. Overexploitation of resources has exacerbated the situation. Thus, parrots of the Lesser Antilles have been hunted to extinction in Guadeloupe and Martinique, even though these two islands once had the largest number of these iconic species and even though mountain forests, one of their preferred habitats, yet remain.

Invasive alien species are another threat, in both natural and cultivated ecosystems:

- Rodents (black rat, brown rat, gray mouse) have been impacting agricultural production for several centuries.
- More recently, the 'cassava ant' has run over all of Guadeloupe within a few decades, causing significant damage to crops and gardens. Pesticides that are used to control it are known for their toxicity and persistence.
- The giant African snail (*Achatina sp.*) amazed residents by the speed of its colonization in the 1990s and the damage it caused. However, as on other oceanic islands, a relative equilibrium has been established, with a strong overall decrease in population and damage that is now localized and/or episodic.
- Many plant species have been introduced, some of which have become a problem for crops.
- Emerging pests threaten crops or livestock: *Ralstonia solanacearum*, tomato bacterium; black cercosporiosis, fungus attacking banana; Senegalese tick, vector of cowdriosis, etc.

In addition, some current agricultural practices threaten biodiversity: excessive use of fertilizers and pesticides, limited water resources, land clearing, etc.

'Biological' agriculture is a promising path to diversification and is partly already being practiced in Creole gardens. The horticultural sectors (fruit and vegetables) and some major crops such as sugar cane are gradually taking this path by adopting the so-called 'organic' practices of agroecology. In close collaboration with research, most sectors, in particular the banana sector, have adopted programmes for sustainable production, especially through the use of functions of agricultural biodiversity as a whole[*]. These paths are explored in detail in this book's later chapters.

[*]As part of the 'Antilles Sustainable Banana Plan', launched in 2008 by the Ministry of Agriculture at the initiative of the Union of Groups of Banana Producers (UGPBAN) and banana producer groups, the Tropical Technical Institute (IT^2), Cemagref and CIRAD are developing solutions to combat diseases of the banana and develop tools for sustainable banana production in the French Antilles.

For further information: Feldmann et al. (2007).

3.4 Effects of the 'Modernization' of Agriculture on Biodiversity

3.4.1 Effects of Agricultural Practices

According to a collective INRA study (Le Roux et al. 2008), there exist no statistics or suitable indicators to assess the environmental costs of agricultural practices, especially on the interactions between organisms. The few indicators that do exist are limited to the extent of the plot or are for periods that are too short. This group of experts referred to some two thousand bibliographical references— concerning mainly temperate crops—for analyzing existing knowledge on the relationships between agriculture and biodiversity.

To estimate the effects of agriculture on agroecosystems, experts have to study mechanisms at various levels: the entirety of agricultural practices at the plot scale; the impact of agriculture on the agroecosystem (cultivated areas, field edges, woods, ditches, etc.); and cohabitation between agroecosystems and natural ecosystems across the landscape or even region. As far as the effects are concerned, they distinguish three categories of biodiversity: alpha diversity, i.e., richness of species of the plot; beta diversity, which reflects changes in alpha diversity between habitats across the agroecosystem; and gamma biodiversity, considered at the landscape, region or country scale.

Studies at the *plot level* highlight a number of general factors that have an impact on biodiversity.

In *annual crops*, material flows (inputs, harvests) are very large and the disturbances are severe: destruction by pesticides, massive export of biomass, modification of the soil by tillage, and of the biocenosis by pesticides or indirect trophic effects. The result is a decline in the richness and abundance of many species: microorganisms, soil flora and fauna, insects, amphibians, birds, etc. Deep ploughing, for example, affects macrofauna, especially earthworms. Depending on their modalities and application frequency, synthetic pesticides used to combat insect pests can have dramatic effects on arthropod life cycles. Fungicides are even more toxic to soil organisms. Herbicides have an effect on a number of plant species, but also on species that are functionally associated with the latter. Finally, the development of species resistant to the molecules used causes significant imbalance in the ecosystem. The use of transgenic plants carrying the Bt toxin carries the same type of risk, to which can be added, over the long term, the transfer of genes into other species. Synthetic fertilizers, which have strong positive effects on the growth of plants and soil organisms, significantly modify the physical chemistry of the soil environment and affect trophic chains. They are also responsible for the disappearance of species better adapted to poor or fragile environments and significantly alters aquatic and terrestrial ecosystems (eutrophication, etc.).

Most of these impacts can be estimated by observing the effects of stopping treatment, but in the process, there will be irreversible loss of biodiversity.

Some loss of diversity can also be reversed by changing the production mode to organic farming, eco-agriculture, conservation agriculture, etc. Properly planned and executed rotations can lead to a reduction in pesticide use by disrupting the pathogen cycles.

Permanent grasslands are usually not subjected to pesticide applications. Even though they can be highly fertilized and intensively exploited, they have a much higher biodiversity than do cultivated monoculture plots. However, heavy grazing has, in general, a negative effect on the wealth of flora, arthropods and soil fauna. Moderate grazing, on the other hand, has a beneficial effect on the richness of many species groups. Finally, hayfields are generally richer in plant species than are grazed grasslands. But other factors come into play and have to be considered, such as the impact of different herbivore species, of the products they excrete, etc.

Pesticides are repeatedly applied to *perennial crops* to fight the always present pests and diseases. This is the primary factor to impact biodiversity. Pesticide use has a significant negative impact, for example, on the functional entomological diversity. It is clear that the presence of several exploitable vegetation strata and use of cover crops are conducive to maintaining trophic networks of species. Agroforestry is thus a possible route to diversification.

The results of *abandoning agricultural practices* on plots previously farmed depends on their initial state: the cultivated plots evolve positively for all groups of organisms in the early years. However, in the case of permanent grasslands, abandonment leads to a systematic decrease in the plant species richness. In all cases, when the abandonment time increases, species richness tends to decrease, especially when woody species start growing there. In functional terms, the short-lived plant species, dispersed by the wind and able to acquire resources, are replaced by long-lived woody species which are dispersed by birds. Soil fauna, mainly earthworms, evolves with these woody species.

Organic farming has a positive effect on biodiversity. The richness of plant species, soil microorganisms, vertebrates and arthropods all increase as does the abundance of invertebrate predators. But the structuring of the landscape also affects species richness and should be tuned to the agricultural practices if rare species have to be restored.

The *use of transgenic plants* is part of the technological intensification of agriculture. In 2011, according to estimates by the International Service for the Acquisition of Agri-Biotech Applications (ISAAA) (James 2011), which promotes GMOs, in particular in developing countries, 160 million hectares of transgenic varieties were cultivated in 60 countries. This represented an increase of 8 % over the previous year, proving the continuing strong growth of these varieties. The most common transgenic varieties are maize, cotton, soybean and potato, and the main feature that is disseminated is herbicide tolerance (59 % of surface area). ISAAA figures are disputed by the NGO Friends of the Earth, which also believes that in 2007 'nearly 90 % of all GM varieties marketed worldwide contained Monsanto traits'[2] (most generic GMO patents are American).

The impact of these transgenic crops on biodiversity is primarly due to their wide dissemination. Today, in the United States, 85 % of all maize grown, 91 % of soybeans, 88 % of cotton and nearly 95 % of the beet is genetically modified. Like all elite commercial varieties, they have a narrow genetic base, and their expansion, based on aggressive marketing, is mainly at the expense of crop diversity.

The impact of the transferred genes on the diversity of insects and plants in and around cultivated fields has been extensively studied for both herbicide-resistant and Bt GMOs. Their transfer to plots cultivated without GMO is sufficiently proven, at least for cross-pollinated crops such as maize and rapeseed, for the Scientific Committee of the High Council on Biotechnology in France to issue a notice[3] regarding the coexistence of genetically modified (GM) crops and non-GM crops. In spite of the arguments for and against, controversies and intellectual property litigation, there is no clear unequivocal link between the use of transgenic plants and biodiversity since the results depend so much on the climatic contexts, cultivated species, changes in practices of pesticide use, target species analyzed, etc. The risks associated with the spread of transgenes into wild plants—and thus the modification of wild biodiversity—are not negligible, but the potential damage of these transfers remains controversial.

3.4.2 Effects of the Pressure on Land and the Degradation of Natural Resources

The expansion of cultivated land in tropical and sub-tropical regions during the past five decades has been at the expense of areas of high biodiversity. Population pressures, depletion of cultivated soils and the need to increase industrial production are its main causes. Intensifying agricultural production in these countries without compromising soil fertility or tropical forests remains a major challenge.

Environmental damage and soil degradation turn 5–10 million hectares of land each year unsuitable for crops. Industrialization and urbanization result in a further 19.5 million hectares becoming unavailable[4] (De Schutter 2010). Restoring these

---

[2] Friends of the Earth, 2007. Qui tire profit des cultures GM? Monsanto et la « révolution biotechnologique » de l'agriculture menée par les multinationales, 20 p., http://www.foei.org/fr/publications/pdfs/gmocrops2006execsummaryfr.pdf (retrieved: 6 April 2013).

[3] Haut Conseil des biotechnologies, comité scientifique, 2011. Avis en réponse à la saisine 100506-coexistence sur la définition des conditions techniques relatives à la mise en culture, la récolte, le stockage et le transport des végétaux génétiquement modifiés, 46 p., http://www.hautconseildesbiotechnologies.fr/IMG/pdf/120117_Coexistence_Avis_CS_HCB.pdf (retrieved: 6 April 2013).

[4] FAO, *Land Policy and Planning,* http://www.fao.org/nr/land/land-policy-and-planning/en/ (retrieved: 6 April 2013).

areas is a major issue in some regions where it is not possible to expand the area available for cultivation.

Some areas suffer from acute water scarcity. The withdrawal of water from lakes and rivers, of which 70 % is used in agriculture, has doubled since 1960. Deforestation itself leads to a decrease in regional precipitation. Yet, irrigated crops have yields that are, on average, double those of the rainfed ones. We must find ways to improve the capacity of existing systems, especially of crop cultivation, to use water while limiting irrigation's negative aspects, in particular the impacts on natural ecosystems and their diversity.

3.4.3 Effects of Changes in Landscapes

Agricultural intensification very often homogenizes the structuring of the landscape. There is, however, little information on biodiversity in the literature: heterogeneity is measured as a percentage of semi-natural elements though sometimes the level of fragmentation and connectivity between habitats is measured instead or also. But the average size of the different surface areas and the diversity of productions are rarely taken into account. Nevertheless, it is clear that increasing areas of cultivated open spaces—at the expense of semi-natural ones—have led to a decline in inter- and intra-specific biodiversity. The MEA thus recognizes landscape diversity and 'ruggedness' as one of the services provided by ecosystems.

Finally, we can report that the effects of farming practices and landscape structuring on species depends on the latters' mobility. Mobile species are the ones most sensitive to landscape fragmentation, whereas sessile or sedentary species are particularly sensitive to farming practices on the plot and their migrations will occur over much longer timeframes.

Various in-depth studies have compared different options for structuring the landscape for an ecological intensification of agriculture. Should areas designated for intensive and highly productive agriculture but low in biodiversity be separated in space from protected natural areas interconnected between themselves (land sparing)? Or, on the other hand, should biodiversity in crops be maintained (land sharing)? The first solution is recommended in intensive agricultural systems (Franklin and Mortensen 2011) for the maintenance of plant biodiversity and of species with low populations (Phalan et al. 2011). This requires that incentivizing public policy be implemented to preserve spaces for biodiversity and its connectivity.

In countries of the South, this type of choice is closely linked to development policies. How to compensate the shortfall in income of people faced with pressure from and proposals of powerful economic groups (examples of rubber and oil palm)? Any workable solution will have to perforce include payments for environmental services (PES) as one of its components, but these are not problem-free themselves: how to calculate payments, how to assess changes, etc.

The second solution, land sharing, is the basis of many development programmes for sustainable development of agriculture, especially in areas where maintenance of agricultural biodiversity requires a real know-how and where the maintenance of populations in rural areas is a priority for food security (De Schutter 2010).

3.4.4 Effects of Climate Change

In the long term, climate change—in particular global warming—could affect agriculture and biodiversity in many different ways. Climate change will notably lead to an increased frequency of extreme weather events (floods and droughts, for example). Rainfall variability already makes it difficult to plan agricultural operations, and reduced rainfall threatens regions that rely on rainfed agriculture. Parts of the world are more susceptible to this variability than others, for example, the Sahel, north-eastern Brazil, Central Asia and Mexico. Global warming has already resulted in changes in agricultural calendars, such as earlier harvest dates. It also results in an increase in net primary production in temperate zones and in a decrease in hot and mountainous regions (Feldmann 2008b).

Furthermore, it is possible that the climatic and ecological zones will shift geographically, disbalancing natural vegetation and wildlife and forcing farmers to scramble to adapt. Some species have already started moving, for example, pests and vector-borne diseases spreading into areas where they were previously unknown.

Rising sea levels lead to water salinity, rendering some coastal land unsuitable for farming, particularly in small low-lying islands. Biodiversity of some very fragile environments—mangroves are the prime example—finds itself under threat.

If agriculture is reeling under the impacts of climate change, one must not forget that it itself is also responsible for 14 % of global greenhouse gas emissions. But it also has the potential of becoming an important part of the solution by reducing and/or eliminating a significant amount of global emissions (see below 'Coping with new hazards caused by global changes, especially climatic ones'). Traditional farming is inherently resilient, a quality it retains due to its agrobiodiversity. By using practices such as conservation agriculture, integrated management and agroforestry, this resilience can be used to improve the management of natural resources such as water, soil and genetic resources.

4 Why 'Cultivate' Biodiversity?

Biodiversity is the undeniable basis of food security for mankind. We have tried to show how far it has been part of mankind's history and how it has provided all that is necessary for feeding man and for his sustenance (clothing, medicines, habitat, energy, etc.).

4.1 Building Up Ecosystem Services and Food Security

Our planet will have to accommodate and feed an additional three billion people over the next 50 years. More than 85 % of them will be added to the populations of developing countries, in an unpredictable context of poverty and access to resources. With such an increase in population, human societies will be and are being forced to draw increasingly on natural resources. Thus, the Global Footprint Network announced on 27 September 2012 that the quota of resources produced by the planet in 2012 had already been consumed by the world population on that date. According to scientists, mankind's global ecological footprint has exceeded the Earth's biological capacity to produce resources and absorb our waste ever since the mid-1980s. The countries that consume the most are, of course, the United States and those in Europe, but emerging ones like China, India and Brazil, are fast catching up, at least in total consumption.

An increase in global agricultural production remains, despite everything, an imperative necessity. This goal has to be pursued agressively but only through an optimum use of current resources. Mankind will also need to limit waste and adopt lifestyles that consume less material and less energy. How can humanity preserve, adapt and mobilize all the know-how, technologies, cultures and lifestyles to *transform agriculture* in order to make this increase possible, while still limiting the impacts on ecosystems to acceptable levels? What useful knowledge will be required on the functional interactions between species regarding efficiency of water use, control of pests and diseases, soil conservation, fertilization, etc.? How to structure agricultural landscapes to promote interaction between species?

Thanks to photosynthesis, agriculture is one of the few human activities that produces renewable biomass. However, its intensification generates externalities that can be very burdensome. The choice of the path to intensification of agricultural production, the burden of fossil fuel dependency and the use of synthetic inputs determine these externalities to a large extent. An improved understanding of the functioning of ecosystems and interactions that will allow us to make the best use of biodiversity is necessary in order to increase production and, at the same time, preserve our planet for future generations.

Ecological intensification of agriculture can provide sustainable solutions to the issues of environmental impact and the finiteness of resources. But the path to follow becomes increasingly complex as our planet deteriorates. The erosion of natural biodiversity is accelerating day after inexorable day. The species extinction rate is 100–1,000 times greater than its average for the past hundreds of millions of years (MEA 2005). Between now and 2025, 10 % of flowering plants will be gone, and with them, a whole population of associated species and their services: pharmacopoeia (40–70 % of medicines are derived from natural substances, especially plants), fibres, genetic resources of cultivated species, auxiliary fauna and flora, fresh water, large biogcochemical cycles, crop values, etc.

Furthermore, we know that man needs a balanced food intake to maintain good health. Food should not only be sufficient in quantity, but must also be diversified.

The extraordinary variety of edible species, culinary know-how and nutrients in all their forms are the basis of diets around in the world. This variety is, in some ways, under threat by the homogenization and industrialization of production and with consumer preferences and diets undergoing profound changes. Nevertheless, at the level of a low-income family or a village, the diversity of agricultural production and food preparation know-how is also a treasure to be preserved.

We must act quickly and avoid mistakes; nature itself can serve as a guide. Ecology and agronomy researchers must build innovative methods and approaches in collaboration with farmers and local communities. This issue is inextricably linked to agricultural development in the countries of the South.

4.2 Overcoming the Finiteness of Resources

4.2.1 The Finiteness of Land

According to work carried out by IIASA (International Institute for Applied Systems Analysis) and the FAO, there are 2.9 billion hectares of arable land in the world, of which 90 % is located in sub-Saharan Africa and Latin America.

The *Agrimonde 1* scenario in the Agrimonde foresight (Paillard et al. 2010) attempts to minimize agriculture's externalities. This scenario suggests that is should be possible to increase the acreage of cultivated land by 25–40 % by 2050 with minimal impact on forests to meet these production requirements. The expansion of arable lands into new areas of high potential is possible mainly in sub-Saharan Africa and South America. Other parts of the world, such as regions of the former USSR, Asia, North Africa and the Middle East will be forced to cultivate land with a much lower potential, some of it even marginal. In some cases, recourse will have to be taken to remediation. But how to control the land rush, especially when it comes to industrial crops requiring large swathes of natural forests to be cleared and which are presented to local populations as important sources of income? (Box 6).

Box 6. Rubber in Laos, Thailand and Côte d'Ivoire

The rubber tree is a veritable natural factory. The latex that it produces has technological qualities not found in any chemical equivalent. At present, China's growing demand has led to an expansion of rubber plantations at the expense of natural forests and their biodiversity (Abel 2007). Thus, in Laos, where plantations today cover 14,000 ha, the authorities are planning for an additional 200,000 ha in 3 years, to be managed mainly by private Chinese firms. Twenty-seven Chinese companies own rubber plantations in Laos. They provide seedlings, technology and chemical fertilizers, train farmers, build refining factories and roads to China for transporting the production. In exchange, they have rights over 40–80 % of the crop for 30 years. Thus, in Bokeo province in northern Laos, not far from the Chinese border, a primary rainforest of great ecological wealth and one of the best preserved in the world was completely destroyed to make way for these plantations. Local farmers agree to manage these plantations because they are allowed to plant rice between the trees for 2 years.

Other approaches are nevertheless possible. In Thailand[*], rubber plantations were set up in the north–east of the country as part of the reforestation movement launched in 1990. This is an impoverished region, where decades of sugar cane and cassava cultivation have led to land degradation. The introduction of rubber is seen as an economic opportunity for farmers[**] as well as an ecological opportunity to maintain or even restore the soil's physical and chemical properties. Unfortunately, rainfall has proven insufficient for the requirements of growing rubber. Extensive studies on the effects of this reforestation on soil fertility, productivity and hydrogeology are underway, as are agroforestry trials.

The approach towards these large monoculture plantations needs to be further improved to orient them towards more diverse systems, such as agroforestry. In Côte d'Ivoire, a 17-year study compared the monoculture production of rubber with when it is associated with other tree crops (Snoeck et al. 2013). It shows that the combination of rubber with coffee or cocoa is quite comparable to that of monocultures, even more profitable in the medium term (10 years). And this without even counting the benefits of an improved use of cultivated land and a better distribution of labour seasons and incomes throughout the year. Furthermore, since the producer grows a wider range of products, he has a greater resilience against market fluctuations.

[*]See http://www.thailand.ird.fr/research-and-missions/research-projects/ecosystems-and-natural-resources/evaluation-of-agro-environmental-impacts-of-rubber-plantations (retrieved: 7 April 2013).

[**]Programme undertaken by IRD (France), Khon Kaen University (KKU), Mahidol University (MU), Land Development Department (LDD), Rubber Research Institute of Thailand (RRIT) (Thailand).

4.2.2 The Finiteness of Mineral Natural Resources

The complete depletion of phosphorus deposits, a large part of which is biogenic—i.e., resulting from a detrital accumulation of living organisms over geological eras—is estimated to take place between 2110 and 2350. This mineral fertilizer, essential for high crop yields, has no substitute. Similarly, the nitrogen supply to crops comes from the conversion of fossil fuels, primarily natural gas, which is, of course, not going to last forever. And yet, feeding the world's growing population will require large amounts of these inputs. What, if any, will be the alternatives found?

4.2.3 The Finiteness of the Water Resource

Even though called renewable, the planet's water resource is becoming increasingly less so. Water withdrawal from lakes and rivers, of which 70 % is used in agriculture worldwide, has doubled since 1960. Some areas are worse off than others, such as the Mahgreb and the Middle East, where non-renewable aquifers are today being exploited. In addition, there is the issue of the quality of water discharged from agriculture.

Fighting against desertification and implementing systems adapted to prolonged droughts is a major challenge in the Maghreb. We now begin to see systems being set up there for the collection and efficient use of water. The imperative challenge is to avoid the depletion of non-renewable fossil water. Another concern remains water potability. It will be necessary to develop ecosystems that can play a purifying or depolluting role or that are resistant to salinity.

4.2.4 The Finiteness of Energy Resources

Biomass is an important source of energy in developing countries in the form of firewood or charcoal. It is naturally abundant in the humid tropics, but its supply is now insufficient around major urban areas. Grown or recycled, it can contribute to population needs and even be a source of income under certain conditions through the emergence of new sectors.

4.3 Coping with New Hazards Caused by Global Changes, Especially Climatic Ones

In 2007, the Intergovernmental Panel on Climate Change (IPCC) estimated that, under certain conditions, agriculture could contribute significantly to sequester greenhouse gases, mainly through biological soil activity. The total stock of organic carbon in the soil is, in fact, at least double that in the atmosphere. There are large variations between ecological zones—the amount of carbon stock varies from about 4 kg/m^3 in arid areas to 8 kg/m^3 in the tropics to as high as 24 kg/m^3 in some polar regions (Batjes 1999)—but we know of agricultural practices that can increase these stocks. The amounts involved can be phenomenally large: a very tiny change in the stock contained in the first 30 cm of soil could either cancel the terrestrial carbon sink or allow it to absorb the annual increases (Bernoux 2011). However, this contribution towards mitigating climate change can be fully effective only when practices that respect soil life are adopted.

Moreover, only chlorophyll production is capable of capturing atmospheric carbon and transforming the inexhaustible energy of the sun into usable biomass.

Biomass production in large quantities for various uses (food, energy, materials, soil fertility, environmental services) should therefore be explored from all angles.

The biodiversity of agroecosystems has a proven impact on their resilience to hazards related to climate change: fight against soil erosion and loss of soil fertility, balance of auxiliary flora and fauna, large biogeochemical cycles, resources for responsiveness to shocks, etc.

Biodiversity should also be explored to develop innovative techniques to counter environmental hazards that remain a constant threat to it: biological invasions, pollution, etc.

Some cultivated *varieties or species* from one climatic region can meet the future needs of another region (drought, rainfall, seasons, etc.). Some wild species can be domesticated. Here, too, the adaptability of producers and populations remains the driver for innovation.

5 What is the Best Way of Understanding the Extraordinary Complexity of Living Organisms and Agroecosystems?

Different paths can be taken to design and evaluate the effects of various approaches to ecological intensification. They must be compared at different spatial scales (in particular the plot, the farm and the landscape) and over various temporal scales.

The *study of functional relationships* within a particular compartment of diversity is very important. It allows the analyses of nutrient cycling, nitrogen conversion, trophic antagonism between species, the chemistry and biochemistry regulating populations and processes, soil structure, and interactions between auxiliaries and pathogens or pests. Some aspects of these relationships have been studied since decades, but others are only now beginning to be documented. Thus, for example:

- *Evolutionary genetics* has been studying, since the early twentieth century, *gene flows between populations of the same species* in time and space in relation to history human (history of cultivated species, origins, domestication, diversification). It accompanies the genetic improvement of cultivated species and their pathogens, with the help of disciplines such as ethnobotany as well as anthropology and, nowadays, linguistics.
- *Functional ecology* has been dealing, since the 1960s, with the functions of organisms and ecosystems in interaction with their environment. It studies, for example, *relationships that connect individuals from a mixture of different species* in a given environment (functional groups of species), with respect to different modes of farming. This branch of ecology has proven especially useful for studying the dynamics of natural forests and grasslands. However, the *functions of soil organisms* are still poorly understood, and the domain of crop mixtures is rarely addressed: nutrient cycling, nitrogen conversion, chemistry

and biochemistry regulating populations and processes, interactions (mutualism, commensalism, competition, pathogenesis).

- *Ecophysiology* addresses the behavioural and physiological responses of organisms to their environment (temperature, altitude, oxygen, food availability, etc.). This discipline also covers *matter and energy flows* between the different compartments of a plot, ranging from the bedrock to the atmosphere and to the climate through plant and animal populations (plantation, grassland, annual crop, agroforestry, natural cover).

Even more integrative scientific approaches have been developed:

- *Agroecology* was born in the 1990s from the convergence of agronomy and scientific ecology. It is considered an approach that combines agricultural development, participatory methods and protection or regeneration of the natural environment. Agroecology is the basis of a multifunctional and sustainable agriculture, which valorises agroecosystems, optimizes production and minimizes input use.
- Various alternative forms of agriculture have been explored, whose impact on the increase of biodiversity and of production can be evaluated only retrospectively: organic farming, high environmental value (HEV) agriculture, conservation agriculture (François et al 2011), eco-agriculture, etc.
- The *study of landscapes*, especially of their structuring between cultivated areas and 'semi-natural' protected areas, is a new area of research. Its goal is to understand what forms of landscape structuring are the most suitable for agrobiodiversity.
- *Associations with civil society and its informed amateurs* (Demeulenaere and Goulet 2012) help human communities share their observations on biodiversity and their know-how for evaluating it, understanding its functions, and managing and restoring it. An example of one such such association is participatory botany,[5] which mobilizes citizens in making observations in time and space. These collected data are then integrated into searchable Internet databases. Other examples include seed exchange networks, such as the Farmers' Seed Network in France[6] (Box 7).

Strategies for agronomy, integrated pest management, improvement of varieties or varietal mixtures, and agroecosystem management can all benefit from the knowledge acquired and methods developed in all these disciplines. But these strategies can be deployed only in processes of innovation that, above all, involve rural communities.

---

[5] Tela Botanica network, http://www.tela-botanica.org/site:accueil?langue=en/, Pl@ntnet initiative, http://www.plantnet-project.org/papyrus.php?langue=en/ (retrieved: 7 April 2013).

[6] Réseau des semences paysannes (Farmers' Seed Network), http://www.semencespaysannes.org/ (retrieved: 7 April 2013).

Box 7. Pl@ntNet and Tela Botanica: tools for collaborative research

Bringing together botanical specialists and amateur enthusiasts is one of the objectives of Pl@ntNet[*], a collaborative network of more than 300 people organized around a software platform. The idea behind Pl@ntNet is simple: to assist observers in identifying plants they find in the field, to share these observations using simple tools and to allow managers and scientists to valorise these observations through their studies. For example, identifying tree species in the flora of metropolitan France; estimating the distribution of tropical plants from heterogeneous occurrence data; gaining a better understanding of the different grape varieties of French vines; identifying and monitoring plants that have invaded natural habitats or weeds in crops; and a better understanding of the endemic flora of Reunion.

These studies have benefitted from 12 years of experience in managing citizen science projects of the Tela Botanica collaborative network. Tela Botanica now has more than 18,000 members worldwide, including 15,500 in France and 1,150 in the Maghreb. This network provides access to more than 200,000 field observations concerning around 6,700 plant species!

A section of the Pl@ntNet platform relies on user-contributors to develop collaborative software for data management and sharing, and to evaluate its features. Thus:

- Pl@ntNet-Identify is a visual search engine which compares photos submitted as a query to a set of stored and identified images.
- Pl@ntNet-Datamanager can manage a wide variety of botanical data, on a fully configurable system, in an individual or collective basis, online or offline.
- IDAO allows users to make a 'composite picture' of a plant by using a fully graphical interface, thus overcoming the constraints of language and specialized vocabulary. Applications exist for different flora from around the world (West Africa, Reunion, India, Southeast Asia, etc.).
- the online Carnet (notebook) allows everyone to enter and manage his or her field observations on an online system, to illustrate them with images and share them with the community.

Communities can be created around common projects through Pl@ntnet. It is thus a powerful tool for promoting citizen science, and a useful vector for accumulating new data on plant biodiversity.

[*]The Pl@ntNet project (2009–2013) is an initiative that brings together the JRU AMAP, Botany and bioinformatics of the architecture of plants (Cirad-Cnrs-Inra-IRD-UM2), Inria (Imedia team) and the Tela Botanica network. It is funded by the Agropolis Foundation.

6 Agrobiodiversity: A Development Issue?

With these few points of reference, we have tried to show the importance that agrobiodiversity had in the history of agriculture and on economic development. The history of tropical plants reflects the issues, power relationships, colonization and violent conflicts that have concerned the great powers. International trade in major crop species is still a very important economic issue for countries that produce them. This is often the reason why swathes of rainforest are still being cleared for plantations and highly profitable crops, such as rubber and oil palm. Aware of these problems, companies that manage large plantations are today conceiving and implementing best practices, certifications, and sometimes investing in the conservation of ecosystems. Nevertheless, the expansion of cultivated areas, depleted in biodiversity, seems unavoidable.

Since the Rio Summit, the right of access to genetic resources, formerly considered by Westerners as a public good, has changed. The role of small farmers in the South in the maintenance and diversification of traditional varieties has been recognized. At the same time, advances in biotechnology and massive private investments have led to the recognition by the CBD of the patentability of living organisms. These two views embody the current confrontation between private and public interests (Bonneuil and Fenzi 2011).

Family farming is considered an 'antithesis' to agricultural industrialization by the many ongoing experiments and by the constant cross-pollination between science and traditional knowledge in ever-evolving contexts. Thus, agroecological practices, supported by research, use beneficial biological synergies between the various components of a given agroecosystem: on-site recycling of nutrients and energy, integration of crops and livestock farming, and diversification and association of species and genetic resources in space and time. Emphasis is placed on interactions and productivity at the scale of the entire agricultural system. Biodiversity provides an opportunity to small producers to adjust and optimize their material and resources and even to take advantage of marginal and difficult lands (Altieri et al. 2011).

There are many different routes to sustainable agricultural intensification. Not only do they depend on farmer expertise and capacity for innovation but also on the institutional and policy environment. Based on assessments of past activities, agronomic and economic studies have shown that production yields of diversified systems can exceed those of conventional intensive monocultures, especially in regions where malnutrition is rampant. Some studies have shown that peasant systems are the most effective in terms of workdays or of energy balance (energy supplied/energy extracted) (Altieri et al. 2011). Thus, a study of the results of 286 sustainable agriculture projects in 57 poor countries reveals an increase in production of 80 %, with African projects having an even higher average of 116 % (Pretty et al. 2006). Recent projects have led to the doubling of harvests over a period of three to 10 years in over twenty African countries. But this intensification must also be evaluated through criteria other than solely of production.

Producer incomes, dependence on technology or synthetic inputs, risk management and resilience are all essential criteria in a context of increasing uncertainty.

Research sometimes lags these innovations. Producer organizations, NGOs, governments and production and consumption networks are playing an increasingly important role in the creation and dissemination of knowledge, know-how and innovations. In West Africa, producer organizations do not hesitate to query experts in this field and information flows freely through meetings, radio, telephone, and farmer field schools (FFS). In Burkina Faso, youth groups specialized in the traditional methods of land reclamation move from village to village to help farmers, some of whom go so far as to buy up degraded land to be able to farm it again (Pretty et al. 2011).

7 Conclusion

These examples show the extraordinary diversity of agroecosystems. Even if we can discern some major trends, we cannot predict their effects on specific local contexts or how these systems will adapt. They also show to what extent diversity is an unavoidable issue for food security at the global level.

As an engine of all the mechanisms at work in cultivated ecosystems, biodiversity is a key resource available to farmers in developing countries to improve their production and increase their incomes. There are choices to be made, ones that depend on the diversity of agriculture and societies. They are implemented at the level of the plot, but act at the scale of landscapes, markets and policy incentives.

Current global changes also show societies in developed companies that they are dependent on the future of countries in the South and on these countries' ability to manage their natural wealth. The expansion of mankind into the landscape is accelerating as is the erosion of our collective resources. Biodiversity will have to be cultivated in an ever increasing measure in order to intensify and transform agriculture systems and enable them to meet the challenge of feeding humanity and fulfilling its needs.

References

Abel, S. (2007). Le Laos soumis à la dictature de l'hévéa chinois. *Libération,* 22 May 2007. Retrieved April 5, 2013 from http://www.liberation.fr/economie/0101102825-le-laos-soumis-a-la-dictature-de-l-hevea-chinois

Altieri, M. A., Funes-Monzote, F. R., & Petersen, P. (2011). Agroecologically efficient agricultural systems for smallholder farmers: contribution to food sovereignty. *Agronomy for a Sustainable Development, 32*(1), 1–13.

Atlan, H. (1999). *La fin du « tout génétique » ? Vers de nouveaux paradigmes en biologie* (91 p), Inra, Sciences en question, Éditions Quae

Atlan, H. (2011). *Le vivant post-génomique. Ou qu'est-ce que l'auto-organisation?* Paris: Odile Jacob.

Batjes, N. (1999). *Management options for reducing CO_2 concentration in the atmosphere by increasing carbon sequestration in the soil.* NRP Report no. 410 200 031.

Bernoux, M. (2011). Le stockage de carbone dans les sols: quels processus ? Comment le mesurer ? *Séminaire « Sols et politiques publiques »* , 20 October 2011, Lyon. Retrieved April 5, 2013 from http://www.gessol.fr/content/sol-et-politiques-publiques

Bonneuil, C., Fenzi, M. (2011). Des ressources génétiques à la biodiversité cultivée. La carrière d'un problème public mondial. *SAC Revue d'anthropologie des connaissances,* 5(2), 206–233.

CBD (Convention on Biological Diversity) (1992). United nations (p. 3). Retrieved April 1, 2013 from www.cbd.int/doc/legal/cbd-en.pdf

Cohen, I.R., Atlan, H., Efroni, S. (2009). Genetics as explanation: limits to the human genome project. *Encyclopedia of Life Sciences* [online], December 2009. doi:10.1002/9780470015902.a0005881.pub2.

Commission on Genetic Resources for Food and Agriculture (2010). *The second report on the state of the world's plant genetic resources for food and agriculture.* Rome: FAO. Retrieved April 5, 2013 from http://www.fao.org/docrep/013/i1500e/i1500e00.htm

Conference of the Parties (1996). *COP3 Decision 3/11: Conservation and sustainable use of agricultural biological diversity,* 4–15 November, Buenos Aires, Argentina. Retrieved April 5, 2013, from http://www.cbd.int/decision/cop/?id=7107

Conference of the Parties (2000). *COP5 Decision V/5: Agricultural biological diversity: Review of phase I of the program of work and adoption of a multi-year work program,* 15–26 May, Nairobi, Kenya. Retrieved April 5, 2013, from http://www.cbd.int/decision/cop/?id=7147

Conklin, H. C. (1957). *Hanunóo agriculture: A report on an integral system of shifting cultivation in the philippine.* FAO: Rome.

De Schutter, O. (2010). *Promotion and protection of all human rights, civil, political, economic, social and cultural rights, including the right to development.* Report submitted to the Human Rights Council of the United Nations, 16th session, 17 December 2010 (special rapporteur on the right to food).

Demeulenaere, E., Goulet, F. (2012). Du singulier au collectif. Agriculteurs et objets de la nature dans les réseaux d'agricultures alternatives. ENS Cachan. *Terrains et travaux,* 1(20), 121–138. Retrieved April 5, 2013, from http://www.cairn.info/revue-terrains-et-travaux-2012-1-page-121.htm

Ecological Footprint Atlas (2009). *Global footprint network, research and standards department.* Retrieved April 5, 2013, from http://www.footprintnetwork.org/

FAO (2008a). *New light on a hidden treasure. International year of the potato 2008.* End-of-year report, 148 p. (p. 14).

FAO (2008b). *The state of the world's animal genetic resources for food and agriculture.* Commission on Genetic Resources for Food and Agriculture. Retrieved April 5, 2013, from http://www.fao.org/docrep/010/a1260e/a1260e00.htm

FAO (2010). *Biodiversity.* Biodiversity for a world without hunger. Retrieved April 5, 2013, from http://www.fao.org/biodiversity/components/animals/en/

FAO (Food and Agriculture Organization of the United Nations) (1996). *Lessons from the green revolution: Towards a new green revolution.* Technical background document. World Food Summit, 13–17 November, Rome, Italy.

Feldmann, P. (2008a). Interactions between human activities and biodiversity in the heart of overseas sustainable development: Stakes for research in managed ecosystems. In: *L'Union européenne et l'outre-mer. Stratégies face au changement climatique et à l'érosion de la biodiversité,* IUCN/région Réunion/ONERC/État français, la Réunion.

Feldmann, P. (2008b). Biodiversité et agriculture: Services écologiques et impacts des changements globaux. In: *Cycle de conférences 2008. Relever le défi de la biodiversité: l'agriculture durable* (Ifore, éd.), Ifore-MNHN, Paris, France.

Feldmann, P., Côte, F., Fernandes, P., Jannoyer, M., Langlais, C. (2007). Biodiversité et agriculture aux Antilles. *Antilles Agriculture.*

François, J.-L., Tissier, J., Legoupil, J.-C., Maraux, F. (2011). *Agriculture de conservation et intensification écologique des exploitations familiales tropicales. Quel partenariat entre recherche et développement?* (4 p). Cirad-AFD, September 2011.

Franklin, J., & Mortensen, D. A. (2011). A comparison of land-sharing and land-sparing strategies for plant richness conservation in agricultural landscapes. *Ecological Applications, 22*(2), 459–471.

Griffon, M. (2006). *Nourrir la planète: pour une révolution doublement verte* (456 p). Odile Jacob.

Griffon, M. (2011). *Pour des agricultures écologiquement intensives* (144 p). L'Aube, Poche Essai.

Heams, T. (2012). Mettons du désordre dans nos idées. *Le Monde, tribune Science et Techno* (p. 8), 22 September.

Ipcc (2007). *Climate change: Impacts, adaptation and vulnerability.* Contribution of working group II. IPCC Fourth Assessment Report, Cambridge University Press, Chapter 9.

James, C. (2011). *Brief 43: Global status of commercialized biotech/GM crops: 2011.* ISAAA Brief no. 43. Ithaca, NY: ISAAA. Retrieved April 5, 2013, from http://www.isaaa.org/resources/publications/briefs/43/

Le Roux, X., Barbault, R., Baudry, J., Burel, F., Doussan, I., Garnier, E., Herzog, F., Lavorel, S., Lifran, R., Roger-Estrade, J., Sarthou, J.-P., Trommetter, M. (2008). *Agriculture et biodiversité. Valoriser les synergies* (178 p). Expertise scientifique collective Inra, Editions Quae.

Mazoyer, M., Roudart, L. (2002). *Histoire des agricultures du monde: du néolithique à la crise contemporaine* (705 p), Seuil.

MEA (Millenium Ecosystems Assessment) (2005). *Ecosystems and human well-being: Biodiversity synthesis,* MA. Retrieved April 5, 2013, from http://www.millenniumassessment.org/documents/document.354.aspx.pdf; see also http://www.maweb.org

Mittermeier, R. A., Goettsch Mittermeier, C. (2005). *Megadiversity: Earth's biologically wealthiest nations* (501 p), Cemex.

Paillard, S., Treyer, S., Dorin, B. (2010). *Agrimonde. Scénarios et défis pour nourrir le monde en 2050* (296 p), Editions Quae.

Phalan, B., Onial, M., Balmford, A., & Green, R. E. (2011). Reconciling food production and biodiversity conservation: land sharing and land sparing compared. *Science, 333*(6047), 1289–1291. doi:10.1126/science.1208742.

Pretty, J. N., Noble, A. D., Bossio, D., Dixon, J., Hine, R. E., Penning de Vries, F. W. T., Morison, J. I. L. (2006). Resource-conserving agriculture increases yields in developing countries. *Environmental Science and Technology, 40*(4), 1114–1119. doi:10.1021/es051670d.

Pretty, J. N., Toulmin, C., & Williams, S. (2011). Sustainable intensification in African agriculture. *International Journal of Agricultural Sustainability, 9*(1), 5–24.

Ruault, M., Dubarry, M., & Taddei, A. (2008). Re-positioning genes to the nuclear envelope in mammalian cells: Impact on transcription. *Trends in Genetics, 24*, 574–581.

Sastre, C., Breuil, A., Bernard, J.-F., Feldmann, P., Fournet, J. (2007). Les causes de régression de la flore. In: *Plantes, milieux et paysages des Antilles françaises: écologie, biologie, identification, protection et usages* (pp. 615–620). Mèze: Biotope.

Snoeck, D., Lacote, R., Kéli, J., Doumbia, A., Chapuset, T., Jagoret, P., Gohet, E. (2013). Association of hevea with other tree crops can be more profitable than hevea monocrop during first 12 years. *Industrial Crops and Products, 43*, 578–586. Retrieved April 5, 2013, from http://www.sciencedirect.com/science/article/pii/S0926669012004311

Volper, S. (2011). *Du cacao à la vanille, une histoire des plantes coloniales* (144 p). Éditions Quae.

WCMC (World Conservation Monitoring Centre). (1992). *Global Biodiversity Assessment.* United Nations: Chapman and Hall.

World Bank (2008). *World development report 2008: Agriculture for development* (30 p). Abridged.

Chapter 3
From Artificialization to the Ecologization of Cropping Systems

Florent Maraux, Éric Malézieux and Christian Gary

In 2050, agriculture will have to feed 9 billion people on Earth, 2 billion more than it does now. Climate change is and will be unconducive to increases in productivity in most parts of the world that are even now deficient in production. This deficiency is the direct result of increasing urbanization and the scarcities of agricultural essentials: arable land, water resources, energy, phosphorus, and basic mineral fertilizers (Kristjanson et al. 2009; Godfray et al. 2010; Brussaard et al. 2010). Cultivated ecosystems are expected to meet this goal of fulfilling food and non-food needs by reversing or overcoming these shortages. In addition, they are also expected to provide a range of ecosystem services. These services benefit agriculture itself and determine its sustainability, in terms of the medium (maintenance of the physical and biological components of soil fertility, recycling of water and nutrients) and regulation (balance between communities of pests and auxiliaries, pollination). Cultivated ecosystems also benefit other sectors of society: they conserve water resources and maintain its quality, they increase biodiversity and improve the quality of life, and they help overcome and/or mitigate the effects of climate change.

F. Maraux (✉)
Director of the SCA Internal Research Unit (Annual Cropping Systems),
Cirad Department Persyst, Montpellier, France
e-mail: florent.maraux@cirad.fr

É. Malézieux
Director of the Hortsys Internal Research Unit (Agroecological Functioning and
Performances of Horticultural Cropping Systems), Cirad Department Persyst,
Montpellier, France
e-mail: eric.malezieux@cirad.fr

C. Gary
Director of the System Joint Research Unit (Tropical and Mediterranean
Cropping System Functioning and Management), Inra Department
Environment and Agronomy, Montpellier, France
e-mail: christian.gary@supagro.inra.fr

É. Hainzelin (ed.), *Cultivating Biodiversity to Transform Agriculture*,
DOI: 10.1007/978-94-007-7984-6_3, © Éditions Quæ 2013

1 The Impasses in the Artificialization of Cropping Systems

Agriculture's evolutionary stages are now well known and can be characterized by a few markers: physical productivity, biological efficiency, division of labour, labour productivity, market access and integration, and recognition of other services provided by agriculture. Certain environmental indicators, such as carbon footprint, pollution or a posteriori analyses of life cycles, can also be used to characterize farming systems (van der Werf et al. 2011).

These changes have been studied less and delineated with less clarity in the tropics. Their important milestones are often shifted in time due to the differences in the dynamics of civilizations, nature of and access to resources, and social organization. Mazoyer and Roudart (1997) offer an overview of these developments with accompanying technical, social and environmental factors. They provide a series of snapshots as well as a view of the changing patterns of agriculture from around the world. The caloric yield per hectare per day (from the physical point of view) and the number of hectares per human labour unit illustrate the improvements in performance of man's various agricultural systems (Paillard et al. 2010). These improvements take place as technical expertise is gained and through social organization. Rising levels of productivity allow increases in populations and settlements, and lead to the division of labour and its specialization. An example is the water management systems in the lower Nile Valley.

Each successive innovation in the Nile valley's water supply and drainage system solved an existing problem in the system. The water provided by the flooding of the Nile was used to irrigate a given area, whose extent depended on the topography, canals, and also methods (other than gravity) to bring water to wider areas. At the same time, a sometimes vicious cycle was created in which the developments and innovations implemented reduced the system's efficiency, and therefore required new innovations. In the case of the Nile system, these undesirable effects took the form of riverbed deepening, scouring of tributaries and canals, concomitant silting up, etc.

The authors thus explain the history of agriculture as a succession of crises and resolutions of crises which are accompanied by, caused by or provide solutions to demographic growth (Mazoyer and Roudart 1997; Griffon 2006).

Agriculture in temperate countries is better documented and formal theorizing becomes easier, as illustrated by the notion of 'agricultural revolutions'. We can note that the productivity of agricultural systems has gone through three phases since the Middle Ages:

- Phase I: a very gradual increase in productivity of agricultural systems from the early Middle Ages to the beginning of the nineteenth century.
- Phase II: a rapid and continuous acceleration of productivity leading up to the Second World War.

- Phase III: a sudden acceleration of productivity (example of wheat in France: a trend of + 0.1 T.year^{-1} since the war[1]).

The system for maintaining soil fertility by biennial or triennial fallow does help reconstruct and maintain fertility, but with an input/output ratio that throttles and limits the system's performance.

During Phase I (Middle Ages), the technology changed little. The biennial or triennial system prevailed and production depended mainly on labour availability (variable over time due to migrations, invasions, epidemics, wars). Productivity increased primarily through gradual genetic progress resulting from the selections made locally by farmers (Bloch 1976; Feyt 2007). A fundamental change also took place between the sixteenth and eighteenth centuries with the introduction of forage crops (Mazoyer and Roudart 1997). This development laid the grounds for the first agricultural revolution.[2]

During Phase II—the first agricultural revolution—the cultivation on fallow land of forage crops including legumes led to both an increase in land under cultivation and an increase in its fertility. There was a big jump in cereal production in France coupled with a marked reduction in surface areas dedicated to cereal crops. The number of animals increased (in absolute terms and by farm acreage) and output too (quality and quantity). The massive injection of nitrogen into the systems by legumes and the introduction of animal manure shifted the thresholds and levels of fertility, with consequent beneficial effects on physical productivity and production.

The experts all agree that during this phase the major environmental quantities (carbon, minerals) remained in balanced and kept pace with demographic changes of men as well as of cattle (Mazoyer and Roudart 1997). Yet we note that the implementation of large projects profoundly changed the rules as far as the water resource was concerned. Irrigated areas around the world began representing an ever-increasing proportion of global production, especially in Asia (Molle and Maraux 2008). So the conclusions on 'environmentally harmonious' development and the accompanying changes in productivity, population and fertility have to be put into a realistic perspective.

During Phase III, characterized by the rapid artificialization of agricultural production (mechanization, massive recourse to chemistry), the trends steepened. Thresholds moved upwards and there seemed to be no upper limit to increases in productivity. In a book written as far back as 1984 (Gay 1984), even before the

---

[1] However in recent years, the curve is inflecting and the trend of higher yields is stalling (Brisson et al. 2010).

[2] The first agricultural revolution of modern times should not overshadow the intermediate events. These included the introduction of animal traction in its various forms and mastery over water supply, which led to leaps in productivity and simultaneous social reorganizations in some areas.

upheavals caused by modern genetics, a very serious AGPM[3] predicted an amazing performance from maize (a yield of over 20 T.ha^{-1}). But by 2011, a yield of 30 tonnes had already been achieved (Crovetto Lamarca 2007–2008). According to the proponents of these methods of production, such performances are the result of genetic progress, advances in crop protection, tillage (or sometimes non-tillage) of soil and chemical fertilizer. They claim that the latter can fulfil (and even exceed) the system's requirements and compensate its exports.

In the South, the general principles remained the same, but the dynamics and milestones were different. Thus, access to water, mastery of its use and the implementation of irrigation were determining factors in arid regions (especially the Middle East, usually believed to be the most likely cradle of civilization) and mountainous areas (mainly in Asia).

Another difference between the North and the South is that the decisive phase consisting of the forage revolution, farming of fallow land, and animal traction—amply described in the evolution of agriculture in the North—seems to have been less prominent and widespread in the South. Productivity increases in the South did not necessarily pass through this stage. The second agricultural revolution (called 'Phase III' above) was exported directly after the Second World War, often as-is, without any attempt at customizing it for the destination. It abruptly displaced existing systems without any preliminary lead-in phases which would have provided some continuity with existing sustainable production models.

The second agricultural revolution—and the artificialization that characterized it—therefore made agriculture production methods converge into a single model (Doré and Maraux 2010). This happened both in the North as well as in the South, often through very different routes. The features of this agricultural production model are (Doré and Maraux 2010):

- A monospecific plant crop is planted and then managed with the help of tillage and cultivation practices.
- Mineral exports that accompany the harvested products are compensated for by massive use of industrially produced fertilizers.
- Bio-aggressors (weeds, insects, diseases) are controlled by the use of industrially manufactured chemical products.

> **Box 1. The Doubly Green revolution**
>
> To go from feeding 3 billion people in the 1950s to feeding 6 billion at the turn of the twenty first century, agriculture had to follow the route of the Green Revolution to transform itself radically, especially in the tropics. To accommodate the additional 3 billion people who will arrive in the next 50 years, the same methods cannot be used as they are neither ecologically

---

[3] French General Association of Maize Producers (*Association générale des producteurs de maïs*).

sound nor economically viable. We have to invent a new Green Revolution in agriculture that pollutes the Earth to the least possible extent and yet helps reduce poverty. It will thus have to be 'Doubly Green' because it will have to combine agricultural productivity with ecological, economic and social sustainability.

Agricultural research will probably never have to do as much—or acquire as much importance—as in the next 50 years. Until the early twentieth century, it was mainly the learning that accumulated over long centuries which allowed human societies to exploit ecosystems and natural resources to feed 3 billion people. During the last 50 years, the population has grown by an additional 3 billion and research has had to come up with a packaged technological solution to cope with the new requirements. This solution—it acquired the name of the Green Revolution in 1970s—was very successful but is now outdated. Its limits have been reached and it harbours dangers. So the model has to be changed in order to accommodate an additional population of 3 billion during the next 50 years. It is this alternative model, the Doubly Green Revolution, that we explore here.

The reasoning behind it is based on two requirements:

- Population growth leads to an ever increasing exploitation of the biosphere, the thin spherical shell of life that includes all living things and their terrestrial and aquatic environments. Mankind extracts from it food, energy, industrial and handicraft products for clothing, building, health and all the other human activities. However, in doing so mankind is abusing the biosphere in very many places on the planet and thus causing major environment crises. The invention of new technologies thus becomes a necessity. These technologies have to ensure the sustainability of the biosphere's ecosystems while helping meet increasing human needs in a context of limited productive space.
- Poverty can only be reduced gradually, as history has shown. It is hereditary and is often passed on for several generations. Three-quarters of the poor live in rural areas. Therefore, the future of the biosphere depends largely on the behaviour of farmers, a section of mankind with the least amount of capital. They often cannot avoid using production techniques that use natural resources without renewing them (soil fertility, firewood, wildlife, etc.). In the process, they have a negative impact on the environment. It is therefore imperative that new technologies also form part of models designed specifically to reduce rural poverty.

Derived from Griffon (2002).

For further information: Conway Conway (1997), Conway et al. (1994), Griffon (1995).

Given this situation, it is necessary to examine the sustainability of these systems of agricultural production. Many authors who have carried out just such an examination conclude their analyses with a call for a radical transformation of agriculture. Examples include Conway and Griffon who advocate the 'Doubly Green' revolution (Box 1). In this chapter, we will only address environmental sustainability, since we know that these modes of production also had an enormous impact on mankind. By making far-reaching changes in farming systems and greatly increasing labour productivity, they encouraged large-scale migration to the cities. The more fortunate migrants found employment which led them to new cycles of economic growth. On some continents, the less fortunate, with no real job prospects, ended up merely bloating megacity populations. The question of sustainability brings to the fore the interlinked scales of space and time, and natural processes and those introduced or accelerated by artificialization. There exists no single solution.

Before addressing this issue, we cast an eye on some details on sustainability by viewing it through fertility and biotic perspectives.

1.1 Managing Fertility

In the North, it is believed that the first agricultural revolution took place under the auspices of a general rise in fertility. Manure recycling and atmospheric nitrogen fixation by forage crops increased chemical fertility (bioavailability of minerals). In addition, the increase at the same time of the forage area/agricultural area ratio led the system into a virtuous cycle in terms of organic matter stored in soils. Indeed, grasslands—both permanent as well as temporary—have the ability to store organic matter at an average rate of 0.5 tonnes per hectare per year, though this rate decreases over time and with a rise in the absolute level of storage (Gastal and Lemaire 2002). There is therefore no doubt that the agricultural systems of the first agricultural revolution had a tendency to maintain or improve fertility in countries of the North.

It is difficult to review the wide gamut of agricultural systems in countries of the South. Let us limit ourselves for the time being to fallow-based systems (Floret and Pontanier 2000). In tropical Africa, a traditional system of land use starts with a crop phase which is 5–15 years long. It is followed by abandonment of cropping (fallow) as soon as yields decline, a drop in fertility is observed or an invasion of weeds or pests takes place. The fallow phase which follows the crop phase is ten to 30 years long depending on the climate. As the land returns to a bushy or tree savannah, fertility recovers. This crop-fallow system worked well until recently. At present, however, population growth and the trend of settling down have led to a significant increase in cultivated acreages and to a proportionate reduction in fallow surfaces.

Increasing population density leads to a reduction in fallow, even to its complete disappearance. Fertility diminishes and soil organic matter declines. This latter is an essential component of fertility in these environments. Organic resources normally used to restore organic fertility (crop residues, sewage powder and manure) are often not available due to strong competition for these resources.

Box 2. The Sahelian fallow systems

Francis Ganry

Current forecasts of the loss of soil fertility capital in Sudano-Sahelian Africa, except in cotton-growing areas, are alarming (Ganry et al. 2011). In the cotton zones, we are witnessing the development of animal traction, and the consequent use of manure. However, fodder is lacking and requires the development of *improved fallow* and its selective exploitation by animal grazing (Sanogo 1997).

The only realistic way forward is organic intensification based on integrated management. This involves the phosphating of land (phosphate is the main soil deficiency of the Sudano-Sahelian region), livestock farming (fodder requirements mentioned above), growing trees (e.g., parklands), development of lowlands (e.g., by *Sesbania* in waterlogged soils), and suitable plot management (e.g., by using ridges).

Trees are part of the agricultural landscape in Sudano-Sahelian Africa with their many uses, their essential products (human food and animal feed, medicinal products) and their beneficial effects on the microclimate and soil fertility. In the Sudano-Sahelian and North-Guinean regions, parklands consist of more or less spontaneously selected species (*Faidherbia aldiba, Vitellaria paradoxa, Lannea microcarpa, Cordyla pinnata, Parkia biglobosa, Ficus* sp.).

In Senegal, the rate of soil organic matter (SOM) and the total nitrogen under *F. albida* can reach as high as 1.5 % and 0.08 % respectively and, under cultivation, come down to as low as 0.5 % and 0.03 % respectively (Oliver et al. in Peltier 1996). In Burkina Faso, Depommier (Peltier 1996) shows that a tree improves soil fertility up to the periphery of its crown. It increases levels of SOM, phosphorus, potassium, calcium and magnesium by 50 %. This effect increases with the size of the tree and is less pronounced in the most fertile soils, but on all sites, a positive effect is observed on cattle under the tree during the dry season.

Shea (*Vitellaria paradoxa*) tree parks are common in all parts of the Sudanese savannah. They are also found in the savannahs of northern Guinea, east of the Senegal River up to Uganda. Fertilization with tree litter is significant, compared to other fertilizer inputs in the low-intensity farming system. In 2002, at the bottom of the slope, with a density of 24 trees.ha^{-1}, the trees restored through their leaves 35 kg.ha^{-1} of CaO, 8 kg.ha^{-1} of MgO, 4.5 kg.ha^{-1} of K$_2$O, 9 kg.ha^{-1} of N and 1.2 kg.ha^{-1} of P$_2$O$_5$. These restorations originate from deep-horizon minerals which are inaccessible to annual crops. Under the trees, MOS levels are significantly increased (Albrecht and Kandji 2003). Using isotopic techniques, it has been shown that under the crown, 70 % of C of the MOS of the 0–20 cm horizon comes from the tree. This proportion gradually decreases to 40 % at a distance of 2.5 times the radius of the crown (Traoré 2003).

In addition, these trees have food (source of fat), cosmetics and pharmaceutical functions. Shea parks are a type of vegetation characteristic of Mali and Burkina Faso.

Watershed erosion control through the use of ridges[*] has a positive impact on crop production (maintenance and/or increase of crop yields) as well as on the maintenance and regeneration of trees (emergence of seedlings and their growth, and nutrition of existing trees) (Traoré 2003).

[*]An earth bank along the contour, made permanent using earthen ridges, quickly covered with vegetation, either natural or planted (*Andropogon*, etc.).

It has therefore become necessary to put in place appropriate management of natural fallow or substitution methods or to rely on compensation by chemical fertilizer with concomitant and unpredictable long-term effects. Recourse to fallow, especially tree fallow (such as shea parks, see Box 2), is a route towards potentially balanced fertility, which, however, is easily disbalanced (by shortening of the cycles). But, even when used optimally, fallow presents a productivity ceiling. We will return to the topic later.

In the savannahs of West Africa, and in cotton growing regions in particular, the shortening or abandonment of fallow and the sedentation of agriculture in general clearly marked the abrupt arrival of the second agricultural revolution (Dufumier 2004). Long-term observations of these systems (Box 3), combined with modelling (Kintche et al. 2011), show that we can maintain a virtuous cycle of organic matter and mineral fertility. To do so, we have to put appropriate cropping systems in place, in which we apply a dose of manure and mineral fertilizer to offset the mineralization of organic matter and mineral exports that result from harvests. We have to realize though that this system is finely balanced and should we be unable to maintain these conditions, the soil will start degrading. With the implementation of structural adjustment policies over the last 30 years—especially accelerated over the past decade—rising fertilizer prices have led to lower consumption and a subsequent lowered contribution to soil fertility. The precarious balance maintained by the monospecific intensive model has been broken and the soil has suffered as a result.

Another anachronistic situation that requires sustainable management of soil fertility arises due to the specialization of production (crops or livestock) in the large agricultural plains (North America, Brazil, Australia). While the first agricultural revolution had generated a virtuous cycle of fertility by combining cropping and livestock activities, the second has created a paradox, an impasse even. In large livestock farming areas, animal waste is considered a nuisance while, at the same time, the level of organic matter content in agricultural soils inexorably declines in cereal growing areas.

Box 3. The model for linking carbon, and stock and bioavailability of minerals

Hervé Guibert and Michel Crétenet

Long-term trials comparing cropping systems in the cotton areas of sub-Saharan Africa clearly show a close relationship between the performance of crops in rotation and soil carbon content (SCC) levels. This relationship can be interpreted in one of two complementary ways depending on whether one's outlook is 'short term', i.e., of the growing season, or 'long term', i.e., of a decade:

- Considered over the growing season: the most fertile soils provide the highest yields and have the highest SCC levels. SCC can therefore be considered a good indicator of soil fertility.
- Considered over a decade: a balance is established of C flows between the various compartments of the soil–plant-atmosphere system. The most productive cropping systems owe their productivity to higher C flows. This increased flow is caused by a more efficient photosynthetic activity. It induces accumulation of C in soils and better soil cover which reduces the mineralization of SCC. Therefore, the soil's organic status can be maintained by an intensification of cropping systems (Fig. 1).

Moreover, these tests show that crops respond differently to mineral fertilization before and after a period of degradation of the soil's organic status. This corresponds to a phenomenon of hysteresis in restoring soil fertility in close relationship with the relative importance of an 'inert' SCC compartment.

For further information: Crétenet and Tittonell (2010).

Fig. 1 Farmer cotton field at Ngong (northern Cameroon, 11 October 2012) with two cropping histories

- **Left:** previous clearing of *Acacia senegal*, regenerated land, soil completely covered most of the season (lower soil temperature, minimal mineralization of soil carbon), high production of aboveground and root biomass (higher C restorations to the soil).
- **Right:** previous continuous crop, degraded land, part of the soil left bare for most of the season (higher soil temperature, higher mineralization of soil carbon) low production of aboveground and root biomass (smaller C restorations to the soil).

1.2 Acid Soils

Soil acidity is known for its negative effects on plant growth. It induces biological effects such as the blocking of phosphorus and, more directly, it causes the release of aluminium toxicity.

Soil acidification is an unfortunately inexorable process, closely tied to the biological activity induced by photosynthesis. It affects natural forests as well as cultivated areas. When soils are not geologically acid (soils formed on sedimentary bedrock), a good reserve of alkalinity and natural acidification pose no problem. But for soils formed on crystalline massifs, the problem is serious. Such soils represent more than 30 % of cultivated land in the world,[4] more or less evenly distributed on all continents.

If we consider soil acidification independently of large biogeochemical cycles (which, over a geological time scale, redistribute alkaline resources on the planet that counteract acidification processes), we could be expected to believe that the world is on a path to an acidic sterility. What then are the processes that govern this phenomenon, and how does farming contribute to it? We know relatively little about the speed of the phenomenon, in any case too little to start sounding the alarm just yet. Nevertheless, we know (Calba et al. 1999) that the acidification of a given soil horizon results from the balance of nitrate and ammonium (nitrogen cycle), the accumulation or export of organic acids (carbon cycle), direct inputs of acids or bases, or the leaching of H^+, OH^- and HCO_3^- ions from outside the cultivated horizon. At this stage, we can assume this approximation to be correct when the phenomena related to mineral weathering and salt precipitation can be ignored.[5]

The limits of the nitrogen cycle in the soil, organic acids exported into plants or accumulated into organic matter, and atmospheric inputs can all be measured directly. To these phenomena which cycle in a loop in the cultivated horizon, we

---

[4] http://www.fao.org/geonetwork/ (retrieved: 12 December 2012).

[5] We do not cover here research initiatives currently underway that are working on stimulating mineral weathering of the bedrock. They contemplate using biological methods in order to introduce massive amounts of elements into the biogeochemical cycles to counteract the phenomena described.

must add the lesser known (and less mastered) phenomenon of acidification by leaching of nitrates or other ions. In fact, even for soils subjected to rainfall below 1000 mm annually, acidification resulting from the leaching of nitrates can be significant, in the same order of magnitude as the phenomena described above.

Over the short term, we have no problems with this process. We have reliable formulas to correct the instantaneous soil acidity (initial conditions) and to rectify the dynamics of acidification. For major crops, such as in Brazil, huge amounts of limestone/gypsum are transported and applied each year to correct the problem. However, projections over the longer term lead us to an impasse (exhaustion of economically exploitable basic deposits) similar to that facing phosphate use in the world.

In summary, as far as acidification is concerned, we can cultivate a third of the world's soils with monocultures and still maintain an acceptable level of acidity and avoid reaching toxicity thresholds. But this can only be done through massive recourse to basic biogenic products, whose local stocks are far from inexhaustible. Given that humanity has no time to wait for geological cycles to redistribute materials, we are well and truly at an impasse—at least over the longer term. Use of other agricultural systems can result in a slowing down of the dynamics and effects of soil acidification. A conventional solution is to use nitrogen fertilizers that are less acidifying than others, or live to with the acidity by cultivating tolerant varieties.[6] But there do exist better approaches. As we shall see, the services of biodiversity can be used to regulate biogeochemical equilibriums, in particular by increasing the stock of organic matter in soils.

1.3 Degradation of Soils Through Irrigation

Man has known for a long time that the phenomenon of land degradation is a major threat to his survival. As far back as a 1,000 years ago, farmers in Lower Mesopotamia had observed some of their fertile land become gradually sterile as a result of salinization caused by irrigation (Cheverry 2010). Currently, with 275 million hectares of irrigated land, the phenomenon is widespread. Worldwide, the salinization of irrigated lands results in an annual reduction of 1–2 % of irrigated cultivated land. This is because of the simple fact that irrigation water, even in low concentrations, deposits chemical elements it carries in solution on agricultural lands. These chemicals accumulate and thus salinize/sodiumize these soils. There is no known sustainable solution to the problem (Marlet et al. 1998). Salinization is a visible and recognized phenomenon but there are others that are less known because the concentrations involved are tiny. Some of them, in spite of their low concentrations, have a very direct effect on human health, such as arsenic

---

[6] However, some experts wonder whether cultivating acidity-tolerant varieties will not lead to further acidification of soils.

in Asia or mercury[7] and fluorine in America. Attempts to counter these phenomena by limiting water supplies or through surface drainage provide only temporary solutions. They may even mask phenomena that may become dominant over the short term. Later on in this chapter, we will explore possibilities of using bioremediation or service plants to fight against these phenomena, but in the meantime, it is clear that agriculture is at an impasse.

1.4 Bio-Aggressors

Even within a pest population, individuals react differently to insecticides. If less than the recommended dose of insecticide is applied, some of the individuals will certainly die, but others will survive. If an insecticide is applied incorrectly, it may kill weak or sensitive individuals, while unintentionally favouring more resistant ones. If this pattern is repeated, the whole of the population, obeying the laws of natural selection and of the survival of the fittest, will begin to develop resistance.[8]

Cropping patterns of intensive production methods obviously encourage all these excesses and lead to long-term impasses. Chapter 5 will address the fundamental issues involved but we illustrate possible and dramatic effects here in Box 4.

> **Box 4. The virtual disappearance of cotton farming in Nicaragua**
>
> Cotton monoculture in Central America, and in Nicaragua in particular, developed under favourable and concurrent factors: the boom in commodity prices, sympathetic political environment for the massive expropriation of peasant communities, and North American support for infrastructure development (Wheelock 1980). The exceptional climate and soil conditions enabled Nicaragua to achieve the best yields in the world for rainfed cultivation (Maraux 1994). All the area that could be mechanized in the western part of the country was used for cotton cultivation. And, indeed, cotton made the fortune of a few dozen *algodoneros*. The main problem encountered in the process was *Anthonomus grandis* Boheman, a beetle causing devastating damage to the cotton crop. This pest had been identified as a national enemy as far back as in 1930s, with eradication measures being enacted as legislation.[*]
>
> In the post-war period, merciless applications of methyl parathion, with increasing doses and increasingly frequent aerial dusting (as many as twenty

---

[7] Recent studies have shown that, in the Brazilian Amazon, mercury pollution due to gold mining activities and deforestation in the last 30 years has contributed less than 3 % of cumulative mercury concentrations in surface soils (Roulet et al. 1999).

[8] We talk about resistance to a product when we find that it is unable to fight pest infestation when used in recommended quantities.

applications per season), ensured the system's continuing profitability. But in 1987 (Laboucheix 1987) doubts arose: 'The increasing doses of methyl parathion they are being forced to use has become a major cause for concern for Nicaraguan cotton farmers. The recent introduction of specific control strategies against the *picudo* has not yielded the expected results. And there is a real possibility of *Anthonomus grandis* developing resistance to methyl parathion. In fact, the evaluation of the effectiveness of methyl parathion by the LD50 method shows that the Nicaraguan populations studied are ten to thirty times less sensitive than the strains tested in the United States and El Salvador'.

Five years later, because of the *picudo* not a single hectare of cotton was being grown in Nicaragua. (This disaster story was recounted by the Commission Chairman at the closing conference of the French Écophyto 2018 programme.)[**]

[*] Decree of measures to eliminate the *picudo* (*Anthonomus grandis*), adopted on 7 July 1936.

[**] Recently, after a 'waiting period' of 20 years and with numerous precautions, Nicaragua is trying to revive cotton cultivation but this time on agroecological bases.

1.5 Intensive Model with Transgenic Crops: An Impasse?

The route to pest control through transgenesis in intensive monoculture systems deserves exploring, especially when agriculture is confronted with disasters such as the one described in Box 4. We are referring to the introduction into plants of pest-resistant genes through genetic means, not through natural cross-breeding. This path deserves investigation since transgenic crops offer the possibility of relatively high productivity while maintaining fertility levels.

The topic of transgenic crops has always been controversial and their pros and cons have been fiercely debated. However, we can now step back and actually analyze the cultivation of these varieties. At first, it seems undeniable that the effects achieved are exactly as expected, i.e., reduced need for pesticide application (Fok 2011; Tabashnik et al. 2009). But when the observation period is extended and/or economic aspects are introduced, the outcome seems less clear-cut. Thus, in an analysis that spans 16 years of GMO use, Benbrook (2012) shows that because of uncontrolled side effects of cropping practices of GM crops, the overall use of pesticides in US agriculture is actually increasing. This is partly due to the fact that new pests, hitherto secondary, are becoming dominant.

1.6 The Need to Find Other Paths to Intensification

We decided to analyze the facts and mechanisms that are behind warnings of the actual or expected consequences of production methods inherited from the post-war period. These systems are still being used in large parts of the world but they only work through constant interventions, inputs, material transfers, artefacts, etc., which obviously are not sustainable. Worse, some transformations of the physical, chemical or biological environments caused by the 'intensive monoculture' model inherited from the second agricultural revolution now seem irreversible.

Agriculture is facing all these challenges at a time when the precepts followed since the Second World War to increase productivity now only lead to a dead end. Genetic breeding of a limited number of species (including through GMOs) and the massive use of chemical inputs to ensure high productivity have led to impacts on the environment (and human health) now considered intolerable in many situations. Genetic erosion of crop species, chemical and organic pollution, and the impact on natural biodiversity have already left a mark on areas of intensive agriculture and continue to be a major environmental risk in the most vulnerable regions (FAO 2009). The loss of natural habitats and intensification of agriculture are associated with the replacement of traditional species by a limited number of high-yield productive species. This poses a particular threat to biodiversity, both natural and domesticated (Jackson et al. 2005; Sachs et al. 2009). Furthermore, these solutions have failed to ensure food security and economic development in the world's poorest regions, especially in Africa. Many parts of the world—again, especially in Africa—suffer from hidden hunger, characterized by high mineral and vitamin deficiencies due to a diet based almost exclusively on a few high-calorie species.

Intensive cultivation, of cereals in particular, has given us huge gains in productivity over the last 20 years via the Green Revolution. These crops—often monocultures—are now a mainstay of global agricultural trade and food security. But their environmental footprint is large and the efficiency of inputs decreases with increasing intensification.

In a far-reaching and oft-cited article, Tilman et al. (2002) established a relationship between changes in input use (nitrogen and phosphorus fertilizers, various pesticides) and the evolution of cereal production. As a first approximation, this relationship initially remained linear. As input use increased between 1960 and 1980 so did cereal yields. But, this trend started to falter in the 1980–2000 period, with an observed decrease in nitrogen efficiency. Agricultural production, already expensive in terms of non-renewable resources and driven by increasing input use, arrived at a situation with worsening material balance. It required ever greater quantities of inputs for the same, or even smaller, production. This shift masks the fact that in some regions, particularly in Asia (intensification of rice) and even in Europe, yields have started stagnating (Brisson et al. 2010). So, in other words, as cereal yields increase, the marginal efficiency of additional nitrogen decreases.[9]

Environmental contamination by fertilizers and pesticides, loss of local cultivars, or at least of their genetic diversity, dependence on fossil fuels and other mineral resources (phosphates, limestone), and soil erosion have often accompanied a relative increase in productivity. In addition, they sometimes have a negative overall social impact. At the same time, this production method has also oriented tropical agriculture towards plantations based on the export of raw or processed commodities. This is the model of large intensive plantations of perennial crops grown in monocultures, such as oil palm and rubber, or semi-perennial ones, such as banana and pineapple. Faithful to this method based on new high-yield varieties and the massive use of inputs, the Green Revolution was successful in the 1970s in increasing productivity to the level necessary to feed a growing world population. Awareness that productivity was not without limits started dawning in the 1970s in Europe. In Africa, the limitations of systems based on this type of intensification were shown by the failures of the Green Revolution in semi-arid zones. Other factors in the relative lack of success of these systems there were the difficulties in sharing water resources amongst a growing number of users, water pollution by chemical inputs, and soil salinization or acidification (Conway 1997). As practiced in the countries of the North, 'modern' intensive agriculture is a major consumer of energy, mineral fertilizers and pesticides, all characteristics that have been called into question today. In the South, it affected only a marginal fringe of the populations. Approximately 80 % of farmers in Africa and 40–60 % of those in Latin America and Asia employ only hand tools. Only 15–30 % of them use animal traction (Mazoyer and Roudart 1997). Hence the importance of traditional farming systems even today. Some of them, such as slash-and-burn systems[10] are known since Neolithic times and still widely practiced. Today, the ratio of labour productivity between the most intensive agriculture and the most extensive is 500 to 1. It was only 10 to 1 at the beginning of the twentieth century (Mazoyer and Roudart 1997).

Given this situation, there is an urgent need to find new conceptual avenues for building sustainable agroecosystems. Crop diversification in farms is an old issue but one that has always remained relevant. It has suddenly become key to finding the ecological, economic and social sustainability of agroecosystems (Connor

---

[9] Nevertheless, it is worth noting that in tropical Africa, where the mineral balance (external inputs of fertilizers minus exports by the crop production) is negative, the marginal efficiency of fertilization is direct and indisputable (Abuja Declaration, IFDC 2006, http://www.ifdc.org/About/Alliances/Abuja_Declaration). This reality places Africa in variance with respect to global calls for reduced input use. National and international public policies towards Africa often—and rightly so—encourage or subsidize the use of chemical fertilizers.

[10] Characteristic cropping system of the humid tropics based on the planting of crops after slashing and burning a forest, most often secondary. The complete rotation period, including the cultivation period followed by the restoration of forest fallow, varies from 10 to 50 years.

2001). The over 1 billion farmers in developing countries still do not have access to the modern technologies of agriculture. Paradoxically, their traditional systems are based on integrated management of local natural resources and, in many cases, rational management of biodiversity. Can they become the models for the cropping systems of the future? Some agronomists today believe so (Ewel 1999; Altieri 2002; Jackson 2002). Thus, as far back as the 1980s, new models were proposed which were based on more or less radical forms of 'organic agriculture' (Cauderon 1981). Multifunctional agriculture and ecosystem services are two concepts that highlight the different functions that agriculture can fulfil. Prominent among them is the environmental function (Bonnal et al. 2012). The recent debate on biodiversity has strengthened and indeed provided an extra dimension to the necessity of finding new avenues.

2 Opportunities and Limitations of Cropping Systems that Promote Biodiversity

Thus, the difficulties of artificialized cropping systems lead us to re-examine the properties and performance of other cropping systems, often traditional or marginal, which are based on the diversification of plant and animal species—cultivated or wild—in agricultural areas. This diversity can be managed in space and time, within plots or within farms and landscapes. It allows the provision of a wider range of ecosystem services and promotes the resilience of farming systems in the face of vagaries and risk (Malézieux et al. 2009).

These multispecies and multifunctional cropping systems are common in the countries of the South. A majority of farmers there farm small areas by combining different plant and animal species for the purposes of food, energy and materials for their families and the market (Boyce 2004; Devendra and Thomas 2002; Kumar and Nair 2004; Morton 2007). Imparting value to biodiversity, however, is not sufficient to ensure the development—or the permanence—of these farming systems, which suffer from certain limitations.

2.1 Diversity within Plots, Productivity and Supply of Ecosystem Services

2.1.1 Many Models of Multispecies Cropping Systems

When several species are intentionally installed in the same space, it is called planned biodiversity (Swift et al. 2004). Planned biodiversity can take a variety of forms depending on the number of species involved, their respective types (annual/ perennial, herbaceous/woody, grassy/leguminous, etc.), their density, and their spatial and temporal arrangement (Malézieux et al. 2009). Annual crops can be associations of varieties or species, mixed, in alternate rows, or in rotation. These species may have a production function or a service function (soil protection, nitrogen trapping, repulsion of pests or attraction of auxiliaries, etc.). Perennial crops can be grassy and be associated with each other, for example for fodder production. They may be woody and associated with each other or with herbaceous crops in agroforestry systems. These latter have a wide variety of structures, ranging from alley cropping which alternates tree rows and strips of annual crops to tropical agroforests involving a large number of species. Finally silvopastoral systems associate trees with forage species grazed by livestock.[11]

The diversity of crop species tends to be linked to an associated biodiversity (Swift et al. 2004). In a monospecific cultivated plot, there is no room for spontaneous vegetation except at its periphery, as long as it is not weeded. In a multispecies plot, there are many areas of transition from one type of crop to another. This promotes increased plant biodiversity with the growth of additional vegetation which may have specific unplanned functions. This plant biodiversity as a whole generates a diversity of habitats for various arthropod and vertebrate communities (Vandermeer et al. 1998). This is especially true when woody species (trees, hedges) are included (Söderström et al. 2001). Similarly, there is an overall correlation between aboveground biodiversity and belowground biodiversity, even though the mechanisms and the quality of this relationship are very context-dependent (Hooper et al. 2000; Wardle et al. 2004).

2.1.2 Multispecificity and Productivity of Cropping Systems

A negative trade-off between plant diversity and productivity is generally observed when one considers a single productive species (e.g., Deheuvels et al. 2012, comparing different structures of cocoa-based agroforests). But by calculating the combined productivity of multispecies systems—the most common indicator is land equivalent ratio (LER)—one can show that their overall production is generally higher than that of monoculture controls (Dupraz and Liagre 2008; Snoek et al. 2013). The LER is the surface area of monocultures that would be necessary to obtain the same output as of a unit of production of the multispecies combination. If LER is greater than one, then the combination of crops uses space more efficiently than its components when grown as monocultures.

Various types of processes can be involved. They can be based on the spatial complementarity of using the light resource or on the complementary use of the soil's resources if the different species explore different soil compartments, for example, woody species accessing deep horizons (or groundwater) which are inaccessible to herbaceous species (Celette et al. 2008). They may also pertain to temporal complementarity when phenological shifts allow different species to

---

[11] Crop-livestock systems will be discussed in Chap. 5.

access the light resource or soil resources at different times of the year. They can depend on facilitation, for example, when the presence of legumes improves the availability of nitrogenous resources for all components of the multispecies combination (Rivest et al. 2010).

2.1.3 Multispecificity and Ecosystem Services of Cropping Systems

In addition to improving agricultural production, multispecificity also tends to favour various types of ecosystem services for agriculture and society. Whenever the overall aboveground production is stimulated, all the root components undergo greater overall development, with attendant beneficial consequences to the physical, chemical and biological soil properties. Thus intercropping promotes nitrogen trapping which would otherwise be leached in bare soil (Maltas et al. 2009). Cover crops favour soil organic matter and soil biological activity (Lienhard et al. 2012; Steenwerth and Belina 2008). Carbon sequestration is stimulated in the tissues of perennial species, in the soils of agroforestry systems (Albrecht and Kandji 2003) and in the soils planted with intercropping species (Metay et al. 2007). However, the carbon content and soil organic matter is less dependent on species diversity per se than on the amount and composition of plant tissues returned to the soil (Russell 2002).

Soil erosion becomes limited if, during the rainy periods, the spatial and temporal arrangement of crops provides good soil coverage by plants or their residues deposited on the ground (Meylan 2012). Ground cover and improved soil porosity associated with a high root density and/or development of soil macrofauna promote infiltration and reduce runoff. This helps prevent particles and pollutants on the surface of the soil from draining away (Gaudin et al. 2010).

Another benefit of multispecies cropping systems is the control of weed communities, pathogenic microorganisms, and arthropod and nematode pests (Jose 2009; Malézieux et al. 2009). The more efficient use of light and soil resources by intercropping leaves fewer resources available for weeds. Examples of allelopathy at the expense of weeds have also been reported (Liebman and Dick 1993). However, an increase in plant biodiversity appears to be linked to an increase in the biodiversity of belowground and aboveground communities of micro- and macro-organisms (Stamps and Linit 1998).

Does the balance between communities become more favourable to farmers, i.e., is there a strengthening of auxiliary species? Several mechanisms may contribute to the process. Some pertain to the structure of the multispecies population, for example, the dilution effect of host species sought by a pest or the barrier effect of other species. Others may cause changes in conditions of development of pests, such as changes in their habitat, emissions of attractive or repulsive compounds by some plant species, and changes in their trophic networks (Djigal et al. 2012; Box 5).

> **Box 5. Inserting maize plants in cucurbitaceae-based agroecosystems in Reunion**
>
> Jean-Philippe Deguine, Serge Quilici and Bernard Reynaud
>
> Fruit flies (*Diptera, Tephritidae*) are major pests in the tropics. In Reunion, flies attacking cucurbits are considered as the major pests in horticultural agroecosystemss. Chemical protection, practiced for many years through massive insecticide use, has shown its limitations: inefficiency, high cost, risks to the environment and human health. Studies nowadays focus on agroecological management of fly populations based on the insertion of maize plants in the vegetable agroecosystem.
>
> Indeed, maize plants are so attractive to these flies (Atiama-Nurbel et al. 2012) that they act as trap plants by concentrating fly populations. It is then possible to regulate them, for example by using adulticide bait (Deguine et al. 2012a). When this is done, there is no need to spray insecticide on the cultivated cucurbits and production losses are minimized. This technique of using maize as trap plants to control vegetable flies has now been successfully adopted in commercial farms (Deguine et al. 2012b).
>
> In addition, the presence of maize trap plants in the agroecosystem (borders around plots, patches or bands in the plots) provides a complementary method for assessing fly populations and for studying their communities. The in situ counting of flies on maize makes it possible to obtain an accurate estimate of the population of flies really present in the agroecosystem. This method is far more reliable than conventional methods of parapheromone-based sexual trapping and of counting pupae and adults obtained from fruits collected in the crops. It is thus possible to characterize certain fly-community parameters such as the relative abundance or sex ratio of different fly species (Deguine et al. 2012c).
>
> Finally, maize plants are home to beneficial insects. Prominent among them are useful Diptera such as syrphids, which are, at the same time, pollinators, predators and indicators of a properly functioning agroecosystem (Duhautois 2010).

However, different responses of the various pathogen communities to changes in cropping systems can head off in opposite directions. In addition, there exist threshold effects. For example, the presence of shade trees in Central American coffee plantations seems to reduce coffee rust disease (*Hemileia vastatrix*), but encourages the spread of American leaf spot disease (*Mycena citricolor*) and of the coffee berry borer (*Hypothenemus hampei*) (Avelino et al. 2011). Only an integrated analysis of ecosystem effects of plant diversity on other living communities can help predict resulting benefits for and threats to agricultural production (Duyck et al. 2011; Staver et al. 2001).

2.2 Diversity within Landscapes and Farms

The use of biodiversity and the services that are expected from it are not limited to the scale of agricultural plots. Biodiversity resulting from the arrangement of different agricultural systems and ecosystems in a landscape, in a territory or on a farm obviously plays a role in the processes that are observed at these supra-plot scales. These include, for example, the movement of organisms and their forms of dissemination, the surface and underground water transfers, and the associated transfers of sediment and pollutants. This biodiversity also affects the corresponding services such as landscape quality, pollination, crop protection, water quality, erosion and flood control, etc.

The dimensions of the landscape and farm provide additional flexibility and leeway, through the spatial and temporal distribution of agricultural activities and practices and the creation and maintenance of landscape infrastructure (fences, ecological corridors, hydrological networks, etc.). Scaling up from the plot to the landscape confers a new dimension to the relationships between biodiversity and ecosystem services. Thus, it seems that plant and arthropod biodiversity depends more on the diversity of farm habitats than on the type of farming practices (organic vs. conventional, intensive vs. extensive) (Weibull et al. 2003). Schroth and Harvey (2007) note that the conservation of plant and animal biodiversity in tropical areas is best achieved in landscapes composed of a complex mosaic of agroforestry and native forests. Such an environment maintains a high diversity of habitats and a high connectivity. In a review of several studies conducted on cereal cultivation in temperate regions, Rusch et al. (2010) note that landscapes with a high proportion of semi-natural habitats favour auxiliary populations in 83 % of cases, while in 50 % of cases the type of cropping system has no effect.

As far as the conservation of biodiversity is concerned, the complexity of the landscape structure can offset the negative effects of locally intensive cropping systems. In other words, the effects of the type of cropping system are more pronounced in landscapes with simple structures than in landscapes with complex structures (Tscharntke et al. 2005). For example, the average level of agricultural intensification in a landscape limits both the diversity of pollinator species and bee species (Batáry et al. 2010). Yet and Ricketts (2004) observes that the presence of tropical forest fragments helps pollinate neighbouring coffee plantations by hosting a variety of bee species. Another example: a complex landscape structure can mask the differences in expected pest regulation between monoculture plots and agroforestry plots (Smits et al. 2012).

These observations suggest that a farmer can use one or more of the several levers mentioned above to manage biodiversity and associated ecosystem services. He can do so at the scale of his individual plots, his entire farm and even larger territories when managed through collective organizations. From an economic point of view, combining species with different production cycles (at an annual scale and/or for production durations for perennial species) can accelerate the return on investment. Thus the farmer gets regular and periodic income from the various crops and productions. Snoek et al. (2013) provide an example of the intercropping of rubber with coffee and cocoa. More generally, in small farms in the South, a variety of activities helps address several concerns: seasonal distribution of production, work and income; ex ante and ex post hazard management, difficulties in obtaining credit, etc. (Ellis 2000). This diversity is also known to impart more resilience to farms in dealing with climate change and its attendant hazards (Lin 2011). Finally, from a nutritional point of view in the context of subsistence agriculture, a link can be established between the species diversity of cultivated species in a territory and the diversity of the diet and the health of the population (Johns and Eyzaguirre 2006).

2.3 Limitations of Multispecies Systems

Are the many identified or theoretical benefits of cropping systems that valorise biodiversity sufficient to convince farmers to continue using or even to disseminate them? Various authors have analyzed situations where multispecies cropping systems have been abandoned in favour of monoculture systems. Feintrenie et al. (2010) show that the profitability of cash crops (coffee, cocoa, rubber) and the opportunities they offer (income, infrastructure modernization and end to isolation) have led farmers in various Indonesian regions to abandon agroforestry. And this despite the cultural attachment they profess for this traditional form of production. Ruf (2011) observes the same trend in Ghana where cocoa cultivation is moving towards monoculture. He identifies several determinants: technical progress (new hybrids offering better returns when grown as a monoculture), land laws, the low value that can be imparted to timber trees, and the advent of cheap labour because of migrations.

Furthermore, biodiversity-promoting cropping systems can sometimes fail. Affholder et al. (2010) analyze the non-adoption of direct-seeding on plant cover (Box 6) by Vietnamese farmers as a result of its increased labour requirements. They estimate the payment for environmental services (PES) which would be necessary to overcome this obstacle to be very high, almost as much as the gross margin. Similarly, Giller et al. (2009, 2011) identify a wide range of technical, economic, land-right and institutional barriers to explain the poor development of conservation agriculture in sub-Saharan Africa, despite a very determined push by various development agencies and NGOs.

In some spatial and temporal configurations, the possibilities of mechanizing multispecies cropping systems is limited and therefore labour productivity remains low. This is the case, for example, of complex agroforests where agricultural machinery cannot enter. To work around this restriction, a planting scheme consisting of different species in alternating bands—already in vogue in alley cropping in temperate regions—would need to be adopted. There are fewer obstacles to mechanization when biodiversity is managed through sequences of different crops. The abandonment of tillage in direct-seeding cropping systems reduces the traction power required and stimulates the invention of new agricultural machinery (Friedrich et al. 2009).

Thus the maintenance or the introduction of cropping systems based on valorising biodiversity are only possible if, on the one hand, they are economically viable and socially acceptable, and, on the other hand, if the know-how of farmers themselves is mobilized in evaluating, designing and modifying these systems (Cerdán et al. 2012).

It is necessary to examine the widespread expansion of no-tillage agriculture, which is fast becoming the norm in large mechanized agricultural regions: the North American plains, South America (Brazil, Argentina, Paraguay), and Australia. This method of agricultural production originated in the United States in response to wind soil erosion (dust storms that scared the farmers and city residents). It started with 'chemical tillage' by contact herbicides (Paraquat is the best known example). With only a few operations, these herbicides could fulfil chemically one of the functions of tillage, i.e., weeding. Other technological developments followed the adoption of this practice, such as the use of plants to protect the soil (service plants). Farmers who have invested in these methods (which now cover about 115 million hectares) rightly wonder whether they are the pioneers of a new agricultural revolution.

> **Box 6. An example of conservation agriculture: direct-seeding on plant cover to protect soils and maintain agricultural production**
>
> The humid tropics usually have fragile soils and the aggressive climate in these regions can lead to degradation of soil fertility and a drop in crop productivity. Conservation agriculture is a technical response to this threat. It relies on three factors: reduced tillage, soil protection by cover crops or organic residues, and diversification of cropping rotations or associations. Within a few decades, conservation agriculture's popularity has grown considerably, particularly in North and South America (Scopel et al. 2012). Direct-seeding on plant cover is part of this approach. It consists of growing a cover crop between two cash crops, the second being sown directly after weeding the cover crop. Several agroecological and socio-economic assessments of conservation agriculture have been conducted in different regions of the world.
>
> A wide range of ecosystem services provided by direct-seeding on plant cover has been observed in different contexts. Most are related to the protection and improvement of the physical and biological functioning of soils. Intercrop vegetation or mulch dissipates rainfall energy and promotes water infiltration, thus reducing soil erosion. Non-tillage promotes the development of macrofauna that maintains or even increases soil porosity (Blanchart et al. 2007). The carbon cycle in soils is activated, with higher CO_2 emissions. However, there is also increased carbon sequestration when the biomass productivity of the vegetation cover is high (Metay et al. 2007). The presence of mulch leads to improved water infiltration and reduced soil evaporation, thus promoting the maintenance of a water reserve (green water) that can act as a buffer during droughts (Scopel et al. 2005). Higher soil humidity stimulates biological activity and the degradation of organic

matter. The cover crop captures the remnants of nitrogen after the previous crop and returns it to the next crop through the mineralization of its residues (Maltas et al. 2009). Finally, after the introduction of an intermediate plant cover which stimulates the production of biomass and the accumulation of organic matter in soils, there is an increase in both the diversity and the abundance of bacterial and fungal soil communities (Lienhard et al. 2012).

While direct-seeding on plant cover favours the productivity of annual cropping systems in general, we find that it is particularly developed on the large mechanized farms (especially in Brazil, Fig. 2) rather than on small family ones (Scopel et al. 2012). On the former, it can reduce production costs by replacing tillage by a herbicide treatment to destroy the plant cover. While on the latter, the additional work necessary to plant the cover crop and herbicide costs can be major barriers to the adoption of the technique (Affholder et al. 2010). An integrated approach to introducing this innovation, in particular one linking crops and livestock, is necessary for the sustainable adoption of this innovation, as is a participatory approach that brings farmers and R&D mechanisms together.

Fig. 2 **a** Direct-seeding on a large Brazilian farm (*source* Scopel et al. 2012). **b** Cotton being grown on *Bracharia* mulch. **c** Manual direct-seeding on a small Brazilian farm. **d** Maize with pigeon-pea relay cropping

3 Towards New 'Ecologically Innovative' Cropping Systems

The world faces a major challenge today of transitioning to a multifunctional agriculture that will be able to feed 9 billion people while protecting natural resources and maintaining the health and the well-being of populations (Millennium Ecosystem Assessment, MEA 2005; IAASTD 2008). It is now widely accepted that the conservation of biodiversity must be a major consideration in the development of all human activities. In addition, biodiversity can also be seen as a key resource for inventing new forms of agriculture for the future. This is a belief—based on the original foundations of agriculture and which forms the basis of the concept of ecological intensification—that we espouse (Fig. 3).

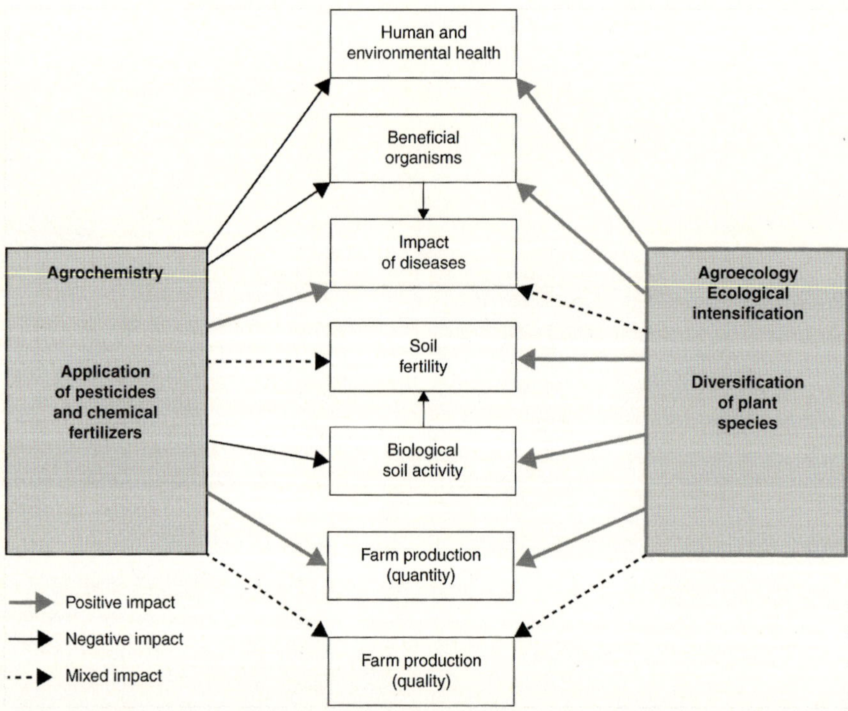

Fig. 3 Plant diversification as one of the two pillars of the agroecological approach, of which ecological intensification is a part. The grey arrows indicate positive impacts on the criteria shown in the central boxes, black arrows represent negative impacts and dotted arrows, mixed impacts (according to Ratnadass 2011)

3.1 Agrobiodiversity and the Design of Innovative Cropping Systems

Agrobiodiversity is an essential component of a multifunctional agriculture, and plays a major role at various levels of organization to provide ecosystem services (MEA 2005). It is the basis of increased productivity, pest and disease control, and recycling of water and minerals, and provides many cultural functions (Tscharntke et al. 2005; Jackson et al. 2007; Jarvis et al. 2011). Today it is clear that the maintenance of intra- and interspecific diversity in agroecosystems is key to building more resilient systems (Box 7). However, the form that this biodiversity should take and the level of organization from which it must be expressed remain topics of intense and unresolved scientific debate (Wood and Lenné 1999; Lenné and Wood 2011).

> **Box 7. An example of ecological intensification through the use of a traditional system: the restoration of savannah soils in agroforestry systems in Cameroon**
>
> Cocoa cultivation is often considered one of the leading causes of deforestation in the tropics. In many countries, it is indeed based on an unsustainable technical model of intensive monoculture consisting of the periodic shifting of production areas at the expense of forest areas (Rice and Greenberg 2000).
>
> However, from surveys of more than a thousand farms in south-central Cameroon, Jagoret et al. (2011, 2012) found old cocoa agroforestry farms, often maintained for several generations (70 % are over 40 years old). These multispecies cropping systems thus demonstrate their sustainability at the agroecological level as well as the socio-economic one. In fact, these cocoa farms are subject to continuous regeneration through coppicing or replacement of dead plants, which leads to a stable cocoa density and consistent yields. The farms have an average of 25 tree species, maintained or introduced deliberately by farmers who expect them to provide diverse and clearly identified services (fruit and wood production, shading and control of pests, maintenance of soil fertility, etc.).
>
> The number of pesticide treatments is reduced and no fertilizer is applied. Even then, the biological properties (density and diversity of mycorrhizal fungi) and chemical properties (stock of carbon and other major elements) of their soils remain similar to those observed in neighbouring secondary forests (Snoeck et al. 2010). This model of sustainable cocoa cultivation based on agroforestry cropping is environmentally and economically sound. It represents a credible alternative to the simplification of cocoa cropping systems for farmers in the regions studied (Jagoret et al. 2009).
>
> Cocoa cultivation can therefore be undertaken without contributing to deforestation. Cocoa-based agroforests can even be instrumental in restoring savannah soils and in introducing cocoa production to suboptimal soil and

                                                                F. Maraux et al.

climatic zones (Jagoret et al. 2012). Oil palm or food crops are first grown on *Imperata cylindrica* grasslands to eliminate this grass. Then a mixture of cocoa and fruit trees is planted, while preserving specific forest trees. The density of fruit and forest trees is gradually reduced, while that of cocoa is maintained (Fig. 4).

By monitoring agroforestry cocoa being cultivated on savannahs and those being cultivated on gallery forests in the same region of central Cameroon, Jagoret et al. (2012) found that after a few decades, yield levels become comparable and similar to those for cocoa grown in secondary forests in southern Cameroon. The organic matter content increased from 1.7 % in the savannah soils to 3.1 % in the agroforests installed on these soils, without any fertilizer use—organic or inorganic.

The productive performance of agroforestry cocoa cultivation on savannahs shows that the initial disadvantage of low soil fertility and erratic rainfall can be overcome. This cropping system demonstrates the success of an ecological intensification process implemented by farmers. Moreover, it opens up prospects for the adaptation of farming systems to climate change that could result in reduced rainfall in the region (Tingem et al. 2009).

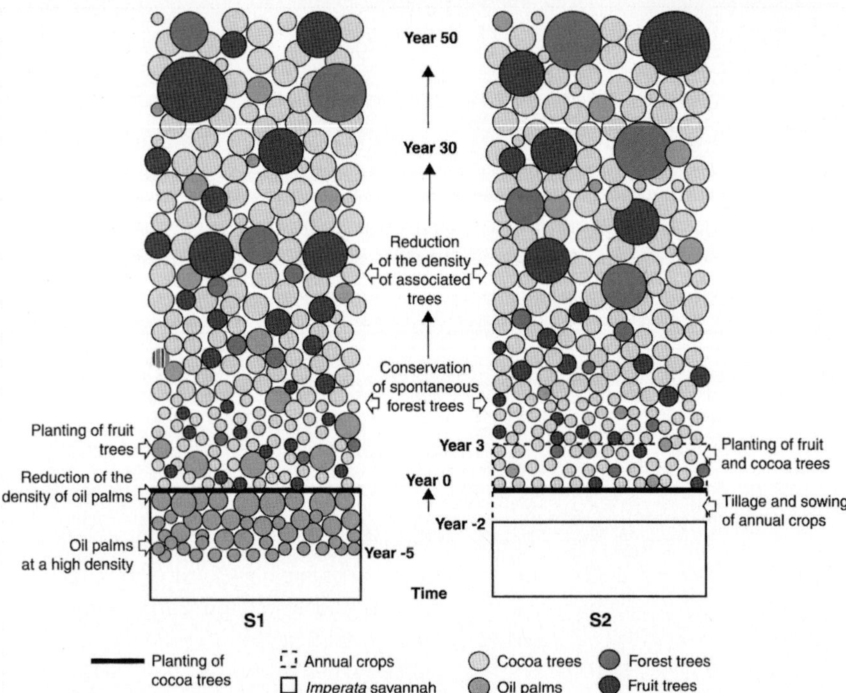

Fig. 4 The dynamics of installing a cocoa-based agroforest on a savannah, beginning with the planting of oil palm (S1) or food crops (S2) (*source* Jagoret et al. 2012)

Biological diversity is put to use in agriculture in a wide variety of practices mainly based on the mixture of cultivated species (intercropping, crop rotations, relay cropping, agroforestry). It can be implemented at the plot level and also across entire cultivated territories, where some specific properties specifically related to landscape mosaics can find expression. The use of biological diversity across a territory for productive ends is garnering increasing attention in the South as well as in the North. In the South, these practices are based on traditional agriculture and are often a reason for its sustainability. In Europe too, the interest is growing, either under the framework of the new common agricultural policy (reducing the size of the plots, the presence of natural elements required in the UAA) or as part of integrated biodiversity-based strategies for defence against pests.

New forms of ecologically intensive agriculture remain to be invented today based on agroecological principles (Altieri 2004). This new agriculture will have to rely on the combined and rational use of ecological principles and agronomic knowledge. It will depend mainly on the interaction between plant and animal species within agroecosystems. The optimal system would be rich in interfaces at the two scales that interest us the most (the plot and the territory[12]) under conditions in which resources are limited in time and space (Passioura 1999).

Among the promising directions for cropping systems is a new strategy based on the principle of imitating natural ecosystems. It relies on the use of a high level of biodiversity, as is often the case in natural ecosystems (Malézieux 2012). For example, the natural grassland ecosystem may appear to be a role model because it protects the soil from erosion, recycles nitrogen thanks to fixing microorganisms, and controls the proliferation and expansion of weeds, pests and diseases (Piper 1999). The grassland model is thus contrary to the single-species 'intensive' model on a large number of criteria such as robustness, resilience, biodiversity, nutrient loss, and energy dependence. This grassland ecosystem's structure relies on a complex mixture of perennial herbaceous species in C3 and C4 and nitrogen-fixing species. Its functions become a model for the design of sustainable agroecosystems (Jackson 2002). But going from the natural prairie to a grain-producing plot, albeit multispecies, is not going to be simple. What ecological continuum can we conceive to create a sustainable—and productive—agricultural system?

3.2 Biodiversity and Functioning of Agroecosystems

Biodiversity's role in the functioning of ecosystems has been and still remains the subject of much research by ecologists. Some recent studies have shown positive correlations between biodiversity and primary productivity, nutrient retention and

---

[12] The plot is our main scope of study. However, the organization of cropping systems in space and their interactions remains a subject for further study in the future.

post-stress resilience, not only in natural ecosystems (Hector et al. 1999; Loreau et al. 2001) but also in cultivated ecosystems (Altieri 1999). Since the time of Darwin, the hypothesis that the stability and sustainability of ecosystems relies on their biodiversity has been the subject of numerous studies (and debates) among ecologists. More recently, Tilman et al. (1996) have evaluated the sustainability of many grassland ecosystems characterized by different levels of biological diversity (number of species present). In this case, the fact that indicators of sustainability—such as the level of mineral recycling or of productivity—rise with biodiversity confirms the general opinion but, more importantly, opens interesting perspectives for grassland management. In reality, the general hypothesis that a complex community is more stable than a community consisting of a limited number of species remains largely to be proven. The confirmation of this hypothesis seems to depend on a large number of factors.

However, the difficulty in predicting the behaviour of a large number of species in very varied situations prevents any clear analysis of biodiversity's role in the performance and stability of systems. One way around this problem is to reduce the diversity of species to a diversity of functions and structures.

For example, in forest ecology, resilience after forest clearing, the exploitation of degraded land, the creation of a multi-layered structure or, more generally, the creation of a favourable living environment for some species, the formation of mineral reserves, and deep-water pumping are all studied and related to biodiversity. Vandermeer et al. (1998) thus formulate different hypotheses about the role of biodiversity in ecosystem functioning: beyond a certain threshold, the number of species has no effect on the functioning of ecosystems. With all functions being performed, the functioning of the system remains stable. Loreau et al. (2001) describe the growing interest in ecology to consider the different species in an ecosystem in terms of their 'function' in it. This is done regarding both the evolution of the vegetation and the relationships between this development and environmental change. This interest has led to the definition of *plant traits* able to translate this functional classification of species. In a given ecosystem, plant traits are thus defined for each species or group of species in the general perspective of linking the ecosystem's composition to its functioning. This approach has been subject to standardization efforts by ecologists at the international level (Cornelissen et al. 2003). It is a matter of understanding the response of vegetation to environmental variations (climate, land use, different disturbance regimes, etc.). Or, conversely, of predicting the impact of vegetation on these different parameters (Lavorel and Garnier 2002). One of the objectives is to ensure that species which present a certain homogeneity in terms of these traits are 'interchangeable' (what is the degree of redundancy present in the ecosystem?) and to qualify and quantify these species' behaviour (search for irreversibility thresholds). The definition of *functional groups* (Gitay and Noble 1997) corresponds to this goal by grouping together species that use the same resources (guilds), and those that respond similarly to a given perturbation (types). Are there many of these groups (new approach to diversity)? Are they the same and, if yes, up to what point in terms of the questions asked? What is their degree of fragility? Can we use them operationally (ranging from the recovery of degraded land to modelling large flows at the biosphere scale) and if yes, at what scales? It is matter also of verifying the type of links existing between specific, functional and intraspecific diversities. Vitousek and Hopper (1993) have established the various possible relationships between functions of an ecosystem (such as its productivity) and the number of species that make up this ecosystem. In some cases, an ecosystem's high species diversity is an obstacle to understanding its functioning and to modelling it. A possible approach consists of clubbing species into functional groups in order to model an ecosystem and thus predict its evolution. This is the approach adopted to understand and simulate the functioning of rainforests which host a large number of species (Gourlet-Fleury et al. 2005).

3.3 Some Examples of the Use of these Concepts by Agronomists

Agronomists today are very interested by these concepts. Injecting complexity into monospecific systems by adding service plants is a transition from theory to practice, and an opportunity for farmers to innovate. There exist many examples already: banana, grapes, orchards or vegetable cultivation (Fig. 5 and Boxes 8 and 9). Analysis of functional traits of service plants has become a major tool for designing systems that combine productive and service plants.

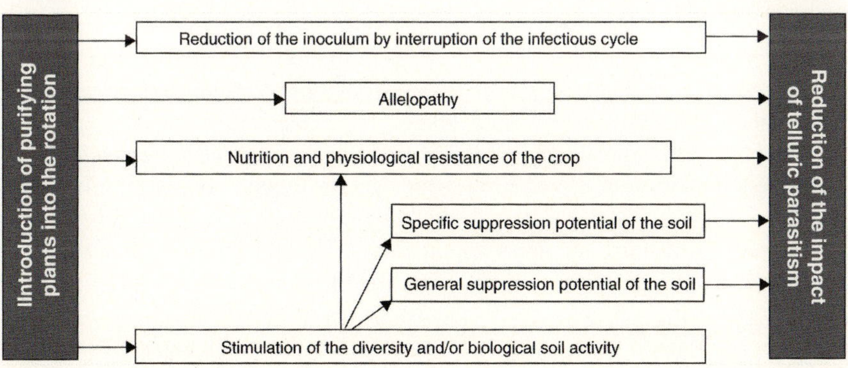

Fig. 5 Principal mobilizable processes on introduction, in a crop rotation, of species with purifying potential on telluric parasites (adapted from Ratnadass et al. 2012)

Box 8. Controlling soil parasitism in vegetable cultivation by introducing purifying plants

Paula Fernandes, Peninna Deberdt and Marie Chave

The development of monocultures in the world and the associated use of chemical inputs not only reduces the diversity of plant communities but is also often accompanied by a reduction in the microbial biodiversity of the cultivated soils. This biological erosion has been associated with an increase in soil parasite populations (Altieri and Nicholls 1999). Across the world, bacterial wilt caused by *Ralstonia solanacearum* is thus responsible for significant economic losses in vegetable crops. In Martinique, where bacterial soil infestation impacts tomato cultivation, research is being conducted into the introduction of plants with purifying properties into the cropping systems. Besides the desired biocidal effects, the introduction of service plants aims to combine the advantages of crop rotation, waste management and organic matter supply (Fig. 5). The increase in microbial soil biodiversity due to the introduction of service plants is also likely to lead to a reduction of the parasitic pressure through other related processes. It will also tend to increase mycorhization of the crop, thus allowing improved bioregulation of telluric pathogens.

Some groups of plants have been the subject of special interest because of their nematicidal properties. Others because of their suppressive effects which lead to the reduction of soil inoculum pressure through the emission of biocidal molecules (Compositae, Fabaceae, Poaceae, Alliaceae, Brassicaceae). A multi-year scheme of multi-criteria selection was used to select four plant species from twenty candidate species which were shortlisted for their multilocal agronomic performance, their host-status, toxicity of their aqueous residues, and their ability to reduce *R. solanacearum* soil infestation in a controlled environment. Three of them exhibit potential to reduce the incidence of bacterial wilt in tomato crop rotations. Three factors come into play: the production of biocidal exudates, increased soil microbial diversity for improved biological control of *R. solanacearum,* and the protective effect via the stimulation of mycorrhizal symbiosis for tomato. The next step is to test the viability of these new systems on a set of plots with infested soil in a real-world situation, i.e., on working farms, by taking into account their organizational constraints and on different soil types.

For further information: Deberdt et al. (2012).

Box 9. The introduction of cover crops in banana plantations and orchards in the West Indies in pursuit of multiple services

Fabrice Le Bellec, Christian Lavigne, Pierre-François Duyck, Raphaël Achard, Philippe Tixier and Marc Dorel.

In the humid tropics, productivity depends to a great extent on being able to control weeds. This is difficult to do in the absence of herbicides, more so in the areas that are not easily mechanizable or when labour is scarce or expensive. This is especially true in the case of West Indian orchards and banana plantations, which are currently major herbicide consumers located in fragile ecosystems. The introduction of cover crops in orchards and banana plantations is being encouraged by CIRAD in partnership with producer groups in order to restrict the use of pesticides and optimize several ecosystem services. Cover crops alter the water and nutrient cycles as well as the interactions between communities of insects and microorganisms. In fact, they are also able to provide multiple ecosystem services through environmental changes brought about in the physical structure and in the chemical state of the soil.

A multi-criteria evaluation grid was created to select an 'optimal' cover crop for citrus orchards in Martinique. The most important criteria include agro-climatic parameters and the use of local seeds to prevent the importation of exotic seeds. Functional groups of candidate plants were identified from more than two hundred shortlisted species using additional criteria such as weed control, the ability to reduce runoff and erosion, competition for water and nutrients, and the ability to control pests and favour auxiliary fauna. A final cut of a limited number of candidates was made from these groups after considering the agroecological characteristics of the target zones and specific objectives of the producer groups. In Guadeloupe, a participatory approach involving scientific databases, expert opinion, experimental measurements and producer objectives led to the selection of nitrogen-fixing plants (*Fabacea*, *Neotonia wightii*, *Stylosanthes hamata*) characterized by their ability to host auxiliaries. In Martinique, *Urochloa mozambicensis* and *Paspalum* grasses were preferred because of their covering ability and their smaller biomass. The multi-criteria evaluation grid created can be used on a generic basis to select service plants for orchards.

A similar approach for banana plantations has shown the important role of the cover plant in controlling major pests such as weevils and nematodes. Integrated methods to manage service plants in orchards and banana plantations are yet to be developed to maximize ecosystem services. The use of mixed cover, combining plants with different properties, remains a promising solution even though such a system would be difficult to master.

For further information: Jannoyer et al. (2011).

3.4 What New Systems can we Design to Meet the Various Economic, Environmental and Social Challenges of Today?

The path we take in the search of new cropping systems depends on first selecting new compromises between different ecosystem services. Among them, the function of production remains, of course, paramount. At the two extremes are cropping systems that we would classify using an agricultural biodiversity index as either intensive monocultures or complex multispecies systems without input use.

Although often left untouched by 'modern agricultural technology', farmers in the South—over 1 billion of them—use traditional practices based on an 'integrated' management of local natural resources and biodiversity. These practices could very well be models for cropping systems of tomorrow, a hypothesis advanced by some agronomists (Ewel 1999; Altieri 2002; Jackson 2002; Malézieux 2012).

Moreover, the various forms of 'alternative' farming systems, such 'organic' ones, are most often based on a rational use of biodiversity. But will these or traditional systems be able to produce as much as intensive systems which use large amounts of chemical inputs to boost productivity?

To address new environmental and societal issues, agronomy has been forced to review its prevailing paradigms (Doré et al. 2011). Two key issues seem central to agronomy today: Which biodiversity to reintroduce into intensive monocultures to limit their well-known ecosystem 'disservices'? Which biodiversity to include in traditional complex systems to improve their productivity while maintaining their ecosystem services?

We will discuss two aspects below, one after another, which seem to require particular attention. First will be the importance of local sources of agrobiodiversity in the design of cropping systems. The second will cover the relationships between biodiversity and pest control, which determines agroecosystems' productivity to a large extent.

3.4.1 Agrobiodiversity, Human Health and Local Resources

The importance of local food production—based on species hitherto often considered minor—is now the subject of renewed international interest (6th Report of the UN Standing Committee for Nutrition). To feed themselves, many human societies still depend on local resources: plant and animal species which have received little or no attention from the scientific community. The role of these species in the nutritional balance and human health—as also in the ecosystem balance—remains little known and promoted, and clearly deserves greater attention from researchers. In contrast, in industrialized regions, interest in local food production is growing, mainly in order to minimize the energy expenditure associated with the transportation of goods. This interest is also driven by a desire to recreate direct social links between producers and consumers. The ability to produce diverse food locally thus takes on a new significance.

There has been recent renewed interest in the importance of local species and varieties for the balance and resilience of traditional agriculture. If the 1970s and 1980s presaged the rapid replacement of hardy varieties with improved ones adapted for intensification, it is now clear that nothing of the kind happened in several situations. The adaptation of traditional varieties to marginal or specific ecosystem conditions in which they were developed—as also their ability to adapt to heterogeneous ecosystems and to climate and soil variability—earns them a particular interest in the context of rapid change (of markets and environmental conditions). In high arid conditions of the Sahel, over 800 species contribute to the nutritional balance of rural human societies (Grivetti and Ogle 2000). We can even conjecture that the possession of a large diversity of species and hardy varieties by poor rural communities is their greatest weapon in combating the vagaries of climate change. The use of a diversity of local species in agroecosystems, themselves diversified, assures food production security in an uncertain environment. It boosts the system's resilience and builds up its *sustainagility*, i.e., the system's ability to identify and deploy appropriate responses to unpredictable hazards (Jackson et al. 2010).

Their conservation depends, however, on their role and their conditions of use in agroecosystems. Despite the fact that the maintenance and use of traditional species and varieties depend heavily on the location, the crop or the species concerned, Jarvis et al. (2011) have proposed a general framework for analysis to address the use and conservation of traditional varieties in production systems.

3.4.2 Biodiversity and the Control of Pests

The search for systems that are not dependent on pesticides constitutes an important part of the effort to improve current agricultural systems. Researchers are focusing on the role biological diversity in cropping systems plays or can play in controlling pests (see Chap. 5), both at the scale of the cropping systems as well as of the territory they belong to. This new interest in biodiversity is driven by its abilities of regulating pest populations biologically and, more broadly, by its potential ability to maintain sustainable systems (biomass production, control of flows, etc.). Furthermore, the reduced use of chemical inputs that 'organic' practices entail is itself a key benefit for the associated biodiversity. A meta-analysis on the subject thus shows a positive effect on species richness and abundance (Bengtsson et al. 2005).

Cropping systems based on little or no use of pesticides (or, more generally, of chemical inputs) will, of course, require fundamental modifications of the crop arrangements and organization, in space and in time. The dynamics of pest populations compel us to also consider the temporal dynamics of the cropping systems in this space and the evolution over time of habitats and biocenoses in these systems. To able to design such systems, we have to ask questions at several levels. At the generic level we have to ask: What biodiversity to introduce into cropping systems to control pests and optimize the production and its quality? And, more specifically: What groups of plants (mixed crops, crops and service plants) to choose? What crop succession/rotation schedules to use? What interstitial areas should be chosen (borders, etc.)? What mosaic of crops and associated areas (grassy strips, etc.) to encourage in the landscape?

As we can see, it is impossible to approach the design of cropping systems without considering the higher levels of organization of which they are part.

3.5 Scales of Study: From the Plot to the Landscape

Ecosystems can be studied at different scales, ranging from a microcosm of earth in a laboratory setting to an entire ocean. The plot is the obvious level of organization that would interest an agronomist. However, the integration of the plot into the wider 'landscape' is necessary to address the supply of ecosystem services (water quality, pest control, mineral recycling) since they all function at this higher level.

3.5.1 Biodiversity and Models of Development: Segregative Versus Integrative

Environmental functions are now recognized as necessary for agriculture, and creating spaces for wildlife in agricultural areas has become an end in itself in many situations (McNeely and Sherr 2003). Different strategies can be adopted in the pursuit of this goal. They include the creation of biodiversity reserves benefitting local communities, the development of a network of habitats in uncultivated areas, and the restriction of the spread of cultivated areas by increasing productivity. But these strategies may be perceived as obstacles to economic development, particularly in developing countries, and will need to be complemented by increasing the ecological value (in terms of habitat) of cultivated areas. This will primarily happen through (McNeely and Sherr 2003) reduced pollution from agriculture; changes in practices to manage soil, water and plant resources; and the search for cropping systems that 'imitate' natural ecosystems. These various issues link the different components of sustainable development closely together and examine them as a whole. They make us question the agricultural production methods from the Green Revolution, highlighting the need to rediscover how some traditional systems function and to develop new cropping systems. For the last few years, biodiversity has thus been the driving force behind an important—and contentious—movement to define new development models. It finds itself at the core of two opposing models proposed for territorial management: segregation and integration. The segregation model—which advocates reserving the most fertile lands for agricultural, specializing agricultural activities therein and producing intensively—often needs defending nowadays in adversarial contexts. It involves the need to manage externalities, environmental ones especially, through explicit compensatory actions, and to reserve other territories for urban or recreational use or strictly for environmental conservation. The integration model, on the other hand, views agriculture as a land-planning tool. This tool can be used by organizing the diversity of its forms to satisfy a wide range of functions while simultaneously preventing harmful or undesirable effects. The debate on the relative merits of these two models accompanies the one regarding conservation: 'sancturizing' nature (segregation) versus managing nature (integration). The two models have very different approaches to food production. The segregation proponents envisage feeding the world with food grown in highly mechanized and intensified areas. Their method includes investing in food distribution programmes, and managing rural exoduses. The integration camp, on the other hand, views agriculture as a factor of local development. Originally limited to the English-speaking world, this debate is now seeing active participation by the French scientific community (Griffon 2006; *Agrimonde* foresight study in Chaumet et al. 2009).

The segregated option, although commonly adopted today, runs strongly counter to the functional agriculture policy that we advocate. Selection of the integrative option involves re-examining the forms of agricultural production, both in their technology as well as their organization. And we have to do so from a viewpoint that does not differentiate between 'natural' and 'cultivated' areas but rather intermingles them both in a landscape mosaic founded on the interpenetration and complementarities of their various separate ecological functions. Such a system would be built around new agricultural models but would also encompass new social and economic models of production. It would rely on a new organization of processing industries and marketing sectors. It will, above all, lay emphasis on diversity and multifunctionality, thus contributing to reducing—rather than exacerbating—crises caused by exclusion and impoverishment (Hubert and Caron 2009).

Our choice of the integrative option prompts us to revisit the work of the agronomist in a wider perspective—that of the territory. The territory, or the landscape, becomes both the subject and the scope of the research and its preoccupations. It becomes the medium for emergent properties often originating from the very heterogeneity that characterizes it. This new approach requires us to embrace the concept of hierarchical organization, i.e., to consider the systems we analyze (cropping systems, farm management systems, landscapes) as a series of interlocking systems and locally interacting processes at varying organizational levels. From an ecological point of view, the landscape is an arena where resources are shared and niches and habitats occupied by plant and animal species. Landscape ecology provides a conceptual and methodological framework to address the role of the landscape's structure in its biological and physical functioning. How should we spatially and temporally structure the species in order to optimize the system's functioning, i.e., to maximize the desired services?

To answer this question, we will have to bring the concepts of agronomy closer to those of landscape ecology.

3.5.2 The Search for Localized Solutions

The decision to promote biodiversity in order to optimize the production of services at various scales reduces the possibility of finding simple and widely replicable recipes. The agro-industrial intensification processes that agriculture has known may have conditioned the agricultural community to expect uncomplicated turnkey solutions (massive use of fertilizers and pesticides to overcome limiting local factors) but no such simple solutions exist here. A new approach must necessarily be implemented, one which endeavours to find local solutions in close partnership with agricultural producers and other stakeholders. To be successful, this new approach also has to take into account economic and local social contexts. Participatory design approaches are thus necessary, at the landscape scale, for the modification of cropping systems. In the case of coffee, for example, intensification has largely led to the elimination of shade trees. They have been replaced with fertilizers and pesticides ignoring the fact that these trees fulfilled functions other than that of shade (Perfecto et al. 1996). Box 10 illustrates a participatory approach that has been implemented in such a context in Costa Rica by CIRAD.

> **Box 10. Participatory design of agroforestry systems for a better balance between ecosystem services: the case of coffee in Costa Rica**
>
> Cropping systems should be designed keeping in mind the diversity of ecosystem services that are expected from them by farmers and other stakeholders. Take, for example, Costa Rican coffee growers located in a watershed characterized by abundant rainfall and steep slopes. They have not only to use their land to produce coffee that will provide them and their cooperative with income but also reduce, through a judiciously chosen cropping system, the risk of landslides and soil erosion to prevent siltation of a downstream hydroelectric dam.
>
> Two major methodological options were explored in designing the cropping systems. The first mobilizes participatory approaches to bring researchers and actors—mainly farmers—together in the shared formulation of the problem, prototype design and experimental evaluation (Rapidel et al. 2009). It takes into account the expectations of all stakeholders, but may only be able to explore a limited number of locally attractive solutions. The second option is to rely on simulation models to explore and evaluate a wide range of scenarios in an effort to optimize the configuration of the cropping system (Bergez et al. 2010). This approach is relatively theoretical but it does open up the field of possibilities. Moreover, simulation models can find links between processes and performance.

These two approaches have recently been combined. The participatory design with farmers of innovative cropping systems relies on virtual experiments to evaluate their performance (Meylan 2012). This combined approach was used for the case study of coffee-based agroforestry systems in Costa Rica for which stakeholders and the ecosystem services expected were clearly identified. This was possible only because a simulation model that could handle the complex interactions between coffee trees, shade trees and the environment was available (van Oijen et al. 2010).

In a first step, the diversity of cropping practices within the watershed was analyzed using surveys to distinguish four types of strategies (Meylan et al. 2013). These were (i) low-intensive systems with low levels of inputs (ii) labour-intensive systems, (iii) systems with a high density of shade trees, and (iv) systems with intensive use of inputs. Each of these types was represented by a conceptual model that could, on the one hand, uncover the links between cropping practices, states of the system (soil, crop, shade trees, pests) and productive and environmental performance and, on the other, identify the strategy's constraints and opportunities (Fig. 6).

In a second step, the digital model was used in a series of participatory workshops with farmers. It allowed participants to explore, for each of the four strategies, scenarios of changes to farming practices and assess progress in the production of expected ecosystem services. The value of combining prototyping and digital simulation could be seen through the ability of farmers to appropriate simulation outputs to enrich the discussion of change scenarios. For example, the farmers appreciated the fact that the model produced information which would otherwise be difficult to obtain, such as the dynamics of soil resources. The value of the approach was also demonstrated by the willingness of the farmers to try out changes suggested by the simulation on their fields.

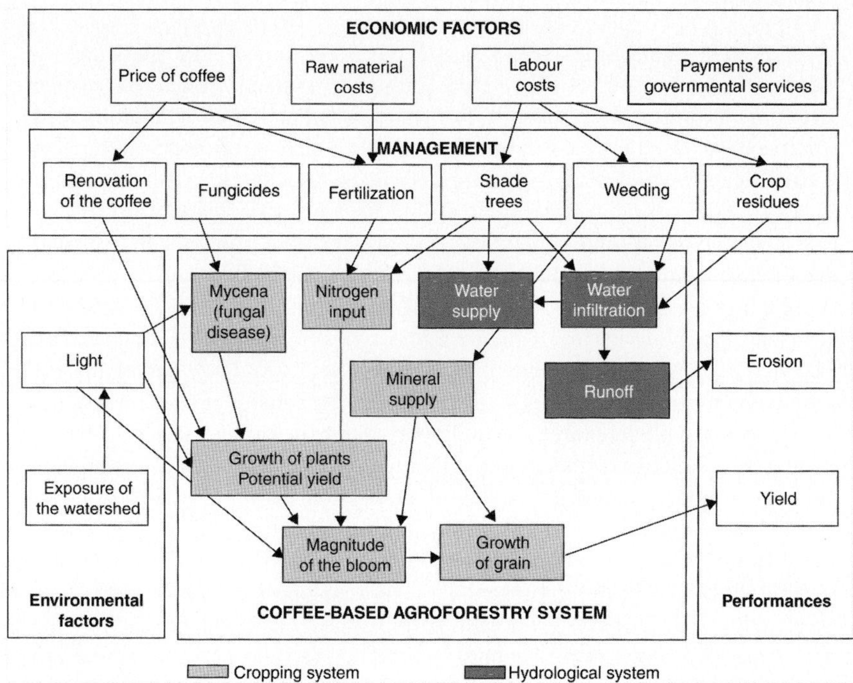

Fig. 6 Conceptual model of relationships between cropping practices, the functioning of the agroforestry system and ecosystem services (coffee production, soil-erosion control)

4 Conclusion

In emphasizing plant and animal production, and especially economic efficiency, the productivist system has long imposed itself at the expense of other ecosystem functions, often ignored or, at best, considered secondary. In response to the almost global demand for productivity, agricultural has followed this system for many decades, especially in Europe, North America and the eastern countries.

Mankind asks a lot from agrosystems. Not only do they have to feed a growing and quality-conscious population and produce and provide multiple services, they also have to be able to respond positively to gradual or sudden climatic or societal changes. Agrobiodiversity is brimming with potential and possibilities and is now recognized as a vital resource in the design of new ecologically innovative cropping systems to meet these requirements. Working with farmers and other stakeholders, researchers—especially agronomists, geneticists and plant breeders—need to exploit this potential in order to develop innovative systems based on new combinations of scientific and local knowledge. 'Cultivating biodiversity' has become a major objective of agronomists around the world in order to develop cropping systems that provide the various ecosystem services required of them.

References

Affholder, F., Jourdain, D., Quang, D. D., Tuong, T. P., Morize, M., & Ricome, A. (2010). Constraints to farmers' adoption of direct-seeding mulch-based cropping systems: a farm scale modeling approach applied to the mountainous slopes of Vietnam. *Agricultural Systems, 103*, 51–62.

Albrecht, A., & Kandji, S. T. (2003). Carbon sequestration in tropical agroforestry systems. *Agriculture, Ecosystems & Environment, 99*, 15–27.

Altieri, M. A. (1999). The ecological role of biodiversity in agroecosystems. *Agriculture, Ecosystems & Environment, 74*, 19–31.

Altieri, M. A. (2002). Agroecology: the science of natural resource management for poor farmers in marginal environments. *Agriculture, Ecosystems & Environment, 93*, 1–24.

Altieri, M. A. (2004). Linking ecologists and traditional farmers in the search for sustainable agriculture. *Frontiers in Ecology and the Environment, 2*(1), 35–42. doi:10.1890/1540-9295(2004)002, [0035:LEATFI] 2.0.CO;2.

Altieri, M. A., & Nicholls, C. I. (1999). Biodiversity, ecosystem function and insect pest management in agricultural systems. In W. W. Collins & C. O. Qualset (Eds.), *Biodiversity in Agroecosystems*. Boca Raton: CRC Press.

Atiama-Nurbel, T., Deguine, J.-P., & Quilici, S. (2012). Maize more attractive than Napier grass as non-host plants for *Bactrocera cucurbitae* and *Dacus demmerezi*. *Arthropod-Plant Interaction,*. doi:10.1007/s11829-012-9185-4.

Avelino, J., ten Hoopen, G. M., & DeClerck, F. (2011). Ecological mechanisms for pest and disease control in coffee and cacao agroecosystems of the neotropics. In B. Rapidel, F. DeClerck, J. F. Le Coq, & J. Beer (Eds.), *Ecosystem Services from Agriculture and Agroforestry* (pp. 91–118). London: Earthcan.

Batáry, P., Báldi, A., Sárospataki, M., Kohler, F., Verhulst, J., Knop, E., et al. (2010). Effect of conservation management on bees and insect-pollinated grassland plant communities in three European countries. *Agriculture, Ecosystems & Environment, 136*, 35–39.

Benbrook, C. M. (2012). Impacts of genetically engineered crops on pesticide use in the US: The first sixteen years. *Environmental Sciences Europe, 24*, 24.

Bengtsson, J., Ahnström, J., & Weibull, A. C. (2005). The effects of organic agriculture on biodiversity and abundance: a meta-analysis. *Journal of Applied Ecology, 42*, 261–269.

Bergez, J.-E., Colbach, N., Crespo, O., Garcia, F., Jeuffroy, M.-H., Justes, E., et al. (2010). Designing crop management systems by simulation. *European Journal of Agronomy, 32*, 3–9.

Blanchart, E., Bernoux, M., Sarda, X., Siqueira, Neto M., Cerri, C. C., Piccolo, M., et al. (2007). Effect of direct seeding mulch-based systems on soil carbon storage and macrofauna in central Brazil. *Agriculturae Conspectus Scientificus, 72*, 81–87.

Bloch, M. (1976). *Les caractères originaux de l'histoire rurale française*, coll. Économies-Sociétés-Civilisations, Paris: Armand Colin.

Bonnal, P., Bonin, M., & Aznar, O. (2012). Les évolutions inversées de la multifonctionnalité de l'agriculture et des services environnementaux. *Vertigo, 12*(3).

Boyce, J. K. (2004). A future for small farms? Biodiversity and sustainable agriculture. In J. K. Boyce, S. Cullenberg, P. K. Pattanaik, & R. Pollin (Eds.), Human development in the era of globalization: essays in honor of Keith B. Griffin (pp. 83–104). Northampton: Edward Elgar.

Brisson, N., Gate, P., Gouache, D., Charmet, D., Oury, F. X., & Huard, F. (2010). Why are wheat yields stagnating in Europe? A comprehensive data analysis for France. *Field Crops Research, 119*(1), 201–212.

Brussaard, L., Caron, P., Campbell, B., Lipper, L., Mainka, S., Rabbinge, R., et al. (2010). Reconciling biodiversity conservation and food security: scientific challenges for a new agriculture. *Current Opinion in Environmental Sustainability, 2*, 34–42.

Calba, H., Cazevieille, P., & Jaillard, B. (1999). Modelling of the dynamics of Al and protons in the rhizosphere of maize cultivated in acid substrate. *Plant and Soil, 209*, 57–69.

Cauderon, A. (1981). Sur les approches écologiques de l'agriculture. *Agronomie, 1*(8), 611–616.

Celette, F., Gaudin, R., & Gary, C. (2008). Spatial and temporal changes to the water regime of a Mediterranean vineyard due to the adoption of cover cropping. *European Journal of Agronomy, 29*, 153–162.

Cerdán, C. R., Rebolledo, M. C., Soto, G., Rapidel, B., & Sinclair, F. L. (2012). Local knowledge of impacts of tree cover on ecosystem services in smallholder coffee production systems. *Agricultural Systems, 110*, 119–130.

Chaumet, J. M., Delpeuch F., Dorin, B., Ghersi, G., Hubert B., & Le Cotty,T., et al. (2009). *Agrimonde. Agriculture et alimentations du monde en 2050: scénarios et défis pour un développement durable.* Note de synthèse, Inra-Cirad: Paris. Retrieved April 21, 2013, from http://www.paris.inra.fr/prospective/projets/agrimonde

Cheverry, C. (2010). Dégradation des terres, comment l'évaluer. In*: Les mots de l'agronomie*, Inra Editions: Paris. Retrieved April 21, 2013, from http://mots-agronomie.inra.fr

Connor, D. J. (2001). Optimizing crop diversification. In J Nosberger, H. H. Geiger, & P. C. Struik (Eds.), *Crop science: Progress and prospects*, (pp. 191–211). Wallingford: CAB International.

Conway G. (1997). *The doubly Green Revolution: Food for all in the 21st Century.* London: Penguin Books.

Conway, G., Carsalade, H., & Griffon, M. (1994). *Une agriculture durable pour la sécurité alimentaire mondiale.* Paris: Cirad-Urpa/Ecopol.

Cornelissen, J. H. C., Lavorel, J. S., Garnier, E., Díaz, S., Buchmann, N., Gurvich, D. E., et al. (2003). A handbook of protocols for standardised and easy measurement of plant functional traits worldwide. *Australian Journal of Botany, 51*, 335–380.

Crétenet, M., & Titonell, P. (2010). Discontinuous fertiliser use affects soil responsiveness and widens yield gaps in cotton-based cropping systems of North Cameroon. In*: Proceedings of Agro 2010: The XIth ESA Congress*, August 29–September 3, 2010 (pp. 347–348). Montpellier: Agropolis International.

Crovetto Lamarc, C. (2007–2008). *Les fondements d'une agriculture durable. 1 et 2*, Panam France, Villemur-sur-Tarn.

Deberdt, P., Perrin, B., Coranson-Beaudu, R., Duyck, P. F., & Wicker, E. (2012). Effect of *Allium fistulosum* extract on *Ralstonia solanacearum* populations and tomato bacterial wilt. *Plant Disease, 96*(5), 687–692.

Deguine, J.-P., Rousse, P., & Atiama-Nurbel, T. (2012a). Agroecological crop protection: concepts and a case study from Reunion. In L. Larramendy & S. Soloneski (Eds.), *Integrated Pest Management and Pest Control: Current and Future Tactics* (pp. 63–76). San Antonio: Intech Publisher.

Deguine, J.-P., Atiama-Nurbel, T., Douraguia, E., Chiroleu, F., & Quilici, S. (2012b). Species diversity within a community of the Cucurbit fruit flies *Bactrocera cucurbitae, Dacus ciliatus* and *Dacus demmerezi* roosting in corn borders near cucurbit production areas of Reunion Island. *Journal of Insect Science, 12*, 1–15.

Deguine, J.-P., Douraguia, E., Atiama-Nurbel, T., Chiroleu, F., & Quilici, S. (2012c). Cage study of Spinosad-based bait efficacy on *Bactrocera cucurbitae, Dacus ciliatus* and *Dacus demmerezi* (*Diptera: Tephritidae*) in Reunion Island. *Journal of Economic Entomology, 105*, 1358–1365.

Deheuvels, O., Avelino, J., Somarriba, E., & Malézieux, E. (2012). Vegetation structure and productivity in cocoa-based agroforestry systems in Talamanca, Costa Rica. *Agriculture, Ecosystems & Environment, 149*, 181–188.

Devendra, C., & Thomas, D. (2002). Smallholder farming systems in Asia. *Agricultural Systems, 71*, 17–25.

Djigal, D., Chabrier, C., Duyck, P. F., Achard, R., Quénéhervé, P., & Tixier, P. (2012). Cover crops alter the soil nematode food web in banana agroecosystems. *Soil Biology & Biochemistry, 48*, 142–150.

Doré, T., & Maraux, F. (2010). Les manières de produire en agriculture, état des lieux et controverses. In T. Doré & O. Réchauchère (Eds.), *La question agricole mondiale: enjeux économiques, sociaux et environnementaux* (pp. 115–134). Paris: Les études de La Documentation française.

Doré, T., Makowski, D., Malézieux, E., Munier-Jolain, N., Tchamitchian, M., & Tittonell, P. (2011). Facing up to the paradigm of ecological intensification in agronomy: Revisiting methods, concepts and knowledge. *European Journal of Agronomy, 34*(4), 197–210.

Dufumier, M. (2004). *Agricultures et paysanneries des tiers-mondes, coll.* Karthala éditions: Hommes et sociétés. 600 p.

Duhautois, S. (2010). Structuration des communautés de Diptères sur le maïs, *Zea mays*, utilisés comme plante piège à la Réunion. Master 2 en Biologie de l'évolution et écologie (BEE), University of Montpellier 2, 37 p.

Dupraz, C., & Liagre, F. (2008). *Agroforesterie, des arbres et des cultures*, éditions (p. 416). Paris: France Agricole.

Duyck, P. F., Lavigne, A., Vinatier, F., Achard, R., Okolle, J. N., & Tixier, P. (2011). Addition of a new resource in agroecosystems: do cover crops alter the trophic positions of generalist predators? *Basic and Applied Ecology, 12*, 47–55.

Ekström, G., & Ekbom, B. (2011). Pest control in agroecosystems: An ecological approach. *Critical Reviews in Plan Sciences, 30*, 74–94.

Ellis, F. (2000). The determinants of rural livelihood diversification in developing countries. *Journal of Agricultural Economics, 51*, 289–302.

Ewel, J. J. (1999). Natural systems as models for the design of sustainable systems of land use. *Agroforestry Systems, 45*, 1–21.

FAO. (2009). *Increasing Crop Production Sustainability: The Perspective of Biological Processes* (p. 36). Rom: FAO.

Feintrenie, L., Kian, Chong W., & Levang, P. (2010). Why do farmers prefer oil palm? Lessons learnt from Bungo District, Indonesia. *Small-Scale Forestry, 9*, 379–396.

Feyt, H. (2007). Évolutions et ruptures en amélioration des plantes. In P. Robin, J. -P. Aeschlimann & C. Feller (Eds.), *Histoire et agronomie. Entre ruptures et durée, coll. Colloques et Séminaires*. Paris: IRD Éditions.

Floret, C., & Pontanier, R. (2000). *La jachère en Afrique tropicale : de la jachère naturelle à la jachère améliorée. 2. Le point des connaissances* (p. 339). Montrouge: John Libbey Eurotext.

Fok, M. (2011). Gone with transgenic cotton cropping in the USA: A perception of the presentations and interactions at the Beltwide Cotton Conferences, New Orleans (Louisiana, USA), January 4–7, 2010. *Biotechnologie, agronomie, société et environnement, 15*(4), 545–552.

Freschet, G., Masse, D., Hien, E., Sall, S., & Chotte, J.-L. (2008). Long-term changes in organic matter and microbial properties resulting from manuring practices in an arid cultivated soil in Burkina Faso. *Agriculture, Ecosystems & Environment, 123*, 175–184.

Friedrich T., Kienzle J., & kassam A. H. (2009). *Conservation agriculture in developing countries: the role of mechanization.* Paper Presented at the Club of Bologna Meeting on Innovation for Sustainable Mechanisation, Hannover, Germany, November 2, 2009. Retrieved April 21, 2013, from http://www.clubofbologna.org/ew/documents/Friedrich_MF.pdf

Ganry, F., & Gueye, F. (1992). La mise en valeur des bas-fonds de la zone soudano-sahélienne par *Sesbania rostrata* est-elle possible ? *L'Agron Trop, 46*(2), 155–159.

Ganry, F., Barthès, B., & Gigou, J. (2011). Les défis du maintien de la fertilité des sols tropicaux: cas de l'Afrique de l'Ouest. In M. C. Girard (Ed.), *Environnement et sols*. Paris: Dunod.

Gastal, F., & Lemaire, G. (2002). N uptake and distribution in crops: an agronomical and ecophysiological perspective. *Journal of Experimental Botany, 53*(370), 789–799.

Gaudin, R., Celette, F., & Gary, C. (2010). Contribution of run-off to incomplete off season soil water refilling in a Mediterranean vineyard. *Agricultural Water Management, 97*, 1534–1540.

Gay, J. P. (1984). *Fabuleux maïs : histoire et avenir d'une plante*, Éditions. Pau: AGPM, Association générale des producteurs de maïs.

Giller, K. E., Witter, E., Corbeels, M., & Tittonell, P. (2009). Conservation agriculture and smallholder farming in Africa: The heretics' view. *Field Crops Research, 114*, 23–34.

Giller, K. E., Corbeels, M., Nyamangara, J., Triomphe, B., Affholder, F., Scopel, E., et al. (2011). A research agenda to explore the role of conservation agriculture in African smallholder farming systems. *Field Crops Research, 124*, 468–472.

Gitay, H., & Noble, I.R. (1997). What are functional types and how should we seek them? In T. M. Smith, H. H. Shugart, & F. I. Woodward (Eds.), *Plant functional types: Their relevance to ecosystem properties and global change* (pp. 3–19). Cambridge: Cambridge University Press.

Godfray, H. C. J., Beddington, J. R., Crute, R. I., Haddad, L., Lawrence, H., Lawrence, D., et al. (2010). Food security: The challenge of feeding 9 billion people. *Science, 327*(5967), 812–818. doi:10.1126/science.1185383.

Gomiero, T., Pimentel, D., & Paoletti, M. G. (2011a). Environmental impact of different agricultural management practices: Conventional vs organic agriculture. *Critical Reviews in Plant Science, 30*(1), 95–124.

Gomiero, T., Pimentel, D., & Paoletti, M. G. (2011b). Is there a need for a more sustainable agriculture? *Critical Reviews in Plant Science, 30*, 6–23.

Gourlet-Fleury, S., Blanc, L., Picard, D., Sist, P., Dick, J., Nasi, R., et al. (2005). Grouping species for predicting mixed tropical forest dynamics: Looking for a strategy. *Annals of Forest Science, 62*, 785–796.

Griffon, M. (1995). Towards a doubly green revolution. In: *Proceedings of a seminar, Poitiers, Futuroscope*. Paris: Cirad-FPI.

Griffon, M. (2002). *Révolution verte, révolution doublement verte: quelles technologies, quelles institutions et quelle recherche pour les agricultures de l'avenir ?* Colloque Nature, sociétés et développement durable, August 29, 2002. Montpellier.

Griffon, M. (2006). *Nourrir la planète* (p. 456). New York: Odile Jacob.

Grivetti, L. E., & Ogle, B. M. (2000). Value of traditional foods in meeting macro and micronutrient needs: The wild plant connection. *Nutrition Research Reviews, 13*(1), 31–46.

Hector, A., Schmid, B., Beierkuhnlein, C., Caldeira, M. C., Diemer, M., Dimitrakopoulos, P. G. et al. (1999). Plant diversity and productivity experiments in European grasslands. *Science, 286*, 1123–1127.

Hien, E., Ganry, F., & Oliver, R. (2006). Carbon sequestration in a savannah soil in Southwestern Burkina as affected by cropping and cultural practices. *Arid Land Research and Management, 20*(2), 133–146.

Hooper, D. U., Bignell, D. E., Brown, V. K., Brussard, L., Dangerfield, J. M., Wall, D. H., et al. (2000). Interaction between aboveground and belowground biodiversity in terrestrial ecosystems: Patterns, mechanisms, and feedbacks. *BioScience, 50*, 1049–1061.

Hubert, B., & Caron, P. (2009). Imaginer l'avenir pour agir aujourd'hui, en alliant prospective et recherche : l'exemple de la prospective Agrimonde. *Natures Sciences Sociétés, 17*, 417–423.

IAASTD. (2008). *International assessment of agricultural knowledge, science and technology for development* Global report. Retrieved April 21, 2013, from http://www.agassessment.org

Jackson, W. (2002). Natural systems agriculture: A truly radical alternative. *Agriculture, Ecosystems & Environment, 88*, 111–117.

Jackson, L. E., Pascual, U., & Hodgkin, T. (2007). Utilizing and conserving agrobiodiversity in agricultural landscapes. *Agriculture, Ecosystems & Environment, 121*(3), 196–210.

Jackson, L., Bawa, K., Pascual, U., & Perrings, C. (2005). *Agrobiodiversity: A new science agenda for biodiversity in support of sustainable agroecosystems*. Diversitas.

Jackson, L., van Noordwijk, M., Bengtsson, J., Foster, W., Lipper, L., Pulleman, M., et al. (2010). Biodiversity and agricultural sustainagility: From assessment to adaptive management. *Current Opinion in Environmental Sustainability, 2*, 80–87.

Jagoret, P., Michel-Dounias, I., & Malézieux, E. (2011). Long-term dynamics of cocoa agroforests: A case study in central Cameroon. *Agroforestry Systems, 81*, 267–278.

Jagoret, P., Michel-Dounias, I., Snoeck, D., Todem, Ngnogue H., & Malézieux, E. (2012). Afforestation of savannah with cocoa agroforestry systems: A small-farmer innovation in central Cameroon. *Agroforestry Systems, 86*(3), 493–504. doi:10.1007/s10457-012-9513-9.

Jagoret, P., Ngogue, H. T., Bouambi, E., Battini, J. L., & Nyasse, S. (2009). Diversification des exploitations agricoles à base de cacaoyer au centre Cameroun: mythe ou réalité ? *Biotechnologie, Agronomie, Societe et Environnement, 13*, 271–280.

Jannoyer-Lesueur, M., Le Bellec, F., Lavigne, C., Achard, R., & Malézieux, E. (2011). Choosing cover crops to enhance ecological services in orchards: A multiple criteria and systemic approach applied to tropical areas. *Procedia Environmental Sciences, 9*, 104–112.

Jarvis, D. I., Hodgkin, T., Sthapit, B. R., Fadda, C., & Lopez-Noriega, I. (2011). An heuristic framework for identifying multiple ways of supporting the conservation and use of traditional crop varieties within the agricultural production system. *Critical Reviews in Plant Sciences, 30*(1–2), 125–176.

Johns, T., & Eyzaguirre, P. B. (2006). Linking biodiversity, diet and health in policy and practice. In*: Proceedings of the Nutrition Society* (Vol. 65, pp. 182–189). DOI: 10.1079/PNS2006494.

Jose, S. (2009). Agroforestry for ecosystem services and environmental benefits: an overview. *Agroforest Systems, 76*, 1–10.

Kintche, K. (2011). Analyse et modélisation de l'évolution des indicateurs de la fertilité des sols cultivés en zone cotonnière du Togo. Thèse en Sciences de la Terre et de l'Environnement, University of Bourgogne, 192 p.

Kristjanson, P., Reid, R. S., Dickson, N., Clark, W. C., Romney, D., & Puskur, R. et al. (2009). Linking international agricultural research knowledge with action for sustainable development. In*: Proceedings of the National Academy of Sciences of the USA* (Vol. 106 (13), pp. 5047–5052).

Kumar, B. M., & Nair, P. K. R. (2004). The enigma of tropical homegardens. *Agroforestry Systems, 61–62*, 135–152.

Laboucheix, J. (1987). Évaluation de l'efficacité du méthyl parathion vis-à-vis d'*Anthonomus grandis* Boheman en culture cotonnière au Nicaragua. *Coton et fibres tropicales, 42*(1), 41–53.

Lavigne Delville, P., Broutin, C., & Castellanet, C. (2004). Jachères, fertilité, dynamiques agraires, innovations paysannes et collaborations chercheurs/paysans. *Coopérer aujourd'hui, 36*, GRET.

Lavorel, S., & Garnier, E. (2002). Predicting changes in community composition and ecosystem functioning from plant traits: Revisiting the Holy Grail. *Functional Ecology, 16*, 545–556.

Le Bellec, F., Dubois, P., Sarthou, J.P., Malézieux, E. (2011). Predicting the nectar provisioning capacity of weeds to beneficial arthropods using an integrative indicator: an application to tropical orchards. In*: Reconception et évaluation des systèmes de culture: le cas de la gestion de l'enherbement en vergers d'agrumes en Guadeloupe* (F. Le Bellec). Thèse, université des Antilles-Guyane, 125–145.

Lenné, J. M., & Wood, D. (2011). Agricultural revolutions and their enemies: lessons for policy makers. In J. M. Lenné & D. Wood (Eds.), *Agrobiodiversity Management for Food Security: A critical review* (pp. 212–227). Wallingford: CABI.

Liebman, M., & Dyck, E. (1993). Crop rotation and intercropping strategies for weed management. *Ecological Applications, 3*, 92–122.

Lienhard, P., Tivet, F., Chabanne, A., Dequiedt, S., Lelièvre, M., Sayphoummie, S., et al. (2012). No-till and cover crops shift soil microbial abundance and diversity in Laos tropical grasslands. *Agronomy for Sustainable Development,*. doi:10.1007/s13593-012-0099-4.

Lin, B. B. (2011). Resilience in agriculture through crop diversification: Adaptive management for environmental change. *BioScience, 61*, 183–193.

Loreau, M., Naeem, S., Inchausti, P., Bengtsson, J., Grime, J.-P., Hector, A., et al. (2001). Biodiversity and ecosystem functioning: Current knowledge and future challenges. *Science, 294*, 804–808.

Mailloux, J., Le Bellec, F., Kreiter, S., Tixier, M. S., & Dubois, P. (2010). Influence of ground cover management on diversity and density of phytosciid mites (*Acari: Phytoseiidae*) in Guadeloupean citrus orchards. *Experimental and Applied Acarology, 52*(3), 275–290. Retrieved April 21, 2013, from http://dx.doi.org/10.1007/s10493-010-9367-7

Malézieux, E. (2012). Designing croping systems from nature. *Agronomy for Sustainable Development, 32*(1), 15–29. doi:10.1007/s13593-011-0027-z.

Malézieux, E., Crozat, Y., Dupraz, C., Laurans, M., Makowski, D., Ozier-Lafontaine, H., et al. (2009). Mixing plant species in cropping systems: concepts, tools and models: A review. *Agronomy for Sustainable Development, 29*, 43–62.

Maltas, A., Corbeels, M., Scopel, E., Macena Da Silva, F.-A., & Wery, J. (2009). Cover crop effects on nitrogen supply and maize productivity in no-tillage systems of the Brazilian Cerrados. *Agronomy Journal, 101*, 1036–1046.

Maraux, F. (1994). Modélisation mécaniste et fonctionnelle du bilan hydrique des cultures : le cas des sols volcaniques du Nicaragua. Thèse de doctorat en Agronomie, Montpellier, Cirad, Paris, INAPG, 285 p.

Marlet, S., Barbiero, L., & Valles, V. (1998). Soil alkalinization and irrigation in the sahelian zone of Niger. 2: Agronomic consequences of alkalinity and sodicity. *Arid Soil Research and Rehabilitation* (now *Arid Land Research and Management*), *12*(2), 139–152.

Mazoyer, M., & Roudart, L. (1997). *Histoire des agricultures du monde. Du Néolithique à la crise contemporaine*, Paris, coll. Points Histoire, Éditions du Seuil, 705 p.

McNeely, J. A., & Scherr, S. J. (2003). *Ecoagriculture: Strategies to Feed the World and Save Wild Biodiversity* (p. 323). London: Island Press.

MEA (Millenium Ecosystems Assessment). (2005). *Ecosystems and Human Well-Being: Biodiversity Synthesis. MA.* Retrieved April 21, 2013 from http://www.unep.org/maweb/documents/document.356.aspx.pdf.

Metay, A., Alves Moreira, J. A., Bernoux, M., Boyer, T., Douzet, J.-M., Feigl, B., et al. (2007). Storage and forms of organic carbon in a no-tillage under cover crops system on clayey Oxisol in dryland rice production (Cerrados, Brazil). *Soil and Tillage Research, 94*(1), 122–132.

Meylan, L. (2012). Design of cropping systems combining production and ecosystem services: developing a methodology combining numerical modeling and participation of farmers. Application to coffee-based agroforestry in Costa *Rica* (p.122). *Ph.D. thesis*, Montpellier SupAgro, Montpellier.

Meylan, L., Mérot, A., Gary, C., & Rapidel, B. (2013). Combining a typology and a conceptual model of cropping system to explore the diversity of relationships between ecosystem services: The case of erosion control in coffee-based agroforestry systems in Costa Rica. *Agricultural Systems, 118*, 52–64.

Molle, F., & Maraux, F. (2008). A-t-on assez d'eau pour nourrir le planète ? *Pour la Science, dossier, 58*, 98–102.

Morton, J. F. (2007). *The impact of climate change on smallholder and subsistence agriculture: Proceedings of the National Academy of Sciences of the United States of America* (Vol. 104, pp. 19680–19685).

Paillard, S., Treyer, S., & Dorin, B. (coord.) (2010). *Agrimonde. Scénarios et défis pour nourrir le monde en 2050,* Éditions Quae, 288 p.

Passioura, J. B. (1999). Can we bring a perennially peopled and productive countryside? *Agroforestry Systems, 45*, 411–421.

Peltier, R. (Éd.). (1996). Les parcs à *Faidherbia albida. Cahiers scientifiques*, 12, Montpellier: Cirad-Forêt, Centre international de Baillarguet.

Perfecto, I., Rice, R. A., Greenberg, R., & van der Voort, M. E. (1996). Shade coffee: A disappearing refuge for biodiversity. *BioScience, 46*(8), 598–608.

Piper, J. K. (1999). Natural systems agriculture. In W. W. Collins & C. O. Qualset (Eds.), *Biodiversity in Agroecosystems* (pp. 167–189). Boca Raton: CRC Press.

Rapidel, B., Traoré, B. S., Sissoko, F., Lançon, J., & Wery, J. (2009). Experiment based prototyping to design and assess cotton management systems in West Africa. *Agronomy for Sustainable Development, 29*(4), 545–546.

Ratnadass, A., Fernandes, P., Avelino, J., & Habib, R. (2012). Plant species diversity for sustainable management of crop pests and diseases in agroecosystems: A review. *Agronomy for Sustainable Development, 32*, 273–303.

Ratnadass, A., Ryckewaert, P., Claude, Z., Nikiema, A., Pasternak, D., Woltering, L., et al. (2011). New ecological options for the management of horticultural crop pests in sudano-sahelian agrosystems of West Africa. *Acta Horticulturae, 917*, 85–91.

Rice, R. A., & Greenberg, R. (2000). Cacao cultivation and the conservation of biological diversity. *Ambio, 29*, 167–173.

Ricketts, T. H. (2004). Tropical forest fragments enhance pollinator activity in nearby coffee crops. *Conservation Biology, 18*, 1262–1271.

Rivest, D., Cogliastro, A., Bradley, R. L., & Olivier, A. (2010). Intercropping hybrid poplar with soybean increases soil microbial biomass, mineral N supply and tree growth. *Agroforestry Systems, 80*, 33–40.

Roulet, M., Lucan, E. M., Fareila, N., Serique, G., Coelho, H., & Sousa Passis, C. J., et al. (1999). Effects of recent human colonization on the presence of mercury in Amazonian ecosystems. *Water, Air, & Soil Pollution, 112*, 297–313.

Ruf, F. O. (2011). The myth of complex cocoa agroforests: The case of Ghana. *Human Ecology, 39*, 373–388.

Rusch, A., Valantin-Morison, M., Sarthou, J. P., & Roger-Estrade, J. (2010). Biological control of insect pests in agroecosystems: Effects of crop management, farming systems, and seminatural habitats at the landscape scale: A review. *Advances in Agronomy, 109*, 219–259.

Russel, A. E. (2002). Relationships between crop-species diversity and soil characteristics in southwest Indian agroecosystems. *Agriculture, Ecosystems & Environment, 92*, 235–249.

Sachs, J. D., Baillie, J. E. M., & Sutherland, W. J. (2009). Biodiversity conservation and the millenium development goals. *Science, 325*, 1502–1503.

Sanogo, Z. J. -L. (1997). Maîtrise de l'azote dans un système cotonnier-sorgho : prévision de la fumure organique et azotée en zone Mali-Sud. Thèse de doctorat, Ensam de Montpellier, France, 72 p. + appendices.

Schroth, G., & Harvey, C. A. (2007). Biodiversity conservation in cocoa production landscapes: An overview. *Biodiversity and Conservation, 16*, 2237–2244.

Scopel, E., Douzet, J. M., Macena Da Silva, F. A., Cardoso, A., Alves Moreira, J. A., Findeling, A., et al. (2005). Impacts des systèmes de culture en semis direct avec couverture végétale (SCV) sur la dynamique de l'eau, de l'azote minéral et du carbone du sol dans les *cerrados* brésiliens. *Cahiers Agricultures, 14*(1), 71–75.

Scopel, E., Triomphe, B., Affholder, F., Macena Da Silva, F. A., Corbeels, M., Valadares Xavier, J. H., et al. (2012). Conservation agriculture cropping systems in temperate and tropical conditions, performances and impacts. *A review. Agronomy for Sustainable Development,*. doi:10.1007/s13593-012-0106-9.

Seguy, L., Bouzinac, S., & Maronezzi, A. C. (2001). *Cropping systems and organic matter dynamics: direct seeding on plant cover, an agricultural revolution*. Retrieved April 21, 2013, from http://agroecologie.cirad.fr/content/download/6886/32983/file/999900043.pdf

Smits, N., Dupraz, C., & Dufour, L. (2012). Surprising lack of influence of tree rows on the dynamics of wheat aphids and their natural enemies in a temperate agroforestry system. *Agroforestry Systems, 85*, 153–164.

Snaddon, J., & Vodouhe, R. (2010). Biodiversity and agricultural sustainagility: from assessment to adaptive management. *Current Opinion in Environmental Sustainability, 2*, 80–87.

Snoeck, D., Abolo, D., & Jagoret, P. (2010). Temporal changes in VAM fungi in the cocoa agroforestry systems of central Cameroon. *Agroforestry Systems, 78*, 323–328.

Snoeck, D., Lacote, R., Kéli, J., Doumbia, A., Chapuset, T., Jagoret, P., et al. (2013). Association of hevea with other tree crops can be more profitable than hevea monocrop during first 12 years. *Industrial Crops and Products, 43*, 578–586.

Söderström, B., Svensson, B., Vessby, K., & Glimskar, A. (2001). Plants, insects and birds in semi-natural pastures in relation to local habitat and landscape factors. *Biodiversity and Conservation, 10*, 1839–1863.

Stamps, W. T., & Linit, M. J. (1998). Plant diversity and arthropod communities: implications for temperate agroforestry. *Agroforestry Systems, 39*, 73–89.

Staver, C., Guharay, F., Monterroso, D., & Muschler, R. G. (2001). Designing pest-suppressive multistrata perennial crop systems: Shade-grown coffee in Central America. *Agroforestry Systems, 53*, 151–170.

Steenwerth, K. L., & Belina, K. M. (2008). Cover crops enhance soil organic matter, carbon dynamics and microbiological function in a vineyard agroecosystem. *Applied Soil Ecology, 40*, 359–369.

Swift, M. J., Izac, A. M., & Van Noordwijk, M. (2004). Biodiversity and ecosystem services in agricultural landscapes: are we asking the right questions ? *Agriculture, Ecosystems & Environment, 104*, 113–114.

Tabashnik, B. E., Van Rensburg, J. B. J., & Carrière, Y. (2009). Field-evolved insect resistance to Bt crops: Definition, theory, and data. *Journal of Economic Entomology, 102*(6), 2011–2025.

Tilman, D., Wedin, D., & Knops, J. (1996). Productivity and sustainability influenced by biodiversity in grassland ecosystems. *Nature, 379*, 718–720.

Tilman, D., Cassman, K. G., Matson, P. A., Naylor, R., & Polasky, S. (2002). Agricultural sustainability and intensive production practices. *Nature, 418* (8).

Tingem, M., Rivington, M., & Bellocchi, G. (2009). Adaptation assessments for crop production in response to climate change in Cameroon. *Agronomy for Sustainable Development, 29*, 247–256.

Tixier, P., Lavigne, C., Alvarez, S., Gauquier, A., Blanchard, M., Ripoche, A., et al. (2011). Model evaluation of cover crops, application to eleven species for banana cropping systems. *European Journal of Agronomy, 34*, 53–61.

Traoré, K. (2003). Le parc à karité, sa contribution à la durabilité de l'agrosystème : cas d'une toposéquence à Konobougou (Mali-Sud). Thèse de doctorat en Sciences du sol, Ensam de Montpellier, France, 188 p. + appendices.

Tscharntke, T., Klein, A. M., Kruess, A., Steffan-Dewente, I., & Thies, C. (2005). Landscape perspectives on agricultural intensification and biodiversity: ecosystem service management. *Ecology Letters, 8*, 857–874.

van der Werf, H. M. G., Kanyarushoki, C., & Corson, M. S. (2011). L'analyse de cycle de vie : un nouveau regard sur les systèmes de production agricole. *Innovations agronomiques, 12*, 121–133.

van Oijen, M., Dauzat, J., Harmand, J.-M., Lawson, G., & Vaast, P. (2010). Coffee agroforestry systems in Central America. 2: Development of a simple process-based model and preliminary results. *Agroforestry Systems, 80*, 361–378.

Vandermeer, J., van Noordwijk, M., Anderson, J., Ong, C., & Perfecto, I. (1998). Global change and multi-species agroecosystems: Concepts and issues. *Agriculture, Ecosystems & Environment, 67*, 1–22.

Vitousek, P. M., & Hooper, D. U. (1993). Biological diversity and terrestrial ecosystem biogeochemistry. In E. D. Schulze & H. A. Mooney (Eds.), *Biodiversity and Ecosystem Function* (pp. 3–14). Berlin: Springer Verlag.

Wardle, D. A., Bardgett, R. D., Klironomos, J. N., Setälä, H., van der Putten, W. H., & Wall, D. H. (2004). Ecological linkages between aboveground and belowground biota. *Science, 304*, 1629–1633.

Weibull, A. C., Ostman, O., & Granqvist, A. (2003). Species richness in agroecosystems: The effect of landscape, habitat and farm management. *Biodiversity and Conservation, 12*, 1335–1355.

Wheelock, J. (1980). *Nicaragua, imperialismo y dictadura*. Editorial de Ciencias Sociales: Ciudad de La Habana.

Wood, D., Lenné, J. M. (1999). *Agrobiodiversity: Characterization, Utilization and Management*. UK: CABI, 490 p.

Chapter 4
Rethinking Plant Breeding

Nourollah Ahmadi, Benoît Bertrand and Jean-Christophe Glaszmann

Plant breeding is the activity of developing diverse plant varieties that can contribute usefully to cropping and production systems. These breeding efforts are directed at plant improvement. But 'improvement' is a subjective and relative goal and it becomes necessary to regularly break up plant breeding objectives and procedures into clearly defined and manageable units.

The goal of ecological intensification is to add the aspect of sustainability to increases in production. Plant breeding must combine this objective with that of adaptation to overall societal and climatic changes. It must integrate diverse objectives and selection criteria. It must accommodate demands made by new stakeholders willing to help define objectives and evaluate breeding results. Indeed, it should come to terms with demands for a completely fresh look at the concept of 'genetic gains'. Such a gain must not only consider the benefits reaped by a farmer using an improved variety at the level of his plot, but also its expected economic, social and environmental impacts on a larger scale in the event of a wider dissemination of this variety. Global change occurs at such scales and speeds that agricultural systems could respond by replacing species rather than by

N. Ahmadi (✉) · J.-C. Glaszmann
Cirad, UMR AGAP (Joint Research Unit Genetic Improvement and Adaptation of Mediterranean and Tropical Plants), Montpellier, France
e-mail: nourollah.ahmadi@cirad.fr

J.-C. Glaszmann
e-mail: jean-christophe.glaszmann@cirad.fr

B. Bertrand
RPB Joint Research Unit (Plant Resistance to Bioaggressors), Department Bios, Cirad, Montpellier, France
e-mail: benoit.bertrand@cirad.fr

J.-C. Glaszmann
Director of the AGAP Joint Research Unit, (Genetic Improvement and Adaptation of Mediterranean and Tropical Plants), Cirad, Montpellier, France

É. Hainzelin (ed.), *Cultivating Biodiversity to Transform Agriculture*,
DOI: 10.1007/978-94-007-7984-6_4, © Éditions Quæ 2013

seeking better adapted varieties of the usual species. Therefore, it is also necessary to foresee the evolution of a 'portfolio' of species used in target regions. The likely increase in diversity and turnover of ecological, agronomic and socio-economic situations for each species raises the question of which varietal deployment strategy to select. Should one select many local genotypes with short lifespans or fewer versatile varieties with longer lifespans?

The key challenge in biological sciences is of integrating knowledge at different scales—of molecules, tissues, organs and whole plants at different phenological stages. This is a prerequisite to understanding the patterns of regulation by genes and assessing their relevance to the spatio-temporal variability of constraints for which an improvement is required. In addition to this scientific challenge, we also need to address the question of which innovation model to implement. Plant breeding is also a business which must ensure a 'return on investment' and produce goods (new varieties) that ensure a convergence of interests of different economic stakeholders.

Before getting into the details of challenges that face plant breeding in the twenty-first century, it may be useful to recall some concepts as well as some lessons learnt from plant breeding practices in the last century. We do so below and also present the levers that are available to us to address these challenges.

1 Plant Breeding: The Past and the Present

1.1 Genetic Improvement: From Empiricism to Science

Plant breeding is the art and science of modifying the plant genotype to achieve a desired phenotype. Man has practiced plant breeding, in an increasingly intentional, organized and efficient manner, from the time he first settled down, more than 10,000 years ago. These activities have witnessed a sharp acceleration in the last century. Domestication was the first result of selection. It involved trying to prevent spontaneous seed shattering (especially in the case of cereals) and also to obtain bigger grains and fruits. Modifications in genomic regions that determine these traits remain the key distinctive signs (domestication syndrome) between cultivated varieties and their wild relatives. With this *directed breeding*, man also reduced the total or *neutral* diversity used, as breeding was carried out in one or more limited-size populations which did not represent the entire diversity of the wild species. This diversity, thus greatly reduced by the domestication bottleneck, underwent further transformation and specialization resulting from the selective pressure exerted by man and the new environments he colonized. Thus were born local varieties or populations, adapted to environmental conditions and to cropping requirements of small agricultural regions.

The need to intensify agriculture in the face of population growth occasionally resulted in a centralized selection of species and varieties to grow. Varieties of the same species, originating from domestication efforts undertaken elsewhere, were then introduced. As early as in 1012, faced with an influx of migrants from the north and a real shortage of arable land, Zhao Heng, the Emperor of China ordered two annual rice crops using a short-cycle rice variety imported from Annam.

Until the early nineteenth century and the discoveries of Darwin and Mendel of biological processes at the origin of genetic diversity (mutation and recombination), plant breeding consisted of identifying individual plants with the best phenotypes in nature or in local varieties to help propagate their progeny. This *massal* selection led to many positive developments. For instance, it was applied to sugar levels in beetroot in 1786, resulting in sugar beet and the setting up, in 1802, of the first sugar factory in Germany. In 1858, Louis de Vilmorin made a first methodological jump by selecting individuals based on the yield of their progeny rather than on their own phenotype.

Man truly started manipulating natural diversity in the late 19th century with the practice of controlled crosses between individual plants with complementary phenotypes. Genetics progressed rapidly in the 1920s with the emergence of quantitative genetics, in particular the model of Fisher (1918) that reflected both Mendelian laws and biometric relationships between related individuals. This model considered an observed phenotype as the sum of a genetic effect, an environmental effect and the effect of the genotype × environment interaction. This led to the development of a wide range of methods for optimizing breeding procedures based on the statistical estimation of genetic parameters obtained from dedicated experiments. The increase in maize yields in the United States, from 65 kg/ha/year between 1925 and 1955 to more than 110 kg/ha/year thereafter, is often attributed to the optimization of breeding procedures based on quantitative genetics. This was undoubtedly made possible by a greater availability of chemical fertilizers.

The accumulation of knowledge in plant genetics was accompanied by the formalization of a plant breeding industry. A seed production sector gradually came into being and necessary regulations were put in place to help the commercial distribution of plant varieties and seeds.

1.2 Agricultural Modernization and the 'Green Revolution'

The term 'Green Revolution' was first coined in 1968 by the director of the U.S. Agency for International Development (USAID) to describe the introduction and rapid increase in wheat yields in Mexico since 1950, and wheat and rice in Asia, especially in the Indian subcontinent since 1970. The phenomenon has been analyzed from various angles ranging from scientific and technological innovation to American foreign policy in the context of the Cold War and containment of the 'red revolution'. It was the result of a combination of several factors, including the implementation of a rural development strategy advocated by the American economist Schultz in his books *Food for the World* (1945) and *Transforming*

Traditional Agriculture (1964), as well as major advances in the genetic improvement of cereals.

The strategy of this revolution was to bring about all the conditions necessary for the adoption of technical innovations by small producers. These pertained to all aspects of agricultural production: not only a 'technology package' (high-yield varieties, chemical fertilizers, pesticides and herbicides, irrigation, etc.) but also subsidies for the purchase of inputs, supply facilities, credit for the purchase of agricultural equipment, support prices, protection against imports, strengthening of agricultural research and extension services, irrigation schemes, etc. This policy resulted in the establishment of international research centres specialized in enhancing the productivity of food crops and of the Consultative Group on International Agricultural Research (CGIAR). Support was also forthcoming from international financial institutions for the setting up and strengthening of national agricultural research systems (NARS) and agricultural extension systems.

The contributions of genetics consisted in developing varieties whose response curve to chemical fertilizers, especially nitrogen, levelled off only at doses that were three to four times higher than those for traditional varieties. This was achieved by identification of dwarfing genes within the existing diversity of each cereal and its wild relatives, and by transferring them to a small number of traditional varieties adapted to a tropical climate. In addition to reducing plant height and increasing the harvest index and the resulting lodging resistance, some of these genes also had a pleiotropic effect on various other yield components. The production potential of wheat and rice varieties rose from 3–4 T/ha to 10 T/ha. Since that first breakthrough, genetics has progressed mainly along two fronts. On the one hand, it has focused on the consolidation of this production potential through the accumulation of resistance genes to diseases and insects. On the other, it has increased productivity per unit of time by shortening the cropping cycle. The transformation of these genetic advances into innovations was achieved through the extensive dissemination of these varieties and cultivation techniques—mainly the use of fertilizers and pesticides required to ensure the expression of the varieties' production potential—via widespread and active extension systems. It was thus that the IR8 rice variety, created by the International Rice Research Institute (IRRI) in 1966, came to be grown over several million hectares by the mid-1970s. By 1980, semi-dwarf varieties of rice were being cultivated in more than 30 % of the rice paddies of the eleven largest Asian rice producing countries (excluding China).

There is no doubt that Green Revolution varieties have contributed in many ways to increasing production and productivity. To this increase in yields was added the effects of allocating larger areas to rice and wheat crops which happened at the expense of other, less profitable, crops. Also notable were the effects of double and triple annual cropping, made possible by shortening plant cycle duration and eliminating photosensitivity, and those of the expansion of rice cultivation to marginal areas that resulted from shorter cycles and better adaptation to abiotic constraints.

Economic, social and environmental impacts of the Green Revolution in the Third World have been subject to the same intense debate as has 'intensive agriculture' in industrialized countries (Evenson and Rosegrant 2003). On the whole, the Green Revolution in Asia ensured that production grew at a faster rate than the population. This led to self-sufficiency in cereals from the 1980s for the continent. This increase in agricultural productivity helped maintain grain prices at affordable levels for the urban poor and thus helped the fight against hunger. At the farm level, while large landowners were the initial beneficiaries—being immediately able to take advantage of new technologies and associated economic measures—the Green Revolution also gradually reached other groups of farmers. However, irrespective of the size of the farm, it was possible to take advantage of new varieties and the associated 'technology package' only under favourable biophysical conditions. Yet, on this front, despite irrigation development efforts, large regional and local disparities still exist. For example, while the average yields of 55 % of the rice paddies of Asia with good irrigation systems increased from 2 T/ha to more than 5 T/ha, the yield in the remaining 45 % exposed to water excess or deficiency, salinity, or other physical or chemical soil problem stagnated around 2 T/ha. The success of the Green Revolution bypassed sub-Saharan Africa where well-irrigated rice paddies represent only 20 % of the total. Thus, differences in biophysical conditions have resulted in increased disparities in farm incomes and wealth. Finally, the Green Revolution was also accompanied by widespread environmental degradation: soils sterilized by salinization in India, water pollution and related diseases linked to excessive insecticide use in Vietnam.

While the Green Revolution was a major phenomenon in Asia, its impact was highly variable across regions, plants and stakeholders. The same period saw Europe and North America undergoing similar changes in agricultural productivity due to agricultural modernization that began in the early twentieth century and accelerated from 1945. In Africa, the principles of the Green Revolution were the basis of the development of peanut cultivation starting in 1918 (Box 1) and cotton cultivation in 1945 (Box 2). The example of rainfed upland rice (Box 3) shows that when genetic innovation is driven by specific needs, it can be disseminated almost spontaneously. Massive efforts were also undertaken to create high-yielding varieties of other food crops, including maize, sorghum, cassava and beans, yet leading to significant dissemination only in the case of maize case only. The pattern of widespread dissemination of a small number of high-yielding varieties in optimal growing conditions was also seen in various non-food crops and perennial plants.

> **Box 1. Genetic improvement of peanuts in Senegal: adaptation to technical, climatic and societal developments**
>
> In his report titled, '*The current state and future of trade in peanuts in Senegal*', Roubaud (1918) states: 'After the abolition of the slave trade in 1815, the processing of peanuts became the primary resource of Senegal. Insurance schemes, cooperative seed stores, etc. were established in most areas directly by the government. Very often, the harvest is booked in

advance by the peanut processor who provides loans to the farmers. The agriculture department has undertaken a process of improving cultivation methods through the use of animal traction and fertilizers. The Senegalese peanut trade has consequently flourished. However, there are areas of concern. The oil yield from seeds has decreased significantly. At the same time, evidence has grown of increasing parasite-induced damages. The damage is all the more marked when drought conditions prevail. One promising control method could be the use of short duration peanut varieties. There are local peanut varieties of shorter duration than the commercial varieties. The variety named Volète deserves a special mention. It can mature in two months. However, Volète's unattractive appearance and low productivity has resulted in its disappearance from the European market, and it now only serves the native diet. Perhaps a hybridization with commercial varieties would lead to more acceptable plants in all respects. Long-term experiments should be undertaken in the colony's experimental stations for selecting seeds and choosing local peanut breeds that are most adapted to different climatic and soil conditions. The recently established research station at Bambey has yet to orient its efforts in a scientific manner in this important direction.'

The work of selection and breeding began at the Bambey research station in 1924. The selection from local material, introductions, especially from the United States, the biparental hybrids and improved populations obtained by recurrent selection, first by the colonial administration and then by the Oil and Oilseeds Research Institute (IRHO), and finally by the Senegalese Institute for Agricultural Research (ISRA) and by CIRAD, have resulted in continuous acquisition and dissemination of new varieties to meet changing needs. These have included an upright shape to facilitate mechanization, a grain size and shape suitable for use (oil mill or confectionery), disease resistance, shortening of the cycle and drought tolerance due to climatic deterioration since the 1970s, and resistance to aflatoxin-producing fungi in line with European standards (Ba et al. 2005). Until the 1980s, most of the production (700,500 T) was destined for industrial oil production for export, as part of a monopolistic model. The government, and subsequently a State-run corporation, fixed prices and oversaw the production and distribution of seeds, while meticulously sticking to the 'varietal map' that was regularly updated by research (Clavel and N'doye 1997). Subsequently, the withdrawal of the State from this process led to a disruption in seed supply and marketing. Production plummeted.

New and more promising perspectives have emerged since 2004. The Senegalese Association for Grassroots Development, with technical support from ISRA and CIRAD, organizes the production of basic seeds by farmer organizations and trains farmers to produce 'farm' seeds (Mayeux and Da Sylva 2008). The drought tolerant varieties used are a result of the breeding program carried out in Bambey (Khalfaoui 1991; Clavel and Annerose 1995; Clavel et al. 2005).

Box 2. Genetic improvement of cotton: the Green Revolution in an agro-industrial sector

Agricultural research was the weak link in the first attempts to develop the cultivation of cotton in sub-Saharan Africa, which began in the 19th century. The establishment of the Research Institute of Cotton and Exotic Textiles (IRCT) in 1946 and of the French Company for the Development of Textiles (CFDT) in 1949 was a turning point. Production increased from 0.1 MT in 1950 to 2.6 MT at its peak in 2004. More than 16 million small farmers made their living from it (Levrat 2009). CFDT and national cotton companies took 'control' of production and enjoyed a monopoly over purchase, collection, ginning and marketing. This 'control' involved providing inputs on credit and dispensing agricultural advice. The choice of inputs, including varieties, and agricultural advice were based on research conducted by the IRCT and subsequently by its national successors associated with CIRAD. This research was conducted in close collaboration with cotton companies, which also provided its funding. It was an integrated system typical of the Green Revolution, with industry ensuring that the farmers met all the conditions necessary for appropriating the results of research. In the case of varieties, this entailed, on the one hand, ensuring production and distribution of the seeds of new varieties, since the variety was selected through research and by the cotton company and not by farmers. And, on the other, the distribution of everything else that was necessary to achieve the full expression of the genetic potential of these varieties to each farmer: fertilizers, a phytosanitary 'umbrella' programme, a guaranteed purchase price and input subsidies. On their part, plant breeders had to pay particular heed to improving the yield of ginned cotton and the quality of fibre, the main determinants of profitability for cotton companies (Collectif 1991).

A network of regional research entities was set up which brought together multidisciplinary teams from each country. Agronomists, entomologists, pathologists and fibre and grain technology specialists contributed to the definition of the plant ideotype and the evaluation of plant material created by geneticists. In the mid-1950s, the focus shifted to improving the species *Gossypium hirsutum*, which currently covers more than 90 % of the world's cultivated area under cotton, without abandoning the use of interspecific crosses. The most tangible genetic advances have focused on plant architecture and the ratio of dry cotton grain to total dry matter; on resistance to diseases and insects; and on fibre characteristics and ginning out-turn, which increased from 35.5 % in 1962 to 41.9 % in 1992. Its potential and yield stability were also boosted, taking the average seed cotton yield in Francophone Africa from 198 to 975 kg/ha between 1962 and 1992. However, it is difficult to pinpoint the exact contribution made by the variety and other production factors in this increase. The deliberate implementation of breeding schemes and seed production that help maintain a residual genetic variability is probably not unrelated to the combination of productivity and hardiness of varieties such as ISA205 and STAM-F. These varieties were cultivated on over 300,000 ha/year in the 1980s. Another significant genetic advance was the creation of *glandless* varieties for Africa. The grain protein of these varieties can be used in human food and animal feed as their gossypol toxins have been removed (Hau et al. 1997). In 2008, varieties that were developed or co-developed by CIRAD covered 83 % of the 1.2 million hectares of cotton grown in eight Francophone African countries.

Box 3. Varietal innovation for a constrained environment: rainfed rice cultivation in the highlands of Madagascar

The highlands of Madagascar (1,200–2,000 m in altitude) have long been confronted by a mismatch between population growth and agricultural productivity. Despite rice being a staple food here, its yields are stagnating and there is an acute shortage of land suitable for irrigated rice cultivation. Initial attempts were made to introduce another form of rice cultivation, called 'rainfed upland' rice, where rice is grown, like other cereals, on non-flooded aerobic soil. These failed because of the region's cold climate. Low night temperatures greatly stretched the sowing-heading cycle, leading to sterility in upland rice varieties usually grown in low-elevation, hot and humid areas.

Madagascar's National Centre for Applied Research on Rural Development (FOFIFA) and CIRAD took up the breeding of rainfed rice varieties tolerant to high-altitude cold temperatures in the mid-1980s. A crossing programme was launched as no suitable varieties could be identified through the introduction and evaluation of numerous traditional and modern varieties from other cold regions of the world. Cold-tolerant irrigated varieties were crossed with recently developed upland rice varieties for the low and medium altitude areas of Madagascar. Pedigree selection based on evaluation of progenies of these crosses at an altitude of 1,500 m and a multi-local evaluation of the best progenies over four years led to the registration of five cold-tolerant rainfed rice varieties in the official catalogue (Dechanet et al. 1997). These varieties have a yield of 5 T/ha for a cycle time of 145–165 days at 1,500 m in experimental conditions. The only downside to this pioneering achievement was the narrow genetic base of the new varieties: the five varieties had the same cold-tolerant parent, a traditional Malagasy variety grown in irrigated rice paddies at altitudes greater than 1,750 m.

The participatory, multi-local and multi-year evaluation of these varieties, conducted in conjunction with seed-producing farmer organizations, quickly led to the adoption of upland rice cultivation by more than 10 % of the farmers (Dzido et al. 2004). A more recent survey of 843 farms in 26 villages in the Vakinankaratra region located above 1,250 m showed that 62 % of these villages and 36 % of the farms cultivated upland rice using one of the varieties created by FOFIFA and CIRAD. The 'new technology' of upland rice cultivation was adopted mainly through informal exchanges of information and seeds between villages and between farmers (Radanielina 2010).

However, since the early 2000s, the large-scale dissemination of upland rice has led to increasingly frequent outbreaks of blast, a disease caused by the fungus *Magnaporthe oryzae*. The resistance of the first released varieties of relatively narrow genetic base, was quickly overcome (Sester et al. 2008). The breeding programme is now focused on building up blast resistance, diversification of the grain quality of upland rice varieties and efficiency in the use of resources (nitrogen, phosphorus). All this without forgetting the necessary cold tolerance. The programme is also paying attention to upland rice adaptation to cover-crop based cropping systems developed and disseminated by agronomists to improve the sustainability of upland cropping systems in the region (Naudin et al. 2010). The effectiveness of the strategy of mixed varieties using diverse sources and types of blast resistance is also being tested (Raboin et al. 2012).

Agricultural modernization in the North and the Green Revolution in the South led to a decrease in the genetic diversity of crops. This was largely due to the substitution of a large number of local varieties by a small number of 'improved' varieties. Subsequently, commodity-based process integration led to additional technological specifications and selection criteria that further narrowed down the range of varietal diversity that could be deployed.

2 Recent Changes and Developments

2.1 Development of Participatory Approaches

One of the key features of agricultural modernization in industrial nations and of the Green Revolution in developing countries has been the separation of the functions of agricultural production, varietal innovation, seed production and germplasm conservation. The latter three functions moved from farmers to public and private professionals.

Plant breeding programmes were carried out in research stations and targeted plant ideotypes maximizing production per unit of area and time under optimal cropping conditions, giving little consideration to specific end-user perspectives. They ignored the spatial variability of biophysical conditions as well as local knowledge and practices for the use of genetic variability. A homogenizing agronomy thus became associated with a unidirectional varietal improvement.

In developed countries, varietal innovation became subjected to a global organizational framework. This started with approval procedures (registration in the official catalogue) based on performance criteria, predictable 'agronomic and technological value' and 'distinctiveness, uniformity and stability'. These criteria were evaluated in highly artificialized conditions (fertilizers, pesticides, etc.) of experimental stations under standard technical processes, eliminating all forms of diversity of the environment. Varieties thus approved were assumed to have wide adaptability, and were recommended by technical-administrative committees for large geographic areas. They were multiplied by public or private seed companies under conditions that ensured the preservation of their conformity, and were distributed to all farmers in the target area. Thus, a small number of inbred lines and hybrid varieties replaced the large number of local and often heterogeneous varieties, each linked to specific biophysical conditions and uses. These varieties, grouped and categorized according to criteria defined by the plant breeder, have become 'genetic resources' conserved ex situ, under conditions that do not leverage their adaptive and evolutionary potential.

This pattern of varietal innovation, that minimizes the impact of genotype $\times$ environment (G $\times$ E) interaction, soon proved, like the whole Green Revolution model, unsuitable for areas with high soil and climatic variability, high diversity of production systems and high inter-annual climatic variability. In his book '*Rural Development: Putting the Last First*', Chambers (1983) highlights the need to mobilize complex knowledge to manage risks. He recommends taking advantage of diversity and combining scientific knowledge with local knowledge in order to optimize research and rural development. These precepts were subsequently gradually integrated into plant breeding programmes. 'Participatory selection' approaches were developed to take local socio-agroecological specificities into account by encouraging farmers to participate in developing new varieties. Formalized as *Participatory Plant Breeding* in the mid-1990s (Hardon 1995) and advocated widely internationally, this new line of research reinstated the farmer at the heart of varietal improvement efforts. While defining the selection criteria, priority is accorded to leveraging of G $\times$ E interactions with limited recourse to expensive inputs and to managing risks that small farmers are subjected to. Emphasis is also placed on the diversity of stakeholders, in particular women, given their important role in production and processing activities. The experiences of participatory breeding, carried out at a small-scale have broadly resulted in an increased adoption of varieties developed by the target communities (Eyzaguirre and Iwanaga 1996). Questions have then arisen on the generalization and sustainability of these actions. Several complementary approaches have attempted to address these issues. Plant breeding activities were somewhat decentralized at the global level by dividing tasks related to *pre-breeding* and actual varietal development between the CGIAR centres and NARS (Box 4). The effectiveness of multi-actor platforms for dialogue and action was tested at the national and regional levels (Box 5). The most ambitious experiments are seeking to transfer the skills and responsibilities of significant parts of the process of varietal creation and dissemination to farmer organizations (Box 6).

Box 4. The decentralization of plant breeding: a new division of tasks between stakeholders of plant breeding

Under the auspices of the FAO, Mahsuri, a semi-dwarf rice variety was widely disseminated in India in the mid-1950s while IR8, arguably the star variety of the Green Revolution, was disseminated throughout Asia (excluding China) and Latin America in the late 1960s. These two varieties boosted the average yield from 2 T/ha to 4.5 T/ha for over 33 million ha of irrigated rice fields. Subsequently, genetic progress focused on stabilizing yields by increasing resistance to diseases and insects, as well as on the 'yield/growth duration' ratio. The lack of progress in genetics with regard to the yield potential was reflected, from the mid-1990s onwards, by stagnating yields and farm production levels.

New breeding patterns and new ways of sharing responsibilities then emerged in response to this stagnation in genetic progress, a result of the shrinking genetic base of disseminated varieties and the inadaptability of Green Revolution varieties in large areas subject to abiotic constraints. This change was also driven by the desire of the NARS, whose capabilities had increased dramatically, to strike out on their own and by the growing interest of private companies in the rice market of developing countries. In Asia, this trend led IRRI to focus on understanding the genetic bases and improvement of specific traits, while leaving to the national systems the responsibility of incorporating them into varieties adapted to the specific conditions of each country and each agricultural region. In Latin America, the division of tasks led to the establishment of a plant improvement programme consisting of decentralized recurrent selection. Synthetic populations with a broad genetic base were thus created, derived from intercrossing dozens of parents. This intercrossing was facilitated by introducing a recessive gene for male sterility into the population. These populations were improved for some major traits by the International Centre for Tropical Agriculture (CIAT) in association with CIRAD. They were then distributed to national research systems to help extract new varieties adapted to each country's specificities as well as to enrich them with local parents. Information from national programmes was, and continues to be, used to orient recurrent breeding at the central level.

Similarly, public–private partnerships were set up in which public research undertakes the pre-breeding phase and passes on the plant material to private firms that are fee-paying members of a consortium. These firms then finalize the breeding for their own geographical areas and target market segments. This is how the Hybrid Rice Development Consortium (established by IRRI) and Fondo Latinoamericano Para Arroz de Riego (established by CIAT) function[*].

It is expected that these changes will promote the expansion of the genetic diversity that is deployed, take advantage of genotype × environment interactions, contributing thus to an increased resilience and productivity of rice systems. However, we currently have few means of quantifying their impact.

[*]Hybrid Rice Development Consortium, http://hrdc.irri.org/; Fondo Latinoamericano Para Arroz de Riego, http://www.flar.org/ (retrieved: 2 May 2013).

Box 5. A platform for varietal innovations of the plantain: a linking of scientific and local knowledge

More than 8 million tonnes of plantain is produced each year in West and Central Africa by small farmers for their own consumption as well as for local and regional markets. In Cameroon, an estimated 600,000 farmers grow plantain as a monoculture or through intercropping.

Several plant breeding programmes based on distinct but complementary genetic concepts have helped breed new varieties since 1987. Their main objective is to improve resistance to diseases and pests such as cercosporiosis, nematodes, and weevils (Tomekpe et al. 2004). Once the improved varieties were obtained, questions arose as to how amenable they were to farmers' choosing criteria, which go beyond mere disease resistance.

To address this issue, the Cameroon-based African Research Centre for Bananas and Plantains (CARBAP) introduced, in 2006 with support from CIRAD, a mechanism called the 'varietal innovation platform'. The objective of this platform is, on the one hand, to get target users to evaluate the varieties developed by research and, on the other, to establish a framework for dialogue between researchers and other stakeholders of the sector.

'Clubs' of users and local experts were set up at the level of small agricultural regions. Members consist of producers, nursery growers, merchants, restaurateurs and consumers. Each club was asked to define a set of constraints concerning their cropping systems and marketing modalities. It then became incumbent upon the CARBAP plant breeders to address these constraints through the development of plant 'ideotype(s)' in terms of on-field performance (robustness, drought resistance, resistance to cold temperatures and leaf diseases, plant height, early cultivation, resprouting ability, etc.), fruit characteristics (size of the plant and stem, fruit size and shape, pulp colour, etc.), leaf characteristics (usability for packaging) and processability (ease of peeling, suitability for different cooking methods, appearance, texture, taste, storability, ability to satiate, etc.). The plant material chosen by the plant breeder was grown and evaluated by the clubs on their plots and under their own real-world conditions. In addition, these clubs have together set up a common steering committee to interface with the research team and public authorities.

This arrangement helps establish links between local knowledge and scientific knowledge, strengthens public–private partnerships in the sector and stimulates the organization of civil society. The closer integration of post-harvest usage criteria in the breeding programme has led researchers to study the relationship between qualitative user preferences and the physicochemical and functional properties of fruits in greater detail (Gibert et al. 2009).

There currently exist three sub-regional platforms for promoting the participatory propagation and evaluation of a dozen hybrid varieties and exotic cultivars in Cameroon, Gabon, Equatorial Guinea, Central African Republic and the Congo. Similar platforms are being set up in Ghana, Togo, Benin and the Democratic Republic of Congo.

While the Green Revolution model was being called into question, other issues on biodiversity and sustainability of production systems arose. This led to the integration of the participatory approach and the in situ management of diversity (Box 6). In addition, the focus of varietal improvement shifted to adapting plants for more sustainable farming systems (Box 7). It is well known and recognized that the reflections on and practical experiences of participatory breeding in Europe originated mainly from proponents of *organic farming* (Wolfe et al. 2008; Ostergard et al. 2009; Dawson and Goldringer 2012).

Box 6. Improvement of sorghum for West Africa: from breeding ideotypes in a research station to the sharing of responsibilities with farmers

Sorghum is a traditional cereal from the African savannah, where it is the staple food of the rural population. It is normally consumed as a thick gruel called *Tô*. The last 50 years have seen a growth in the production of sorghum, due largely to an increase in the acreage devoted to it. In West Africa, while the cultivated area almost doubled, increasing from 5.4 million to 10.5 million hectares between 1961 and 2010, the yields (840 kg/ha) increased by a mere 12 %.

Breeding programmes for sorghum varieties, inspired by Green Revolution ideotypes of other cereals, have focused since the 1960s on modifying plant architecture (short size and single stem) and the panicle (compact).

They have also worked towards eliminating photoperiod sensitivity, something that is considered inconsistent with intensive agriculture. A large number of improved fixed line or F1 hybrid varieties were developed based largely on exotic genetic resources. It was thus that the IRAT204 variety, for example, was released in 1980. It was a *caudatum*-type variety of short height (1.4 m against 2.5 m for local *guinea* ecotypes), with a rather compact panicle and very short cycle, not photosensitive, and with a production potential of 5 T/ha, against 3 T/ha for local varieties (Chantereau et al. 1997). These varieties were only marginally adopted by farmers (Ouedraogo 2005) for at least two reasons: the economic conditions for intensifying sorghum production were not met and most of this material did not possess the qualities required for traditional grain and straw uses.

Recently, CIRAD plant breeders and their partners showed the importance of photosensitivity in the adaption of sorghum to climatic and disease constraints (Vaksmann et al. 1996) and analyzed the biological bases of the grain's technological and organoleptic qualities (Fliedel 1995). They thus reoriented their work towards valorising local genetic diversity, based on methods that balanced productivity, quality in a multi-use context (human food and animal feed) and in situ conservation of agricultural biodiversity. In Mali where intensive cereal cultivation practices are being progressively adopted in cotton-growing areas, recurrent selection scheme using populations consisting of a broad local genetic pool was associated with decentralized participatory breeding. This was done in an attempt to develop photoperiodic varieties with a reduced size and a better yield index, while maintaining grain quality (Vaksmann et al. 2008). In Burkina Faso, ancient local varieties which were conserved ex situ and improved varieties released by research, were evaluated in a participatory process. This participatory evaluation and the participatory development of new varieties led to an improved inclusion of farmers' highly specific selection criteria (Vom Brocke et al. 2008, 2010). This work helped in identifying several varieties with a reasonably wide geographic adaptation and in confirming the capacity of farmers and their organizations to implement important components of a plant breeding programme (Vom Brocke et al. 2011). Supported by the French Global Environment Facility, this programme is currently seeking to build the organizational capacity of the farmers as a prerequisite to a formal transfer of the management of activities of sorghum breeding and evaluation, registration and protection of the new varieties developed, and production and marketing of seeds.

Box 7. Development and dissemination of Arabica varieties for agroforestry production systems

Two species of coffee are grown worldwide: *Coffea arabica* (about 65 % of global production) and *Coffea canephora* (known commercially as 'robusta') representing the remaining 35 %.

Coffea arabica was introduced in Latin America using a very small number of plants (Anthony et al. 2002), a fact that led to a marked founder effect. Nevertheless, this initial low genetic diversity was used judiciously by breeding programmes launched in the 1930s. They gave birth to a dwarf variety that allowed the adoption of full-sun intensive cropping systems similar to the Green Revolution described above, mainly in Brazil, Colombia and Costa Rica. However, the combination of dwarf varieties, high-density cropping and expensive pest control methods never took hold in the rest of Latin America and Africa. Coffee continued to be grown under shade without any major technological innovations, resulting in a stagnation of yields. And yet, it became clear in the 1990s that despite their low productivity, agroforestry systems had positive effects on the maintenance of biodiversity, soil fertility, etc. and the quality of the coffees grown.

CIRAD proposed creating hybrid varieties that were suitable for agroforestry systems. The selection method used was based on intercrossing two pools of genetic material: the American lines and the 'wild' coffee varieties from Ethiopia or Sudan. Following 20 years of experiments in controlled environments and on the producers' farms, it appears that the F1 hybrids produce 30–60 % more in agroforestry systems without the use of added fertilizers (Bertrand et al. 2011). It was possible to select individuals from the hybrid families which demonstrated strong resistance to leaf rust and nematodes. Finally, the overall quality of the hybrids is comparable to that of standard varieties (Bertrand et al. 2006). In some specific environments, we even observe a much higher aromatic quality, for example, floral notes in hybrids grown above an altitude of 1,200 m.

The somatic embryogenesis technique was used to reproduce these hybrids (Etienne et al. 2012). This technology was developed by CIRAD over 20 years and was transferred to the private sector which built two tissue culture laboratories. More than 4 million plants have now been sold in Central America and Mexico. The Nicaraguan experience (pilot laboratory and technology transfer) appears to be a model that can be replicated in other countries around the world. It is estimated that 500 million new plants will be needed to renew all the planet's agroforestry plantations. The enthusiasm for the new hybrids is genuine and the impact of these new varieties is only limited by the method of reproduction used. Using genetic male sterility may help overcome this bottleneck. Preliminary hybrids obtained by this technology are currently being evaluated by farmers and it is hoped that they can be widely distributed to agroforestry systems as early as in 2016.

2.2 The Advent of Molecular Tools and Breeding on Genotype

Most traits of agronomic interest are quantitative and are determined by a large number of genes as well as by their interactions with each other and with environmental factors. Breeding based on the phenotype is inefficient for these traits because of the confusion between the effects of genes and those of the environment. Quantitative genetics improves breeding accuracy by segregating these effects and their interactions. However, it is a statistical approach based on genetic parameters (genetic variance, genetic correlations) of the population subject to breeding. Any increase in the average population performance from one generation to another depends on the accuracy of phenotypic data and trait heritability. Moreover, even setting aside the precision necessary and the associated cost, some traits of some species can be observed only at maturity or in an adult individual. The breeding cycle can then extend over several years and the breeding efficiency per unit time is reduced. Finally, the effect of genes and their interactions is treated arbitrarily as a single block, without taking their distribution on chromosomes into account.

These limitations quickly led to the development of methods for labelling genes with particular agronomically useful traits with the help of markers with a low dependency on the environment and the age of the plant (Sax 1923). However, it was not until the late 1980s that the first genetic map was created using polymorphism found directly in DNA. This map broke down a quantitative trait into discrete Mendelian factors, *quantitative trait loci*, or QTL (Paterson et al. 1988). Genes can be labelled at two levels of precision: with the causal mutation of the phenotypic variation or through nearby markers presumed to be non-functional. The former are currently thought to be almost exclusively related to qualitative traits while the latter to both qualitative and quantitative ones. In recent years, there has been a flood of information for labelling agronomically useful genes. Methods of genetic mapping of progeny derived from model crosses (such as 'good × poor') are being enriched by methods using phenotypic and genotypic diversity of representative populations of the species or of the material from a breeding programme (Jannink and Walsh 2002), as well as by information from the annotation of genomic sequences and different methods of analyzing gene functions in model organisms. This information, combining genetic segregation and functional assumptions, helps define the ideal genotype, i.e., the mosaic of chromosomal segments of the parent population that needs to be combined in a single individual to get the best possible expression of the target trait(s). The integration itself is achieved through a succession of crossing cycles and selection on genotype or marker assisted selection (MAS). One must note, however, the instability of QTL effects in different genetic backgrounds and biophysical environments as well as the erosion of QTL-marker linkage over breeding generations. For these and other reasons, labelling with non-functional markers must be redone for each population, and must be regularly renewed over breeding generations (Dekkers and Hospital 2002).

These methods, already commonly used by private firms, are gradually being adopted in breeding practices by CGIAR centres and NARS in developing countries, both for simple and complex traits (Box 8).

> **Box 8. Creating the ideal genotype: case of marker assisted recurrent selection in sorghum**
>
> Marker assisted recurrent selection (MARS) is part of a set of new approaches that use molecular markers to create varieties. In this approach, molecular markers are used to break down the variation of different quantitative traits which breeders are interested in into single traits (QTL). The originality of the approach lies not only in incorporating the use of markers in the breeding method, but also in working simultaneously with all useful traits and in diverse environments. One or more ideal genotypes can thus be defined as the mosaic of chromosomal segments that carry favourable alleles from the parents for all traits considered. It is theoretically impossible to obtain this ideal genotype using a classical pedigree method and with realistic population sizes when a large number of QTLs are involved. The MARS method—which involves several generations of successive crosses between progenies on the basis of their genotype to molecular markers associated to target QTLs—and the definition of multi-trait breeding indices is used to create material that is close to this ideal genotype. The value of the material evaluated by the breeder is thus optimized. In addition, the multi-trait and multi-environment aspect of this approach helps explore different breeding assumptions and objectives using the same material.
>
> Since 2008, CIRAD and the Institute of Rural Economy (IER) in Mali have successfully implemented the MARS approach for sorghum to obtain photoperiodic varieties which combine productivity with grain quality. In this endeavour, they have received financial support from the Generation Challenge Program and the Syngenta Foundation, and methodological support from Syngenta Seeds.

2.3 Complex Crop Stands

The past decade has seen a spurt in efforts to develop improved varieties for better efficiency in using resources (fertilizers, water and even pesticides) and for resilience and robustness. This was the result of the environmental impacts of the

Green Revolution observed in the South and the development of organic farming and low-input agriculture in the North, which were more exposed to biotic and abiotic constraints than conventional agriculture. Various experiments were conducted, including some on the use of intravarietal diversity of crop stands. Breeders and pathologists started using varietal mixtures as far back as the 1920s to counter a rapid reduction of varietal resistance to diseases (Finckh et al. 2000). This approach is also used to dampen variations in environmental constraints, improve resources (water, fertilizers, etc.), increase usage efficiency, or improve lodging resistance or grain quality (Ostergard and Fontaine 2006). In some ways it is akin to restoring erstwhile uses of intraspecific diversity as it existed within local varieties, and interspecific diversity of crop associations. After all, it still remains the basis of traditional agriculture on hundreds of millions of hectares.

Numerous studies have focused on multi-line stands, particularly for cereals in organic agriculture (Newton et al. 2009; Wolfe et al. 2008; Kiaer et al. 2009; 2012). They are seen as a promising option for general agriculture in developing countries (Faraji 2011). While the modalities of the functional advantage of complex populations still need to be spelt out, techniques of analysis have advanced using molecular procedures. For example, the dynamics of the diversity of a variety of multi-line bean, composed of contrasting root behaviour lines, was studied in the field for different fertilization levels (Henry et al. 2010). Molecular markers were used to quantify the contribution of each line to root stand and grain yield. The test's level of accuracy was, however, not high enough to arrive at any conclusion regarding a potential advantage of the composite population. New methods under development, based on a quantitative DNA analysis, should help quantify the root development of different components of multi-varietal and multi-specific heterogeneous crop stands (Haling et al. 2011). Similarly, the concept of *evolutionary plant breeding* has emerged recently. It envisages the deployment of varietal stands that are capable of adapting to changes in environmental conditions (Döring et al. 2011). A more detailed understanding of ecophysiologic interactions involved in improving the performance (primary production and its stability) of monospecific stands endowed with a functional diversity is necessary to help select the best possible complementary components. Which traits are to be diversified, and how to go about it without adversely impacting the homogeneity desirable for other traits?

3 The Challenges of Ecologically Intensive Agriculture

While it is well known that conventional intensification, as part of the Green Revolution, helped boost yields substantially in some countries, it also had serious negative effects on the environment. This has led to a search for new ways of increasing production. Ecological intensification represents a comprehensive transformation of agriculture through the adoption of production models to 'obtain desired *output yields* from a cultivated ecosystem that are intrinsically high per biosphere unit, while maintaining the functionality and viability of various eco-system functions and without forcing with artificial inputs' (Griffon 2007).

This new model can only be adopted through the use of a new generation of plant material. In addition to producing biomass, this material will have to opti-mize biological interactions to ensure adaptation under various environmental constraints, protection against pathogens, symbiotic fixation of atmospheric nitrogen, greater mobilization and recycling of soil minerals, protection against erosion, and a maintenance of agricultural biodiversity, including intraspecific genetic diversity. The example of rice (Box 9) shows that research in genetics and plant breeding has started down this path. It has endeavoured to create plant material suitable for constrained environments and which uses resources with greater efficiency. Yet, much remains to be done in terms of acquiring knowledge as well as of practices.

Box 9. Adapting rice to ecologically intensive agriculture

One of the central pillars of the Green Revolution has been the combination of semi-dwarf varieties and a high level of mineral fertilizer. But not only is the production and distribution of these fertilizers very energy intensive, existing phosphorus resources are limited and may well run out before the end of the twenty-first century. Furthermore, nitrogen use often leads to water pollution.

Two lines of research are being explored to reduce the use of nitrogen fertilizers: biological nitrogen fixation or BNF (Choudhury and Kennedy 2004), and nitrogen use efficiency or NUE, i.e., the efficiency of nitrogen use by the plant (Peng and Bouman 2007). The genetic programme for the endosymbiotic association between rice roots and nitrogen-fixing bacteria is currently being dissected. It has already been shown that rice possesses the essence of the genetic programme involved in the nodulation process in legumes. Combined with work on the ability of bacteria to associate themselves to cereal roots, this research should lead to operational results with regards to BNF in the next decade. As far as NUE is concerned, such operational results seem to be available already. Transgenic rice varieties that halve the requirement for nitrogen fertilizer while maintaining pro-duction levels are being evaluated on a large scale in China[*]. Patents related to the biological processes involved here have been filed. The possibilities of monetizing the reduction of CO_2 emissions resulting from the use of NUE rice is an additional motivating factor.

As far as reducing phosphate fertilizer requirements is concerned, the most promising line of research is to attempt to improve the plant's ability to use the insoluble fraction of soil phosphorus. The genetic diversity for such ability exists in most major field crops. Efforts to clone the gene (*Pup1*) responsible for this capability in rice are very advanced (Gamuyao et al. 2012). The transgenic approach of transferring microbial genes to increase root excretions that can utilize non-soluble forms of soil phosphorus also shows promise.

In another vein, while the first Green Revolution was based on the combination of the semi-dwarf varieties and mineral fertilization, it is generally accepted that, for rice, any further significant increase in production potential can only come from changing its photosynthetic mechanism from type C3 to type C4. This transformation would enhance the efficiency of rice to convert solar energy into biomass by 50 %, without additional consumption of water or fertilizer. The C4 Rice Consortium is exploring two complementary lines of research: research into wild relatives of rice that have already changed partially to the C4 type, and into transgenic transfer of genes from maize and/or other C4 plants to rice (Sheehy et al. 2007).

* See http://www.arcadiabio.com/nitrogen (retrieved: 4 May 2013).

3.1 Understanding Complexity

3.1.1 Biological Interactions

Biological interactions, defined at the level of the plant breeder, reveal unprecedented dimensions of complexity. To select genotypes that are adapted to a particular use in a given type of environment (one among the target population of environments, or TPE, of agronomists/ecophysiologists) is in itself an important task. The selection power will increase with the number of genotypes sorted, the number and diversity of selection sites, the accuracy of description of sites, and the accuracy of measurement of the target plant traits. It entails manipulating large numbers, optimizing flows and making the best possible use of available capabilities. Taking into account biological interactions with other living entities and variable components of the system leads to an exponential expansion of the conditions that require testing. Using the individual genotype as a factor in these interactions seems impractical.

For example, an extensive experimental effort was needed to determine the effect of maize genotypes on the composition of the associated rhizobacterial community. It was possible to show the influence of the varietal type, but not of the interaction with the genotype within a varietal group (Bouffaud et al. 2012). This kind of result emphasizes the existence of a potentially significant effect of soil biology but does little to indicate how the breeder could use it. Rather, it is an invitation to cultivate an intraspecific and interspecific diversity to promote biological diversity.

The valorisation of biodiversity's ecosystem functions via the use of varieties/populations with large internal diversity raises a whole new set of questions.

Expanding the scope to multispecies populations widens the issue to encompass new traits which correspond to the mutually beneficial functions sought. To what extent should the role of a cereal be promoted as a guide to facilitate the development of the associated legume? What are the implications for harvesting methods and the use of the products?

The challenge of integrating knowledge becomes even greater if we transpose the issue of the valorisation of biodiversity's ecosystem functions from a cultivated plot to a wider canvas: the *terroir* or a landscape. We would then have to consider intraspecies and interspecies diversities, their interactions and their spatio-temporal arrangement. Faced with this complexity, we must identify local and contextualized solutions, while allowing more room for empiricism, local knowledge and its interactions with the knowledge related to mechanisms involved.

3.1.2 Interactions with the Abiotic Environment

A large amount of data has been gathered on different plant species over the past two decades. It pertains to the functioning of genes studied individually or in small groups in contexts of responses to specific stimuli (salinity, drought, cold, etc.). The major scientific challenge is the integration of this data at different scales (of molecules, tissues, organs, entire plants at different phenological stages). This is a prerequisite for understanding the principles of regulation by genes and assessing their relevance in terms of spatio-temporal constraints for which the adaptation is sought. Ongoing research on adaptation to climate change illustrates the scope of the challenge. Genes and gene networks involved in basic biological processes that underpin the growth and development of plants as they interact with different environmental factors (water, minerals, light, temperature, CO_2) are described quite extensively. Significant efforts are underway at the level of the plant to assess the genetic diversity available for traits necessary for adaptation: tolerance to non-optimal temperatures at different development stages, valorisation of higher atmospheric CO_2 concentrations, regulation of nocturnal respiration, etc. The uncertainty of climate change and its variability adds further complexity to the process of defining plant ideotypes (relative weightage of adaptation traits) and their construction (hierarchy of genes and gene networks to be considered). Given the logistical and temporal difficulties that would arise from the evaluation of all new trait combinations in in situ experimental conditions, it becomes essential to make use of *ex ante* methods based on modelling and simulation to evaluate new ideotypes. The example of variation of the effect of drought tolerance QTL in maize, according to target environments (Box 10) shows the importance of such an assessment. There is still a long way to go before we can integrate multidisciplinary knowledge and understand complex biological reality in order to develop innovative production systems which combine varietal innovation and innovative cropping practices (Hammer et al. 2010; Passioura 2012; Tardieu 2012; Parent and Tardieu 2012). This complexity is further multiplied when perennial plants are considered.

Another area where modelling and simulation can guide plant breeding is the characterization of the target population of environments (TPE). Environments can be characterized, not in terms of the variation of classical physical descriptors (temperature, rainfall, solar radiation, soil water retention capacity, etc.), but also in terms of their impact on plant performance (target yields, stress indicators during the cycle, efficiency of resource use, etc.). The characterization of upland rice and maize growth environments in central Brazil using the Sarra-H model (Dingkuhn et al. 2003) is a good example of TPE definition (Heinemann et al. 2008).

Box 10. Modelling gene-phenotype at the stand level

A major challenge in predicting the performance of a genotype (G) in a range of environments (E) is the taking into account, at the level of the entire plant, of a multitude of G × E interactions at more basic levels: of the gene, molecule, cell and organs, throughout the entire growth cycle (Cooper et al. 2009). This is especially significant considering that the analysis of the contributions of 10 genes with two alleles in 10 environments would lead to an examination of $3^{10} \times 10$ combinations! Another approach based on biophysical simulation models has been developed over the last 20 years in order to meet this challenge. Integrating physiological processes and their genetic control, these predictive models can integrate G × E interactions at different levels of the organism (Messina et al. 2009). They also allow us to explore in silico a large number of genotype combinations (alleles at many loci) and environments. However, the simulation of the effect of genes/QTL for complex traits such as the response of the plant's growth and architecture to the environment requires modelling physiological processes that remain stable irrespective of the environment (Tardieu 2003; Hammer et al. 2006).

This approach was recently tested to predict the behaviour of maize vis-à-vis drought (Chenu et al. 2009). The leaf elongation rate (LER) is a major factor in maize's response to water stress. It is the same for anthesis-silking interval (ASI). Several QTL that control LER and ASI were mapped, and Reymond et al. (2003) were able to simulate and validate the effect of these QTL in different environments for new lines defined by their genotypes for these QTL. Welcker et al. (2007) have shown that several LER and ASI QTL co-exist. Chenu et al. (2008a) subsequently incorporated the leaf elongation model (which functions at the organ level with time steps ranging from a minute to an hour) in another model (APSIM-Maize) that works at the level of the stand and incorporates more complex population × environment interactions (Fig. 1). The APSIM model thus complemented was used to simulate the effect of LER on soil water reserves throughout the plant cycle. It was validated by comparing experimental and simulated results of a hybrid maize variety in different environments, for leaf surface area, biomass and grain yield. Finally, Chenu et al. (2009) used this model to simulate the impact of LER QTL on maize yield under different drought conditions. This simulation showed that the two major QTL, having similar effects on LER, affect yield differently under water stress. The authors thus demonstrated that robust gene-phenotype models can be used as decision-making tools to select traits (and associated QTL) based on target environments.

This example outlines a two-step selection process for adaptation to environmental conditions: first, the genetic analysis of traits that boost resilience to environmental conditions, resulting in an agronomic value for alleles, and second, the creation of ideotypes adapted to a given region, based on the combination of these alleles. The first step is based on phenotyping platforms that help the genetic analysis of a large collection of plants under varying conditions. These platforms, whether in greenhouses or fields, involve semi-controlled conditions and continuous measurement of environmental conditions and responses of plant materials. The second step comprises an in silico search of alleles favourable to a given situation (taking the results of the previous step into account), followed by the testing of a limited number of promising combinations in field trials. These trials are accompanied by a frequency study resulting from the simulation of the productivity of genotypes with different allele combinations. This simulation helps predict the frequency at which a combination of alleles is favourable in a given region and cultivation system (Tardieu 2012).

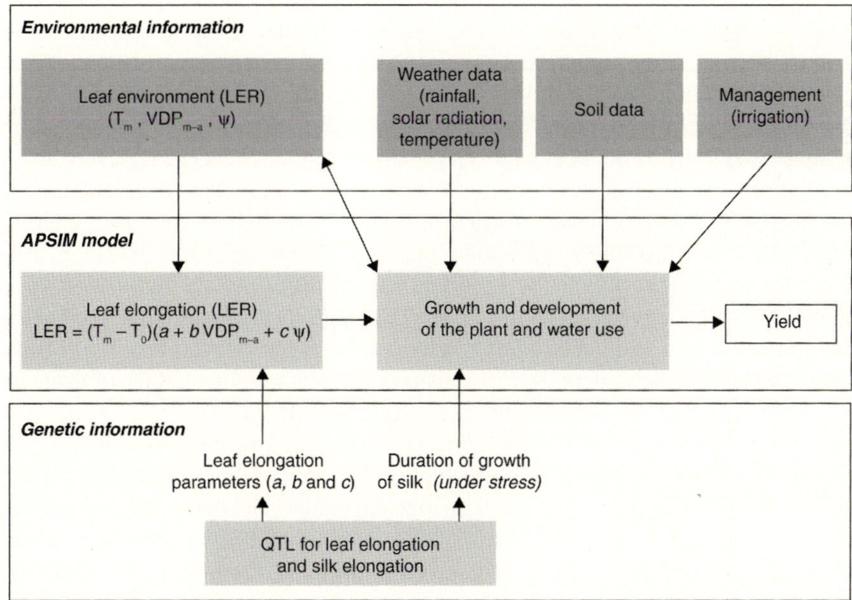

Fig. 1 Schematic view of the 'gene-phenotype' model showing how the leaf elongation module interacts with other components of the APSIM model (adapted from Chenu et al. (2009))

Environmental and genetic information is used as input to simulate leaf growth, plant growth and development, and grain yield of the crop. Leaf elongation (LER) is a function of environmental factors of the leaf: temperature of the meristem (T_m), base temperature (T_0), water vapour pressure deficit between meristem and surface area (VPD_{m-a}), the leaf water potential at dawn (ψ), and leaf elongation parameters (*a:* rate of LER potential; *b:* response of LER rate at VPD_{m-a}; *c:* response of LER rate at ψ).

3.2 Rethinking the Varietal Innovation System

Plant breeding in the twenty-first century has to contend with a context of rapid global changes (climatic, technological, etc.) and a multitude of often divergent requirements of decision makers and users (civil society, governments, farmers, processors, distributors, consumers, etc.). It also has to take into account the increase in the number of actors involved in the production and distribution of varieties and associated knowledge (national and international research institutions, private seed companies, NGOs, farmer organizations, etc.) and the increase in the number of stakeholders who fund the production and dissemination of plant varieties (national governments, international organizations, charitable foundations, producer organizations, etc.). In such a context, the dissemination of genetic progress in the form of new varieties is no longer a linear and unidirectional process, moving from the geneticist-breeder to the farmer via agricultural extension services or seed companies. This dissemination takes place rather within a complex innovation system that requires the convergence of views and interests of a large number of stakeholders. The proper functioning of this system depends on freedom from technological bottlenecks (e.g., incapability/difficulty of adapting processing equipment to traits of the new variety), regulatory impediments (e.g., standards to distinguish homogeneity and stability of varieties which prevents the marketing of varietal mixtures), commercial restrictions (e.g., use of hybrid varieties requiring the purchase of seeds for each cropping cycle) or ideological resistance (e.g., out-of-hand rejection of genetically modified organisms). The transformation of a varietal prototype developed by the breeder into a varietal innovation that is disseminated widely and quickly among farmers thus requires several back-and-forth cycles between stakeholders. Such exchanges allow each stakeholder to evaluate the constraints and the opportunities associated with the adoption of the new variety. Constraints in terms of altering practices (e.g., in order to adopt an F1 hybrid rice whose seed price is much higher than that of line varieties, a farmer must reduce planting density and thus change the seeder) and opportunities in terms of income and organization of work (e.g., the adoption of the F1 hybrid rice variety results in increased productivity and shorter cycle time, facilitating the establishment of the next crop).

If the objectives of ecological intensification are taken into account, the varietal innovation system becomes even more complex. Indeed, the optimization of biological interactions involved in the adoption of this new production model requires a clear contextualization of varietal solutions (taking into account not only soil and climatic conditions, but also production systems, succession and inter-cropping, processing, etc.). We thus have to proceed in stages and/or decentralize the process of creating and evaluating varieties. However, this complexity is also an opportunity to incorporate local knowledge, something that is particularly difficult to formalize in the context of family farming in the South.

The process of breeding and dissemination of varieties has to change from the breeder-extension officer-farmer triptych to a more complex innovation system. This constitutes a challenge for the breeder who must manage and coordinate the mobilization of a wide range of knowledge and skills to design future varieties (ideotype/prototype) and who must then create and evaluate prototypes and pro-mote the resulting innovation. To this end, we would have to study action models proposed by the sciences of analysis and creation of innovation systems in other economic sectors in order to adapt them to varietal innovation (Klerkx et al. 2009; Berthet 2010).

4 Mechanisms to Help Meet the Challenges of Ecological Intensification

Faced with these challenges, we must find more nimble ways to undertake plant breeding for the benefit of farmers and for leveraging agrobiological interactions in different environments and for different plant species.

Plant breeding can be made flexible, quick and responsive, on the one hand, by using the knowledge and ability to manage important genetic factors and, on the other, by producing populations with a large genetic base and at a high enough analytical resolution. This would allow the creation of material that is adapted to a broad range of farming systems.

To be really beneficial to farmers, plant breeding must progress with them through a better understanding of their limitations, aspirations and practices. It must draw on strengthened interactions and partnerships for joint design and joint validation of new material. The material produced by the breeding programme should incorporate characteristics of their preferred material and offer incremental improvements over varieties in use.

In order to valorise agrobiological interactions, plant breeding must be con-sidered in terms of traits that are known/expected/proven to be favourable, and must include regular tests and implementations in interactive situations. In par-ticular, it must strengthen breeding methods for complex crop stands that are composite mono-specific populations or have multispecies associations.

In order to be efficient in different environments, plant breeding will have to incorporate analyses of the diversity of environments and rely on the modelling of methods of local adaptation in the broadest possible overall environmental diversity.

In order to be implemented for large array of plants, plant breeding will have to adapt itself to work in stages and with numerous partners. It will have to include amongst its outcomes a provision of access to genetic resources described in the most relevant way possible, or provide rapid characterization to guide the efforts of stakeholders.

The principal elements of the process include a systematic mobilization of the diversity of genetic resources; appropriate partnerships and innovative ways of collaboration; and shared research objectives and questions to bring different disciplines closer together.

4.1 A Systematic Mobilization of Genetic Diversity

4.1.1 Structured Access to Genetic Diversity

Genetic diversity forms the basis of all genetic progress. Viewed as such, it has become a strategic issue, and sometimes even the source of political tensions. Some international initiatives aim to maximize access to global genetic diversity for a large number of species by offering representative samples of large collections created both on an ecogeographical and a molecular basis (Glaszmann et al. 2010; Billot et al. 2013). These 'core' or 'mini-core' samples are offered as references to different research stakeholders. This allows bringing together various traits in order to better understand the relationships between observed traits and to explore the link between behaviour and molecular polymorphisms. This concentrated diversity is also a gateway to global diversity, which can then be explored on the basis of trends observed in the reference sample (e.g., towards a geographical region rich in diversity or in sources of resistance, etc.). It can also be used to identify an even smaller sample (core sample) which can be used to represent the species in comparative studies or as a starting point for the intermixing of diversity.

It is the understanding of diversity, more than the describing of it, that will guarantee its best use. In the course of domestication, it was—and continues to be—influenced by natural selection and human selection in all their complexities. Also influencing it is the demographic history of cultivated populations, basically linked to the histories of the environment and of mankind. Understanding history helps identify situations which may have led to the emergence of unique genetic factors, and which can be utilized in new contexts through recombination. Sometimes the evolution of cultivated forms is achieved through successive jumps, related, for example, to the emergence of new hybrid forms or even new genomic configurations. This knowledge helps identify successful breeding directions, accompany or accelerate them, or even re-explore them on new bases. Histories of cocoa, coconut or banana all reveal episodes of genetic intermixing slowed down by an extended cycle or the pre-eminence of vegetative propagation (Loor Solorzano et al. 2012; Gunn et al. 2011; Perrier et al. 2011). These structuring episodes help inspire hybridization strategies for crop breeding. The analysis of contemporary situations helps clarify another level of evolution by throwing light on the current processes governing seed management and their diversity in societies (Leclerc and Coppens d'Eeckenbrugge 2011; Pautasso et al. 2013). Such an analysis helps us in the overall understanding of the dynamics of diversity.

4.1.2 Intensive Intermixing in Exploratory Populations

Why practice an intensive intermixing? Because retrospective analysis shows that intermixing has been the source of major advances. Recent studies on the domestication of rice in Asia show that it was carried out in eastern Asia on a local wild species and led to the selection of specific allelic forms for several genes distributed in the genome. It was the recycling of these alleles by spontaneous hybridization that allowed the replication of this domestication from wild rice species from South and South-East Asia (Huang et al. 2012). We can similarly say that cultivated wheat, cotton, bananas and peanuts originate from early crosses that combined different species. Moreover, conventional breeding produces new crosses between distant species, a fact made possible by the technology of in vitro culture to not only get, for example, disease-resistant genes (rice, barley, etc.), but also productivity-boosting genes, as was done in tomato (Gur and Zamir 2004) and in rice (Thalapati et al. 2012).

A detailed knowledge of the structure of diversity helps us create populations that will be able to host new variability traits through intensive intermixing, e.g., as *Multi-parent Advanced Generation Inter-Cross* (MAGIC) populations (Cavanagh et al. 2008), which consists of intercrossing a limited number (usually 4, 8 or 16) of representative genotypes. Another example is NAM-type focused populations, where NAM stands for *Nested Association Mapping* (Yu et al. 2008), made from recombinant lines derived from crosses between a central genotype and a series of diverse genotypes.

4.1.3 Detailed Information on the Genome, its Structure, Diversity and Expression

Genomics is a recent and growing discipline. It involves studying the functioning of organisms at the level of the complete genome. Genes and their arrangement in genomes are decoded and their role is explored by studying the genome's expression at the transcriptomic level, derived directly from this expression and its regulation, and at the proteome or metabolome level which are an indirect result of expression and regulation and are more related to the physiology of organisms.

Technological advances allow the rapid analysis of genomes on an increasing number of plants, even on those considered orphans until recently (Varshney et al. 2010).

The sequences of basic genomes are being decoded more and more rapidly, the latest being tomato and banana (D'Hont et al. 2012). This allows us to reconstruct the evolution of genomes and analyze the dynamics of gene families involved in the most important functions and metabolic pathways.

The extent of advances on the plants most commonly worked on is impressive. Thus, data on the re-sequencing of over 1,500 varieties of rice (446 wild rice species and 1083 cultivars) have recently been made public (Huang et al. 2012). Nearly 8 million polymorphisms are available for analysis, allowing a detailed description of the domestication of rice and a listing of the parts of the genome most responsible. Huang et al. (2010) were able to identify genetic factors that explained about 36 % of the phenotypic variation observed for 14 agronomically useful traits with this type of data.

Knowing the repertory of genes (30,000–50,000 per basic genome) one can analyze their expressions throughout development stages in different environmental conditions in order to identify genes that meet specific conditions and specific breeding objectives.

The analysis of the genome's structure and functioning opens up access to new features such as transposable elements (Kidwell 2005) and small RNAs, whose activity induces changes in the genome's expression, some of which are carried over to the next generation (He and Hannon 2004). This pertains to epigenetics, a field concerned with environment-induced changes that are usually reversible, but can sometimes be inherited. Epigenetics bears a troubled history, its reputation marred by the excesses of Lyssenko, but whose biological basis can shed new light on epistasis, heterosis and reproductive isolation, and thus encourage new paths in plant breeding (Tsaftaris et al. 2008; Durand et al. 2012; Paszkovski and Grossniklaus 2011).

The sheer stream of data generated by genomics poses a massive challenge to existing data processing capabilities and even those of data storage. Phenomics, a technological domain for describing organisms (phenotypes) in various conditions and at different levels (of the cell, tissue, organ, organism, stand, etc.) is also developing. It is also churning up data, albeit not as much as genomics, that is biologically and mathematically very complex. Just as technological changes have resulted in the creation of high-density, high-tech platforms, so too bioinformatics is being influenced by a realignment of strengths beyond the boundaries of traditional disciplines. It often brings together specialists in human health, evolution, animal, plant or microbial biology, much like the Computational Biology Institute in Montpellier, France, is doing (Box 11).

Box 11. Computational Biology Institute in Montpellier, France

Plant genomics projects are increasingly using new and very high-throughput sequencing technologies (*next generation sequencing*, or NGS) and high-throughput phenotyping methods. The application of NGS is not limited to the sequencing of new genomes; it can be used for genome re-sequencing and detection of genomic variations such as *single nucleotide polymorphism* (SNP) or transcriptomics. These projects generate a large quantity of information. Its integration, for association studies at the genome level for example, requires the deployment of innovative approaches for data analysis. The lack of software tools capable of dealing with ever-increasing volumes of data constitutes a major bottleneck.

In this context, the Computational Biology Institute (IBC) in Montpellier, France, is developing methods and software applications to analyze, integrate and contextualize biological data on a large scale in the fields of human health, agronomy and the environment. Several domains are involved: algorithms (combinatorial, numeric, highly parallel, stochastic), modelling (discrete, quantitative, probabilistic) and data and knowledge management (integration, workflow, cloud). The challenges are thrown up by the exponential growth of data, the complexity of the models as well as the heterogeneity and distribution of data and biological knowledge. In order to address these factors and make a concerted approach to solve this well-defined set of problems, the project is divided into five work packages. Each of them pertains to one key aspect of existing biological-data processing techniques: methods for high-throughput sequencing; advancing to the level of evolutionary analysis; structural and functional annotation of proteomes; integration of cell and tissue imaging with omics data; and integration of biological data and knowledge. Concepts (computational methods, mathematical models, etc.) and tools (software applications, platforms and databases, etc.) will be validated mainly with the help of agronomic applications (plant genomics, agriculture of the South) and environmental ones (population dynamics, biodiversity).

This work involves 57 permanent scientists, encompassing a broad spectrum of disciplines drawn from one private company and 13 public institutions in Montpellier, including CIRAD, the National Centre for Scientific Research (CNRS), the National Institute for Agronomic Research (INRA), the National Institute for Research into Computer Science and Automation (INRIA), the Institute of Research for Development (IRD), and the Universities of Montpellier 1 and 2.

4.2 An Integrative Approach to Biology

The phenotypic response of a plant, observed macroscopically (growth, development, transition from vegetative to reproductive phase, etc.), to stimuli or environmental stresses (light, drought, salinity, high or low temperatures, fertilization, disease, etc.) clearly results from the integration of diversity and the expression of a large number of genes. While the plant breeder seeks to identify and recombine the modules segregated in the progeny as best as possible, the major challenge in biology is to understand how metabolic pathways, cell signalling pathways and different developmental processes are linked to the expression of the genome, on the one hand, and to phenotypic expression, on the other. It is a matter of describing relationships that are intrinsic to the system under study by characterizing relationships that connect the system in its entirety to the systems that surround it.

A study group set up by INRA's scientific council concluded in 2004 (Charrier et al. 2005) that integrative biology is the new paradigm that must 'make sense' of the analytical and quantitative approaches of genomics. It must also help integrate the information acquired at different approach levels to elucidate and accord a significance to the processes studied. While the approach is still in the conceptual phase, it is expected to lead to new lines of thinking and to discover original epistemological directions. New methods and concepts are emerging. Investments in bioinformatics, stimulated by the 'mass' of data produced by genomics, together with new methodologies based on artificial intelligence and statistics have resulted in new hypotheses on genomics data and efforts to validate them. Various technological developments help improve the in vivo monitoring of metabolism and the detailed observation of functional ultrastructures of the living cell.

By equating plant communities to mass and energy exchange surfaces, the field of ecophysiology has been able to model the primary production of plant communities without delving into the underlying biological processes and their genetic regulation. By working at the level of the individual plant, this approach takes processes into account which control the distribution of assimilates and thus of the biomass between different parts of the plant that result from morphogenesis. They can only be understood through an architectural approach to plant growth. This has led to the development of the approach that considers communities or stands to be a collection of individual plants that interact among themselves, with their emergent properties reflecting the collective functioning of the population.

In integrative biology, genetics plays a pivotal role in determining the functions of genes in the expression of the phenotype by analyzing natural variation.

Large-scale studies in molecular biology highlight the dynamics of expression and co-expression that characterize interactions between genes as regulatory networks. At the same time, genome-wide association studies (GWAS) present statistical connections between genotypes (in terms of allele combinations) and phenotypes. These two types of studies provide aggregated information that could be put together with the help of mathematical modelling (Nuzhdin et al. 2012).

Plant models are simplified mathematical representations of biological and environmental interactions of the dynamics of growth and development of a plant or community. They represent the preferred tool to understand and predict genotype-phenotype relationships for complex traits such as phenotypic yield or plasticity (adaptability) (Dingkuhn et al. 2003; Hammer et al. 2006). However, most existing models do not contain the required level of detail of biological functioning. It is therefore necessary to emphasize explanatory approaches through an understanding of the dynamics of the processes that underlie the plant's growth and development. To this end, the modelling of morphogenetic processes of the whole plant, based on a source-sink relationship for carbon assimilates (Box 12), is a promising approach. It provides the bases for breaking down physiological functions that underlie changes in the main adaptive traits and for identifying genomic regions involved in the control of these functions.

Box 12. EcoMeristem: model simulating plant growth within the crop stand to support phenotyping and ideotype exploration

The EcoMeristem model (Luquet et al. 2006) formalizes the morphogenesis of the plant (rice, sorghum and other tropical grasses) and its phenotypic plasticity as a response to the abiotic environment in the stand or community. This formalization is done on the basis of equations whose genotypic parameters define a morphogenetic potential (phytomere height, tillering ability, phyllochron, etc.), physiological potential (radiation interception, carbon assimilation, leaf transpiration) and its regulation by the nutritional status (carbon, water) of the plant (threshold parameters and rates of regulation of potentials) according to its photothermal and hydrological conditions. The model thus formalizes hidden morphophysiological parameters that control the functioning of source and sink organs in response to the environment and the potential competition between sink organs for the same pool of resources within the plant. In this way, it formalizes the compensation and regulation of these processes which could result from their physiological and/or genetic connections.

Once calibrated on a panel of varieties representative of the species diversity, the model helps estimate the genotypic parameters mentioned in the model for any new variety studied by using a set of relatively simple experimental data. These genotypic parameters are supposed to be less influenced by the environment than the directly measured variables (Luquet et al. 2012a, b). EcoMeristem has thus been extensively used to analyze the phenotypes for growth and vegetative vigour of rice and sugar cane. In particular, its ability to represent the genetic diversity of behaviours related to vegetative vigour and susceptibility to water stress (stomatal opening) in a diversity panel composed of 200 *japonica* rice accessions has been demonstrated (Luquet et al. 2012a, b).

The parameters of the model are optimized for each of the 200 accessions of *japonica* rice and will now be used to identify the genetic bases of these parameters by genome-wide association analysis. Similarly, the range of values for the parameters representing the diversity of *japonica* rice will be used to define, in silico, varietal ideotypes for tolerance to water stress. The EcoMeristem model will be used in the ANR Grand Emprunt Biomass Crop for the Future project (2012–2019). It will be applied to phenotyping and ideotypaging sorghum—biomass for the production of bioproducts. These future applications involve linking it to 3D representation models of plants (Soulié et al. 2010). There are also plans to deploy a more elaborate photosynthesis model in order to access, depending on the application, additional architectural or physiological traits.

Recently, the fundamental concepts of the EcoMeristem model were simplified and integrated with the Sarra-H model (Dingkuhn et al. 2003) to create a new generation of agricultural models. The Samara model thus developed provides a simplified representation of source-sink relationships for carbon assimilates. It can simulate the plasticity and multiple yields (grain, sugar, and biomass) of tropical grass stands in a very elaborate representation of the cropping systems specific to its flagship species: rice and sorghum.

In the immediate term, the proposed models are assessed by genetic improvement practitioners in terms of their ability to simplify the representation of the inheritance of phenotypic behaviour by identifying hereditary modules with significant effects on the target phenotype.

These approaches should help effectively address not only conventional plant breeding objectives such as 'adaptation to abiotic constraints', but also new and very complex objectives. These can include the competition-mediation process, which underlies the improved performance of heterogeneous stands, or the plant's interaction with elements of its surrounding environment, such as pathogens, pollinator insects, neighbouring plants, seed-dispersing animals, etc.

4.3 Shared Purpose and Research Questions for a Bringing Together of Disciplines

4.3.1 Ideotype

The ideotype is a concept originally proposed by ecophysiologists. It represents an ideal plant endowed with a set of traits that give it the best adaptation to a given agricultural system (of cultivation, of production). It originates from an initial understanding of associations between traits that form the basis of agronomic behaviour and is an intellectual construct similar to a model. Progress in this area helps improve simulations, place ideotypes in perspective based on environmental conditions, incorporate genetic factors identified through logical reasoning, and guides recombination programmes across breeding generations. In a wider sense, beyond that of biologists alone, the ideotype is a virtual object which combines the vision of researchers from different disciplines and of different stakeholders of agriculture, processing and distribution, or even of framers of public policy. It becomes an issue of sharing, participation, co-adaptation, or negotiation between stakeholders, which leads to shared representation and a phase conducive to innovation. This process starts with each stakeholder's objectives and relies on the analytical capabilities of researchers to assess feasibility and scientific opportunity, and on the ability of co-adaptation of different stakeholders. When considered fixed, the ideotype induces a planning process that has little to do with innovation. When considered as a result of iterations, it creates an opportunity for multidisciplinary synthesis and cyclic and programmatic communication and interactions capable of fostering innovation. It is through these interactive processes that the objectives of ecological intensification, their diversity and specificity can best be incorporated into varietal improvement programmes.

4.3.2 Populations Which are Vectors of Genetic Progress and Biological Resolution

The ability to produce new segregating populations based on an understanding of genomic diversity and recombination encourages research to produce populations which will be most likely to lead to genetic progress as well as fulfil the desire to better understand the genetic factors that can be best used by the breeding programme.

This approach leads to different options depending on the biological traits of the species under consideration. Vegetatively propagated species often represent an extreme situation. These species can benefit greatly from the dissemination of allelic diversity because of a strong heterozygosity, of complex and relatively unknown genetics, of a fragmentation of habitats or due to a limitation of exchanges for prophylactic reasons. Such dissemination can be organised by using reference samples that are cleared of their pathogens, or progenies of crosses between germplasms selected for their ability to replicate a broad and new diversity. Thus, DAD (*distribution of allelic diversity*, Lebot et al. 2005) opens up excellent prospects for crops such as cassava, yam, taro, or plantain banana.

Creating nested association mapping (NAM) populations, i.e., focused on a selected genotype, helps mobilize a broad-based genetic pool while maintaining a controlled analytical framework and a genomic interlace with high genetic resolution. The recombinant lines derived from crossing of the same key variety with a group of varieties having different and complementary traits give more control to genetic analyses and help better assess incremental genetic progress. Thus, the NAM populations have helped launch a new generation of highly informative analyses (Cook et al. 2012; Hung et al. 2012). The generalization of this kind of approach holds great promise.

As we move the 'focus' further towards a target genotype, producing introgression lines becomes possible by successive backcrossing assisted and accelerated by the use of molecular markers. Similar to what has been achieved at the level of the entire genome with the development of populations of chromosome segment substitution lines (CSSL) using interspecific hybrids in peanuts (Fonceka et al. 2009; 2012a, b) (Box 13) and in rice (Bocco et al. 2012), the rapid and systematic production of introgression lines on a gene pool with proven agronomic interest allows us to test the effect of the introgressed chromosome segment in the very genetic pool to which it will be transferred. Geneticists are thus able to get rid of epistatic relationships that usually render genetic analysis more complex, and to accurately measure the phenotypic impact of the introgressed factor. Similarly, biologists are able to understand pleiotropy from which a lot can be learnt and farmers are able to assess the usefulness of the applied change. These approaches have the benefit of simultaneously producing information that is directly useful and easily applicable in breeding, and material that could very likely be of use to the actors.

Box 13. Valorisation of wild species related to chromosome segment substitution lines: the case of peanuts

The cultivated peanut, *Arachis hypogaea,* is an allotetraploid derived from a recent hybridization between two wild diploid species, *A. duranensis* and *A. ipaensis.* This tetraploidy isolated it from a reproductive point of view from about 80 known species of the genus *Arachis.* Since the diversity available within the crop is limited, its expansion by crossing with wild relatives constituted an important aspect of the peanut's varietal improvement. The first step was to create synthetic tetraploids by crossing two wild species and by doubling the hybrid's chromosomes. A synthetic amphidiploid was thus developed by Empresa Brasileira de Pesquisa Agropecuária (Embrapa) in Brazil by crossing the two wild parent species, *A. duranensis* and *A. ipaensis,* of the cultivated peanut.

Based on this amphidiploid, CIRAD has embarked on an ambitious programme to create chromosome segment substitution lines (CSSL) with the dual objective of increasing the diversity of the cultivated peanut and of producing a material that allows the identification of genomic regions (QTL) involved in traits of agronomic interest. To this end, the Fleur11 variety that is grown widely in Senegal was selected as recipient parent. A genetic map (Fonceka et al. 2009) was constructed for the BC_1F_1 generation to monitor and control, using markers, the distribution of the introgression of wild genome in that of Fleur11 in subsequent backcross generations. A population of 122 CSSL lines was obtained (Fig. 2) at the end of the BC_4F_3 generation. It represented the entire genome of the wild species in the form of overlapping segments introgressed in a cultivated genetic background (Fonceka et al. 2012b). Most of the lines of this population (62 %) have a unique segment, thus making it possible to directly link observed effects to it through comparison with the cultivated parent.

In the course of development of the CSSL population, an advanced backcross population was used for an early detection of QTL for plant architecture, morphology of seeds and pods as well as yield components (Fonceka et al. 2012a). This revealed the existence of wild alleles having a positive effect on agronomic traits such as the number and size of grains and pods and pod maturity. In addition, some of the QTL identified were unlike any QTL with adverse effects and could, thus, be directly used for breeding purpose.

An initial study of CSSL has confirmed the usefulness of this type of population to dissect the genetic control of useful traits such as plant bearing and height (Fonceka et al. 2012b). In light of this study, plant bearing which hitherto was described as a relatively simple trait controlled by one to four genes, appears to be the result of a greater number of QTL. This leads to 'crawling' or 'erect' phenotypes and to their intermediates as seen in cultivated groundnut.

Developed in the framework of international collaboration and exchange of genetic material and through a major effort to integrate molecular markers in the breeding process, this CSSL population is an important resource for the discovery of favourable wild alleles and the study of the genetic control of agronomically useful traits.

4.3.3 Diversifying the Species Worked On

European history is studded with examples of power plant breeding (Stamp and Visser 2012). It only took a century following the blockade imposed by the British navy on Napoleon 1's French empire to transform fodder beet to sugar beet, a temperate-zone sugar crop to rival tropical sugar cane. In less than 50 years, the quality of rapeseed was radically improved, mainly by reducing its erucic acid content, in response to a directive from the European Union. In less than 40 years, soy has been adapted to the European climate to become an economically viable crop north of the Alps. With the help of tools like molecular markers, reproductive biology and physiology, plant breeders are now able to adapt an annual crop to completely new conditions or requirements in less than 25 years if the necessary research is undertaken, genetic resources are available and the context of the agricultural system is conducive.

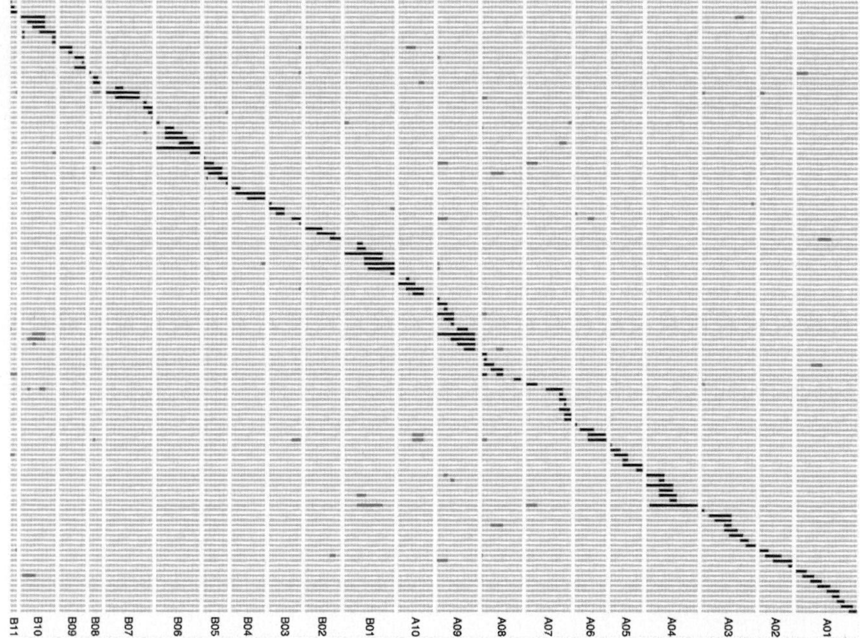

Fig. 2 Graphical genotype of 122 chromosome segment substitution lines (CSSL) in peanut. The 21 chromosomes A01–A10 and B01–B11 are represented on the abscissa. The *122 lines* are represented on the ordinate. The *white* chromosomal areas represent the cultivated gene pool; the *black* areas represent the segments substituted by the wild genome; the *grey* areas represent wild additional chromosomal segments (unintended)

Without seeking to expand the area of distribution or changing uses, an attempt can be made to maximize the value of species that are identified as largely under-utilized. These can be revisited with new analysis methods. Public research institutions should probably undertake a new generation of research on a number of chosen plants, such as fonio (*Digitaria exilis* Stapf, *D. iburua* Stapf), considered to be the oldest cereal in West Africa. Currently grown from Senegal to Lake Chad, it is adapted to dry and low-fertility environments, is not labour-intensive, and can bridge the hunger gap before the harvest of a major cereal with its high nutritional value (Vodouhè and Achigan-Dako 2006). Its rehabilitation raises issues linked not only to genetics, but also of integration into cropping, processing and distribution systems, as also of perception of rural and urban consumers.

Similar tools and approaches can be adapted to numerous plants that are currently neglected, such as service plants that make decisive contributions to the sustainability of cropping systems. Public agricultural research cannot probably handle all the genetic improvement programmes that appear useful. It can, however, share its tools and sign up partners for work that is simple but with very high added value. It can provide these partners with some molecular markers (cytoplasmic and nuclear), facilities for in vitro culture and cytology, seeds conservation capacities, etc. We could soon witness the sequencing of genomes of representative samples of genetic resources being offered as a service.

Agricultural research could also arrange for training by drawing on its experiences—including on work currently under way within ARCAD[1] projects supported by the Agropolis Foundation. This is particularly relevant with regard to strategies for exploring genetic diversity, the associated survey methodologies and modalities for accessing and sharing this diversity, all of which are components that promote processes likely to lead to a real leveraging of local knowledge.

4.3.4 Experiments in Genetic Improvement of Complex Crop Stands

The challenge of breeding for complex crop stands, i.e., mixtures of several genotypes of the same species or of associated crops, also needs addressing. But, as we pointed out, if we take biological interactions into account, we literally flood the matrix of conditions that need to be compared in order to make a choice similar to those in conventional breeding.

Developing near-isogenic material in the context of incremental breeding provides the opportunity to use very similar genotypes. These genotypes differ only in some behavioural aspects that are easy to characterize as the isogenic lines have few genes that differ.

An approach of a progressive and iterative exploration from an initial existing system becomes easier to adopt with the use of such material with discontinuous genetic variation (e.g., inbred lines) and which is limited in scope (e.g., near-isogenic). This facilitates experiments in a space where complexity is reduced to situations that can be identified by the experimenter, thus accelerating a global empirical approach.

Complementation tests can be carried out in mono-species populations for plant shape, root establishment, early vigour or other traits that differentiate the genotypes used. Maintaining a finite number of variable components helps draw conclusions based on genetics and ecophysiology, and reuse them in a simple way in succeeding generations and in models used to explore what is achievable.

The same reasoning applies to multispecies stands. The evaluated parameters have to characterize the quality of co-adaptation between genotypes of associated species. The Fabatropimed[2] project, supported by the Agropolis Foundation, provides an interesting framework to encourage breeding measures for cereal-legume association. It provides a multidisciplinary framework for the functional characterization of the benefits of the system's fertility.

---

[1] Agropolis Resource Centre for Crop Conservation, Adaptation and Diversity, http://www.arcad-project.org/about_arcad (retrieved: 6 May 2013).

[2] http://www.agropolis-fondation.fr/ (retrieved: 6 May 2013).

4.4 Tools and Methods to Accelerate the Creation of Prototype Plants

4.4.1 Fine Control Over Recombination

Methods for controlling the recombination of genomes within populations continue to progress with advances in high-throughput genotyping and statistical methods. Marker-assisted recurrent selection (MARS) has already been discussed (Box 8). An extreme version of the application of genome marking is 'genomic selection', where the value of candidates available for selection is no longer determined on the basis of the genotype with a small number of QTL, but by estimating the effect of thousands, or even hundreds of thousands, of markers on a phenotype. The absence of a priori assumptions about causal relationships between markers and target traits allows the breeding of complex traits whose genetic basis is not completely known (Heffner et al. 2009).

At the same time, marker assisted selection allows massive and rapid transfer of targeted alleles or chromosomal segments from one gene pool to another and to test the transfer's phenotypic impact. It thus provides new insights on the effects of pleiotropy and epistasis and consequently leads to a better understanding of agronomic behaviour and biological processes that underlie them, while producing material of great agronomic utility.

4.4.2 A Parsimonious Use of Genetic Transformation

Altering the genetic makeup of plants by inserting one or more new genes in their genome (transgenics) is a new form of hybridization. It allows the exchange of genetic information between organisms that cannot do so by conventional reproductive means. This brings us emphatically into the realm of GMOs. Few topics in recent times have engendered such fierce debates as GMOs. However, these debates remain inconclusive and unproductive since the concept is as yet vague and the context includes rapidly evolving technologies, high economic stakes, conflicting and changing laws, and public controversy between very active stakeholders.

The first GMOs resulted from using rather primitive technologies, leaving large traces in the genome and integrating copies of the transferred gene in indeterminate numbers and sites. Recent methodological advances have made it possible to envisage gene 'surgery', i.e., to make the desired changes with perfect accuracy. Examples include adjusting the expression of a pre-existing gene or replacing a gene by an allelic version from elsewhere by changing some targeted nucleotides—or even just a single one amongst hundreds of millions or billions (depending on the species) that the genome contains. This is the goal of Genius project, funded in the framework of the "Investment d'avenir" initiative, dedicated to develop new technologies to 'breed varieties that are more resistant, less polluting and better adapted to consumer needs.'[3]

Research in biocellular and biomolecular technology therefore promises a future with far more efficient genetic engineering and opens an almost limitless range of possibilities. Considering this, it is important to translate aims into precise objectives of genetic progress before selecting the technique to achieve these aims. A finer and more comprehensive analysis of the diversity of genetic resources will often lead to sources of high-value diversity that can be mobilized by conventional breeding. Only occasionally will it be found necessary to rely on genetic engineering to help incorporate some new traits into the plant material. Analyzing the issue, the partnership framework, expected benefits and their recipients, risks and their origins—such as biotechnical challenges—will help gauge the ethical relevance of such a path.

4.5 A Decentralized Distribution of Plant Material

Once a new 'improved' variety is created, it has to be maintained and propagated, and disseminated to potential users. The modalities for such maintenance and propagation depend, on the one hand, on the reproductive pattern of the species (sexual self-pollinated, sexual cross-pollinated, asexual) and, on the other, on the variety's genetic structure (homozygous and stable over generations of sexual propagation or heterozygous and unstable). Dissemination modalities also have to include an economic angle, as it is usually through the sale of seeds that the investment made in the plant breeding activity can be recovered.

Self-pollinated crops (wheat, rice, soyabeans, etc.) disseminated as homozygous varieties (pure lines) are the easiest to distribute. Because of their stability over generations, these varieties can be reproduced from year to year on the farm. This is clearly beneficial for the farmer but poses a threat to the breeder of the variety if equitable compensation mechanisms are not put in place. The situation is similar for species with asexual or vegetative reproduction (banana, yam, cassava, etc.). Moreover, vegetative propagation allows the maintenance of the same heterozygous genetic structures.

Cross-pollinated species, on the other hand, often with heterozygous genetic structures, are difficult to maintain and disseminate. Farmers must procure their seeds every year, a fact that translates to an economic opportunity for the breeder but creates a dependency for the farmer. Maintenance and dissemination difficulties increase when varietal improvement of a species (cocoa, teak, mahogany, coffee, rubber, etc.) relies on unique heterozygous genetic entities and on their

---

[3] http://media.enseignementsup-recherche.gouv.fr/file/Fiches_biotech_bioressources_2/93/4/ GENIUS_208934.pdf (retrieved: 6 May 2013).

dissemination by vegetative propagation—which is not their normal way of reproduction.

The choice of the procedure through which new varieties are developed is therefore not inconsequential. Public research must, on the one hand, analyze its impact in terms of access to genetic progress for various farmer categories. On the other, it should strive to reduce seed production costs and prevent monopolistic situations that arise from exclusive control of the production and dissemination of seeds.

Heterozygous genetic structures often present genetic and agronomic advantages: higher and more stable yields, homogeneity, speed of combination of favourable dominant genes in the same genotype, etc. They encourage breeders to increasingly use hybrid varietal formulas, even for self-pollinated species (Gallais 2009). Mastery of apomictic reproduction would help propagate these hybrids at the farm without a risk of altering their genetic structure (Box 14). The use of techniques to produce hybrid seeds in perennial plants that are based on the use of male sterility, already widely used for annual plants, would allow the production of hybrid seeds on a large scale in these species (Box 15). The development of somatic embryogenesis techniques would lead to the massive propagation of exceptional heterozygous genetic entities and the diversification of actors in this propagation, a factor of self-sufficiency conducive to a more intensive and diverse agriculture (Box 16).

Box 14. Apomixis: decentralizing the exploitation of hybrid vigour

Apomixis is the ability to produce seeds that contain only maternal genetic heritage. It can thus be compared to vegetative reproduction, but with a difference—it is disseminated by seeds. This possibility of identical reproduction is especially helpful for the use of hybrid varieties in self-pollinating as well as cross-pollinated species. It avoids the need to continuously renew seeds from parental lines, which would push up seed prices and require the existence of a well-organized seed industry. Apomixis also helps seed producers maintain elite genotypes from cross-pollinated species without pollen isolation. The introduction of the apomixis trait in major food crops (maize, wheat, millet, rice, etc.) would therefore be an effective way for expanding the use of hybrid vigour over a larger cropping area (Hoisington et al. 1999).

The phenomenon has been observed in many botanical families. There are several types of apomixis (apospory, diplospory and adventive embryony), ranging from an embryo formed immediately after the interruption of meiosis in megaspore mother cells to an embryo formed from cells of the nucellus or the ovule.

A good amount of research is needed to understand the molecular mechanisms of apomixis for its use to become more widespread in breeding programmes and seed reproduction (Grimanelli et al. 2001). Currently, such research is mainly the preserve of the private sector, leading to patents, especially for methods aimed at increasing the percentage of apomictic seeds from plants that reproduce sexually or from facultative apomixis. It is important for public research programmes to invest in this field of reproductive biology which promises a 'democratization' of the use of hybrid formulas.

Box 15. Seed production based on self-incompatibility or genetic male sterility

In the reproduction of many cultivated cross-pollinated species, it is almost impossible for male and female gametes produced by the same plant or plants with identical genotypes to produce viable zygotes (Gallais 2009). This is self-incompatibility. This feature is used by breeders of perennial crops, such as coffee and cocoa, to obtain hybrids between complementary clones (Charrier and Eskes 1997). The two clones are planted together in 'seed fields' in order to obtain these 'hybrid clones'. This was how hybrid clones derived from research on cocoa by CIRAD and its partners were disseminated in Côte d'Ivoire, Togo and Cameroon at subsidized rates. More recently, the rootstock variety Nemaya (species *Coffea canephora*) which is multi-resistant to nematodes was reproduced and disseminated as seeds by the Asociación Nacional del Café of Guatemala (Guatemalan National Coffee Association or ANACAFE) (Bertrand et al. 2000).

Another reproductive technique which can be readily popularized is based on the use of male sterility. The plants undergo mutations in the genes involved in the development of reproductive organs, leading to male sterility (no stamen, no viable pollen grain, etc.). Although rare, sterile male mutants have been found in all cultivated species which were subject to such research (Gallais 2009). In perennial plants where vegetative reproduction is also possible, the sterile male mutants can enter the hybrid composition of clones and be used through the system of seed fields described above. CIRAD did just this with a sterile male of the arabica coffee to produce a hybrid variety.

The genetic determinism of male sterility is generally simple, making it possible to rapidly locate and clone the gene responsible, and to ensure the transfer of a variety (or clone) to another with the help of marker-assisted selection or genetic transformation.

Both these techniques of seed production and valorisation of heterosis are relatively simple to use and can be easily implemented in other species and transferred to producer groups.

Box 16. Somatic embryogenesis

Somatic embryogenesis is a technique of in vitro culture based on the ability of some species to develop embryos from one or more somatic cells. These embryos, when placed in suitable culture conditions, grow to plants in a morphogenesis similar to that of the zygotic embryo, which is itself derived from sexual fertilization. Somatic embryogenesis allows the multiplication of clones whose unit production cost by conventional methods is otherwise high, and of genetically transformed individuals which, in some cases, cannot be propagated sexually.

It is possible to carry out in vitro culture in large containers, or bioreactors, and produce over a thousand seedlings per bioreactor. CIRAD has developed such devices for the production of coffee plants. Its plastic bioreactor (Matis) is inexpensive and easy to use, and ensures competitive costs for producing germinated somatic embryos (<0.50 Euro) (Etienne et al. 2012). Once developed in a bioreactor, the seedlings are planted on a horticultural substrate.

The major challenge now is to transfer the tools to producer organizations or to private companies. For this to be possible, the culture media and procedures will have to be simplified further, phases of autotrophic conditions increased and horticultural phases also simplified. In addition, stakeholders will have to be involved in these simplification processes through participatory approaches. For example, to make the technology more accessible, it is envisaged to build laboratories using natural light to provide heating and lighting. This would reduce the time required to obtain autotrophic embryos that germinate directly in the bioreactor.

4.6 A Renewed Varietal Innovation System

Industrial innovation systems are based on two main innovation models: through dissemination or through incentives (Akrich et al. 1988). The dissemination model assumes that the new object can disseminate by itself, by contagion, due to its intrinsic properties. It further assumes that its use presents the same benefits to everyone and in all places. In contrast, the incentive-based model assumes that the fate of the new object depends on the possibilities of simultaneous evolution of the innovation and the social environment that adapts and adopts it. The fate of a project thus depends on the alliances it forms and the interest it evokes, leading to the conclusion that no criteria or algorithm ensures, a priori, its success. Rather than discuss the rationale of decisions, we must speak of the convergence of interests that they can, or cannot, produce. Innovation is the art of evoking interest in a growing number of allies who, in turn, make you stronger (Tidd and Bessant 2011).

This attention to innovation process clearly explains the successes and setbacks of the creation and dissemination of new varieties over the past 50 years. And it is even more necessary today since ecological intensification entails a greater valorisation of G × E × cropping-system interactions. When a wider range of situations is taken into account, the probability that any one variety would present the same advantages and ease of use to everyone and at every place is reduced. Similarly, an adaptation to ecological intensification would also probably lead to disruptions of varietal traits. This would require major changes and/or disruptions in user practices and thus the adoption of innovative models that open doors to numerous fruitful exchanges between developers and users, as well as to adjustments and initiatives that mobilize local knowledge. These issues are explored more extensively in Chap. 7.

It is therefore necessary, more than ever, to establish platforms for dialogue and participatory action with all stakeholders to formalize their expectations and motivations. This will also help define with them the specifications of new varieties and evaluation and feedback methods, including their environmental and social externalities. Given the increasing diversity of scientific disciplines to be mobilized, it is also necessary to formalize, within research teams, the nature and sequence of knowledge to be developed as well as the methods, tools and actions necessary for developing prototypes for varietal innovation.

In such a context, the plant breeder is expected to play the important roles of initiator-coordinator of dialogue and of integrator of knowledge for varietal innovation. By observing and tracking the impact, he will help in the proper recognition of the roles of the various stakeholders in the innovation process and help share the resulting benefits equitably.

The establishment of such renewed varietal innovation systems depends on supportive and incentive-based policies from sponsors as well as from research institutions involved in varietal innovation. Beyond these incentives, it is necessary to ensure the implementation of *ex post* monitoring-evaluation procedures for assessing the impact of new varieties, especially in the context of family farming.

5 Conclusion

For a long time, plant breeding activities have been undertaken in an agricultural context of artificialization and standardization of the crop environment. Up to now, only a limited number of target environments were even considered. Plant breeders optimized the use of resources and practices—population size, selective pressure, etc.—in this configuration. This approach was very effective in applying quantitative genetics and in according limited importance to the biological fundamentals of variation in traits and adaptation. Recent technological and methodological developments in the field of genomics now offer plant breeders new capabilities in analyzing the traits' genetic architecture and biological adaptation mechanisms. They also have a better understanding of the dynamics of diversity and adaptation during domestication. New partnerships are also being explored, incorporating more participatory methods, in order to diversify the environmental frameworks of intervention and fine-tune the adaptation of the final products. Some methods of decentralized dissemination are already available or will be soon.

In the context of ecologically intensive agriculture, plant breeding must also address more diverse needs and take into account more complex biological functions which are in interaction with other organisms of the cropping systems. In some cases, these functions can be explained by specialized research and can be translated into absolute selection criteria (e.g., an intrinsic ability to use mineral resources). In a majority of cases, however, new and multifaceted phenotyping methods of unprecedented complexity will have to be implemented, ones that use biological interactions.

Plant breeding must also expand its scope to include a greater number of species in order to encourage a general expansion of the biological bases that agronomists and farmers rely upon.

The search for continued technological and methodological improvements and their use in sustaining existing dynamics will help address some of these challenges. We must, however, affirm and strengthen plant breeding in some of its reorientations and initiate newer ones.

Genetic diversity should be actively and systematically mobilized, based on a better description and understanding of this diversity and with the help of a rapid and accurate management of genomics. The modelling of biological systems must help translate a greater number of complex biological objectives into traits that can be inherited and selected for breeding.

The systematic creation of populations centred around established varietal types (known as 'quasi-ideotypes') will help provide opportunities for incremental breeding that will not lead to agroecological destabilization. These 'progressive' populations will be accessible to all actors for analysis and qualification.

We will have to expand the range of species we work with to include new ones, especially service species and/or those that have not been—or are as yet little— domesticated. Our range of breeding objectives and conditions under which we undertake breeding should also be expanded. Plant breeders should focus on developing new skills in multigenotypic breeding for using internal complementarities in order to create complex crop stands which are conducive to ecological intensification.

Associations with farmers—in their roles as intermediaries or full partners— must be strengthened and simplified. This will require an analysis of roles of all actors, a translation of methods and a structuring of partnerships in order to optimize the process of innovation as a whole, including the fine-tuning of the innovation to the local context. Dissemination methodologies and approaches will remain important issues and a source of determinant technological options.

References

Akrich, M., Callon, M., & Latour, B. (1988). À quoi tient le succès des innovations? L'art de l'intéressement, gérer et comprendre. *Annales des Mines, 11,* 4–17.

Anthony, F., Combes, M. C., Astorga, C., Bertrand, B., Graziosi, G., & Lashermes, P. (2002). The origin of cultivated *Coffea arabica* L. varieties revealed by AFLP and SSR markers. *Theoretical and Applied Genetics, 104,* 894–900.

Ba, M., Schilling, R., N'doye. O., N'diaye M., & Kan A. (2005). L'arachide. In ISRA-Cirad (Ed.), *Bilan de la recherche agricole et agroalimentaire au Sénégal* (pp. 163–188). ISRA-ITA-Cirad.

Berthet, E. (2010). La conception innovante à l'appui d'une gestion collective des services écosystémiques. Étude d'un cas de mise en œuvre de Natura 2000 en plaine céréalière. Mémoire de master II, Paris West University Nanterre La Défense, Mines ParisTech, ESCP.

Bertrand, B., Peña-Duran, M. X., Anzueto, F., Cilas, C., Etienne, H., Anthony, F., et al. (2000). Genetic study of *Coffea canephora* coffee tree resistance to *Meloidogyne incognita* nematodes in Guatemala and *Meloidogyne* sp. nematodes in El Salvador for selection of rootstock varieties in Central America. *Euphytica, 113*(2), 79–86.

Bertrand, B., Vaast, P., Alpizar, E., Etienne, H., Davrieux, F., & Charmetant, P. (2006). Comparison of bean biochemical composition and beverage quality of Arabica hybrids involving Sudanese–Ethiopian origins with traditional varieties at various elevations in Central America. *Tree Physiology, 26,* 1239–1248.

Bertrand, B., Alpizar, E., Lara, L., SantaCreo, R., Hidalgo, M., Quijano, J. M., et al. (2011). Performance of *Coffea arabica* F1 hybrids in agroforestry and full-sun cropping systems in comparison with American pure line cultivars. *Euphytica,.* doi:10.1007/s10681-011-0372-7.

Billot, C., Ramu, P., Bouchet, S., Chantereau, J., Deu, M., Gardes, L., et al. (2013). Massive sorghum collection genotyped with SSR markers to enhance use of global genetic resources. *PLoS One* (sous presse).

Bocco, R., Lorieux, M., Seck, P. A., Futakuchi, K., Manneh, B., Baimey, H., et al. (2012). Agro-morphological characterization of a population of introgression lines derived from crosses between IR 64 (*Oryza sativa indica*) and TOG 5681 (*Oryza glaberrima*) for drought tolerance. *Plant Science, 183,* 65–76.

Bouffaud, M. L., Kyselkova, M., Gouesnard, B., Grundmann, G., Muller, D., & Moenne-Loccoz, Y. (2012). Is diversification history of maize influencing selection of soil bacteria by roots? *Molecular Ecology, 21,* 195–206.

Cavanagh, C., Morell, M., Mackay, I., & Powell, W. (2008). From mutations to MAGIC: Resources for gene discovery, validation and delivery in crop plants. *Current Opinion in Plant Biology, 11,* 215–221.

Chambers, R. (1983). *Rural development: Putting the last first.* Harlow: Longman. 246 p.

Chantereau, J., Trouche, G., Luce, C., Deu, M., & Hamon, P. (1997). Le sorgho. In A. Charrier, M. Jacquot, S. Hamon, & D. Nicolas (Eds.), *L'amélioration des plantes tropicales* (pp. 565–590). Orstom, Repères: Cirad.

Charrier, A., & Eskes, A. B. (1997). Les caféiers. In A. Charrier, M. Jacquot, S. Hamon, & D. Nicolas (Eds.), *L'amélioration des plantes tropicales* (pp. 171–196). Orstom, Repères: Cirad.

Charrier, A., Boemare, N., Bouchez, D., Glaszmann, J. C., Joyard, J., & Lemaire, G. (2005). La biologie intégrative végétale. Rapport au Conseil scientifique de l'Inra, 43 p.

Chenu, K., Chapman, S. C., Hammer, G. L., Mclean, G., & Ben Haj Salah H. (2008a). Short-term responses of leaf growth rate to water deficit scale up to whole-plant and crop levels: An integrated modeling approach in maize. *Plant Cell Environment, 31,* 378–391.

Chenu, K., Chapman, S. C., Tardieu, F., McLean, G., Welcker, C., & Hammer, G. L. (2009). Simulating the yield impacts of organ-level *quantitative trait loci* associated with drought response in maize—a "gene-to-phenotype" modeling approach. *Genetics, 183,* 1507–1523.

Choudhury, A., & Kennedy, I. R. (2004). Prospects and potentials for systems of biological nitrogen fixation in sustainable rice production. *Biology and Fertility of Soils, 39,* 219–227.

Clavel, D., & Annerose, D. J. M. (1995). Genetic improvement of groundnut adaptation to drought. In S. Risopoulos (Ed.), *Research projects* (pp. 33–35). Summaries of the Final Reports STD2, UE-DG12, Wageningen, The Netherlands.

Clavel, D., & N'doye, O. (1997). La carte variétale de l'arachide au Sénégal. *Agriculture et développement, 14*, 41–46.

Clavel, D., Drame, N. K., Diop, N. D., & Zuily-Fodil, Y. (2005). Adaptation à la sécheresse et création variétale: le cas de l'arachide en zone sahélienne. Première partie: revue bibliographique. *OCL, 13*(3), 246–260.

Collectif. (1991). *Le coton en Afrique de l'Ouest et du Centre,* Editions du ministère de la Coopération et du Développement, 354 p.

Cook, J. P., McMullen, M. D., Holland, J. B., Tian, F., Bradbury, P. J., Ross-Ibarra, J., et al. (2012). Genetic architecture of maize kernel composition in the nested association mapping and inbred association panels. *Plant Physiology, 158*(2), 824–834.

Cooper, M., Van Eeuwijk, F. A., Hammer, G., Podlich, D., & Messina, C. (2009). Modeling QTL for complex traits: Detection and context for plant breeding. *Current Opinion in Plant Biology, 12*, 231–240.

D'Hont, A., Denoeud, F., Aury, J. M., Baurens, F. C., Carreel, F., Garsmeur, O., et al. (2012). The banana (*Musa acuminata*) genome and the evolution of monocotyledonous plants. *Nature, 488*(7410), 213–219.

Dawson, J. C., & Goldringer, I. (2012). Breeding for genetically diverse populations: Variety mixtures and evolutionary populations. In E. T. Lammerts Van Bueren & J. R. Myers (Eds.), *Organic crop breeding* (pp. 77–98). Chichester: Wiley-Blackwell.

Déchanet, R., Razafindrakoto, J., & Valès, M. (1997). Résultats de l'amélioration variétale du riz d'altitude Malgache. In: C. Poisson, & J. Rakotoarisoa (Eds.), *Rice for highlands* (pp. 43–48). *Proceeding of the International Conference on Rice for Highlands*, March 29–April 5, 1996, Antananarivo, Madagascar/Cirad, Montpellier, France.

Dekkers, J. C. M., & Hospital, F. (2002). The use of molecular genetics in the improvement of agricultural populations. *Nature Reviews Genetics, 3*, 22–32.

Dingkuhn, M., Baron, C., Bonnal, V., Maraux, F., Sarr, B., Sultan, B., et al. (2003). Decision support tools for rainfed crops in the Sahel at the plot and regional scales. In TESBaMCS Wopereis (Ed.), *Decision support tools for smallholder agriculture in Sub-Saharian Africa* (pp. 127–139). *A practical Guide*, IFDC-CTA, Wageningen, The Netherlands.

Döring, T. F., Knapp, S., Kovacs, G., Murphy, K., & Wolfe, M. S. (2011). Evolutionary plant breeding in cereals: Into a new era. *Sustainability, 3*, 1944–1971.

Durand, E., Bouchet, S., Bertin, P., Ressayre, A., Jamin, P., Charcosset, A., et al. (2012). Epistasis, pleiotropy and maintenance of polymorphism at a locus associated with flowering time variation in maize inbred lines. *Genetics, 190*, 1547–1562.

Dzido, J. L., Vales, M., Rakotoarisoa, J., Chabanne A., & Ahmadi, N. (2004). Upland rice for highlands: New varieties and sustainable cropping systems for food security. Promising prospects for the global challenges of rice production. *FAO Rice Conference,* February 12–13, 2004, Rome, Italy, 11 p.

Etienne, H., Bertrand, B., Montagnon, C., Dechamp, E., Jourdan, I., Alpizar, E., et al. (2012). Un exemple de transfert technologique réussi en micropropagation: la multiplication de Coffea arabica par embryogenèse somatique. *Cahiers Agriculture, 21*, 115–125.

Evenson, R., & Rosegran, M. (2003). The economic consequences of crop genetic improvement programs. In R. E. Evenson, & D. Gollin (Eds.), *Crop variety improvement and its effect on productivity: The impact of International Agricultural Research.* CABI.

Eyzaguirre, P., & wanaga, M. (1996). Participatory plant breeding. *Proceedings of a Workshop on Participatory Plant Breeding,* July 26–29, 1995, Wageningen, The Netherlands, IPGRI, Rome, Italy.

Faraji, J. (2011). Wheat cultivar blends: A step forward to sustainable agriculture. *African Journal of Agricultural Research, 6*(33), 6780–6789.

Finckh, M. R., Gacek, E. S., Goyeau, H., Lannou C., Merz, U., Mundt, C. C., et al. (2000). Cereal variety and species mixtures in practice, with emphasis on disease resistance. *Agronomie, 20*, 813–837.

Fisher, R. A. (1918). The correlation between relatives on the supposition of Mendelian inheritance. Transactions of the *Royal Society, Edinburgh, 52*, 399-433.

Fliedel, G., 1995. Appraisal of sorghum quality for making tô. *Agriculture et développement, Special Issue, 35–45*.

Fonceka, D., Hodo-Abalo, T., Rivallan, R., Faye, I., Sall, M. N., Ndoye, O., et al. (2009). Genetic mapping of wild introgressions into cultivated peanut: a way toward enlarging the genetic basis of a recent allotetraploid. *BMC Plant Biology, 9*, 103.

Fonceka, D., Tossim, H.-A., Rivallan, R., Vignes, H., Faye, I., Ndoye, O., et al. (2012a). Fostered and left behind alleles in peanut: Interspecific QTL mapping reveals footprints of domestication and useful natural variation for breeding. *BMC Plant Biology, 12*, 26.

Fonceka, D., Tossim, H.-A., Rivallan, R., Vignes, H., Lacut, E., & De Bellis, F. (2012b). Construction of chromosome segment substitution lines in peanut (Arachis hypogaea L.) using a wild synthetic and QTL mapping for plant morphology. *PLoS One, 7*(11), e48642, 11 p.

Gallais, A. (2009). *Hétérosis et variétés hybrides en amélioration des plantes*. Versailles, coll. Synthèses, Éditions Quae, 356 p.

Gamuyao, R., Chin, J. H., Pariasca-Tanaka, J., Pesaresi, P., Catausan, S., Dalid, C., et al. (2012). The protein kinase Pstol1 from traditional rice confers tolerance of phosphorus deficiency. *Nature, 488*, 535–541.

Gibert, O., Dufour, D., Giraldo, A., Sánchez, T., Reynes, M., Pain, J. P., et al. (2009). Differentiation between cooking bananas and dessert bananas. 1. Morphological and compositional characterization of cultivated Colombian *Musaceae* (*Musa* sp.) in relation to consumer preferences. *Journal of Agricultural and Food Chemistry, 57*(17), 7857–7869.

Glaszmann, J. C., Kilian, B., Upadhyaya, H. D., & Varshney, R. K. (2010). Assessing genetic diversity for crop improvement. *Current Opinion in Plant Biology, 13*, 167–173.

Griffon, M., 2007. Pour des agricultures écologiquement intensives. In *Les défis de l'agriculture au xxie siècle*, Leçons inaugurales du Groupe ESA, Angers.

Grimanelli, D., Leblanc, O., Perotti, E., & Grossniklaus, U. (2001). Developmental genetics of gametophytic apomixis. *Trends in Genetics, 17*(10), 597–604.

Gunn, B. F., Baudouin, L., & Olsen, K. M. (2011). Independent origins of cultivated coconut (*Cocos nucifera* L.) in the old world tropics. *PLoS One, 6*(6), e21143.

Gur, A., & Zamir, D. (2004). Unused natural variation can lift yield barriers in plant breeding. *PLoS Biology, 2*(10), 1610–1615.

Haling, R. E., Simpson, R. J., McKay, A. C., Hartley, D., Lambers, H., Ophel-Keller, K., et al. (2011). Direct measurement of roots in soil for single and mixed species using a quantitative DNA-based method. *Plant and Soil, 348*, 123–137.

Hammer, G. L., Cooper, M., Tardieu, F., Welch, S., Walsh, B., Eeuwijk, F., et al. (2006). Models for navigating biological complexity in breeding improved crop plants. *Trends in Plant Science, 11*, 587–593.

Hammer, G. L., van Oosterom, E., McLean, G., Chapman, S. C., Broad, I., Harland, P., et al. (2010). Adapting APSIM to model the physiology and genetics of complex adaptive traits in field crops. *Journal of Experimental Botany, 61*(8), 2185–2202.

Hardon, J. (1995). Participatory plant breeding. The outcome of a workshop on participatory plant breeding. *Issues in Genetics Resources*, 3, IPGRI, Rome, Italy.

Hau, B., Lançon, J., & Dessauw, D. (1997). Les cotonniers. In A. Charrier, M. Jacquot, S. Hamon, & D. Nicolas (Eds.), *L'amélioration des plantes tropicales* (pp. 241–266). Orstom: Cirad.

He, L., & Hannon, G. J. (2004). MicroRNAs: Small RNAs with a big role in gene regulation. *Nature Reviews Genetics, 5*, 522–531.

Heffner, H. L., Sorrells, R. E., & Jannink, J. L. (2009). Genomic selection for crop improvement. *Crop Science, 49*, 1–12.

Heinemann, H. B., Dingkuhn, D., Luquet, D., Combres, J.-C., & Chapman, S. (2008). Characterization of drought stress environments for upland rice and maize in central Brazil. *Euphytica, 162,* 395–410.

Henry, A., Rosas, J. C., Beaver, J. S., & Lynch, J. P. (2010). Multiple stress response and belowground competition in multilines of common bean (*Phaseolus vulgaris* L.). *Field Crops Research, 117*(2–3), 209–218.

Hoisington, D., Khairallah, M., Reeves, T., Ribout, J. M., Skovmand, B., Taba, S., et al. (1999). Plant genetic resources: What can they contribute toward increased crop productivity. *The Proceedings of the National Academy of Sciences (USA), 96,* 5937–5943.

Huang, X., Wei, X., Sang, T., Zhao, Q., Feng, Q., Zhao, Y., et al. (2010). Genome-wide association studies of 14 agronomic traits in rice landraces. *Nature Genetics, 42*(11), 961–969.

Huang, X., Kurata, N., Wei, X., Wang, Z., Wang, A., Zhao, Q., et al. (2012). A map of rice genome variation reveals the origin of cultivated rice. *Nature, 490,* 497–501.

Hung, H.-Y., Shannon, L. M., Tian, F., Bradbury, P. J., Chen, C., Flint Garcia S. et al. (2012). ZmCCT and the genetic basis of day-length adaptation underlying the postdomestication spread of maize. *PNAS,* DOI: 10.1073/pnas.1203189109.

Jannink, J. L., & Walsh, B. (2002). Association mapping in plant populations. In M. S. Kang (Ed.), *Quantitative genetics, genomics and plant breeding* (pp. 59–68). CAB International.

Khalfaoui, J. L. B. (1991). Determination of potential lengths of the crop growing period in semi-arid regions of Senegal. *Agricultural and Forest Meteorology, 55,* 251–263.

Kiaer, L., Skovgaard, I., & Ostergard, H. (2009). Grain yield increase in cereal variety mixtures: a meta-analysis of field trials. *Field Crops Research, 114,* 361–373.

Kiaer, L. P., Skovgaard, I. M., & Ostergard, H. (2012). Effects of inter-varietal diversity, biotic stresses and environmental productivity on grain yield of spring barley variety mixtures. *Euphytica, 185,* 123–138.

Kidwell, M. G. (2005). Transposable elements. In T. R. Gregory (Ed.), *The evolution of the genome* (pp. 165–221). San Diego: Elsevier.

Klerkx, L., Hall, A., & Leeuwis, C. (2009). Strengthening agricultural innovation capacity: Are innovation brokers the answer? UNU-MERIT Working Paper Series #2009-019, United Nations University-Maastricht, Economic and social Research and training centre on Innovation and Technology, Maastricht, The Netherlands.

Lebot, V., Ivancic, A., & Abraham, K. (2005). The geographical distribution of allelic diversity, a practical means of preserving and using minor root crops genetic resources. *Experimental Agriculture, 41,* 475–489.

Leclerc, C., & Coppens d'Eeckenbrugge G. (2011). Social organization of crop genetic diversity. The G × E × S interaction model. *Diversity, 4*(1), 1–32 (2012).

Levrat, R. (2009). *Le coton dans la zone franc depuis 1950. Un succès remis en cause.* L'Harmattan, 256 p.

Loor Solorzano, R. G., Fouet, O., Lemainque, A., Pavek, S., Boccara, M., & Argout, X. (2012). Insight into the wild origin, migration and domestication history of the fine flavour national *Theobroma cacao* L. variety from Ecuador. *PLoS One, 7*(11), e48438.

Luquet, D., Rebolledo, M. C., & Soulié J. C. (2012a). Functional-structural plant modeling to support complex trait phenotyping: Case of rice early vigor and drought tolerance using Ecomeristem model. In IEEE (Ed.), *PMA* Shanghai, China.

Luquet, D., Dingkuhn, M., Kim, H. K., Tambour, L., & Clément-Vidal, A. (2006). Ecomeristem, a model of morphogenesis and competition among sinks in rice. 1. Concept, validation and sensitivity analysis. *Functional Plant Biology, 33,* 309–323.

Luquet, D., Soulié, J. C., Rebolledo, M. C., Rouan, L., Clément-Vidal, A., & Dingkuhn, M. (2012b). Developmental dynamics and early growth vigour in rice. 2. Modelling genetic diversity using Ecomeristem. *Journal of Agronomy and Crop Science, 198*(5). 385 p.

Mayeux, A., & Da Sylva, A. (2008). Guide pratique de production de semences d'arachide de bonne qualité semencière. Document de l'Association sénégalaise pour la promotion du développement à la base (Asprodeb), 46 p.

Messina, C., Hammer, G., Dong, Z., Podlich, D., & Cooper, M. (2009). Modelling crop improvement in a G × E × M framework via gene-trait-phenotype relationships. In V. O. Sadras & D. Calderini (Eds.), *Crop physiology: Applications for genetic improvement and agronomy* (pp. 235–265). The Netherlands: Academic Press, Elsevier.

Naudin, K., Scopel, E., Rakotosolofo, M., Solomalala, A. R. N. R., Andriamalala, H., Domas, R., et al. (2010). Trade-offs between different functions of biomass in conservation agriculture: Examples from smallholders fields of rainfed rice in Madagascar. In *11th congress of the European Society for Agronomy (ESA)*, August 29–September 3, Montpellier, France.

Newton, A. C., Begg, G. S., & Swanston, J. S. (2009). Deployment of diversity for enhanced crop function. *Annals of Applied Biology, 154*(3), 309–322.

Nuzhdin, S. V., Friesen, M. L., & McIntyre, L. M. (2012). Genotype-phenotype mapping in a post-GWAS world. *Trends in Genetics, 28*(9), 421–426.

Ostergard, H., & Fontaine, L. (2006). Cereal crop diversity: Implications for production and product. In *Proceedings of the COST SUSVAR workshop*, June 13–14, La Besse, France, Institut technique de l'agriculture biologique.

Ostergard, H., Finckh, M. R., Fontaine, L., Goldringer, I., Hoad, S. P., Kristensen, J. K., et al. (2009). Time for a shift in crop production: Embracing complexity through diversity at all levels. *Journal of the Science of Food and Agriculture, 89*(9), 1439–1445.

Ouédraogo, S. (2005). *Intensification de l'agriculture dans le plateau central du Burkina Faso: une analyse des possibilités à partir des nouvelles technologies*. Thèse, Groningen University, 322 p.

Parent, B., & Tardieu, F. (2012). Temperature responses of developmental processes have not been affected by breeding in different ecological areas for 17 crop species. *New Phytologist, 194*(3), 760–774.

Passioura, J. B. (2012). Scaling up: The essence of effective agricultural research. *Functional Plant Biology, 37*(7), 585–591.

Paszkowski, J., & Grossniklaus, U. (2011). Selected aspects of transgenerational epigenetic inheritance and resetting in plants. *Current Opinion in Plant Biology, 14*, 195–203.

Paterson, A. H., Lander, E. S., Hewitt, J. D., Peterson, S., Lincoln, S. E., & Tanksley, S. D. (1988). Resolution of quantitative traits into Mendelian factors by using a complete linkage map of restriction fragment length polymorphisms. *Nature, 335*, 721–726.

Pautasso, M., Aistara, G., Barnaud, A., Caillon, S., Clouvel, P., Coomes, O. T., et al. (2013). Seed exchange networks for agrobiodiversity conservation. A review. *Agronomy for Sustainable Development, 33*, 151–175.

Peng, S., & Bouman, B. (2007). Prospects for genetic improvement to increase lowland rice yields with less water and nitrogen. In J. H. J. Spiertz, P. C. Struik, H. H. van Laar (Eds.), *Scale and complexity in plant systems research: Gene-plant-crop relations* (pp. 251–266). Springer.

Perrier, X., De Langhe, E., Donohue, M., Lentfer, C., Vrydaghs, L., Bakry, F., et al. (2011). Multidisciplinary perspectives on banana (*Musa* spp.) domestication. *Proceedings of the National Academy of Sciences of the United States of America, 108*, 11311–11318.

Raboin, L. M., Ramanantsoanirina, A., Dusserre, J., Razasolofonanahary, F., Tharreau, D., & Lannou, C. (2012). Two-components cultivar mixtures reduce rice blast epidemics in an upland agrosystem. *Plant Pathology,*. doi:10.1111/j.1365-3059.2012.02602.x.

Radanielina, T. (2010). Diversité génétique du riz (*Oryza sativa* L.) dans la région de Vakinankaratra, Madagascar. Structuration, distribution écogéographique et gestion in situ. Thèse de doctorat, AgroParisTech, Paris, no. 2010/AGPT/0093.

Reymond, M., Muller, B., Leonardi, A., Charcosset, A., & Tardieu, F. (2003). Combining *quantitative trait loci* analysis and an ecophysiological model to analyze the genetic variability of the responses of maize leaf growth to temperature and water deficit. *Plant Physiology, 131*, 664–675.

Roubaud, E. (1918). L'état actuel et l'avenir du commerce des arachides au Sénégal. *Annales de géographie, 27*, 357–371.

Sax, K. (1923). The association of size differences with seed-coat pattern and pigmentation in *Phaseolus vulgaris*. *Genetics, 8*, 552–560.

Schultz, T. (1945). *Food for the world*. Chicago: University of Chicago Press. 353 p.

Schultz, T. (1964). *Transforming traditional agriculture*. New Haven: Yale University Press. 206 p.

Sester, M., Raboin, L. M., Ramanantsoanirina, A., & Tharreau, D. (2008). Toward an integrated strategy to limit blast disease in upland rice. In *Diversifying crop protection, Endure international conference,* 2008, La Grande Motte, France.

Sheehy, J. E., Mitchell, P. L., & Hardy, B. (2007). *Charting new pathways to C4 Rice,* Los Baños (Philippines): International Rice Research Institute, 422 p.

Soulié, J. C., Pradal, C., Fournier, X., & Luquet, D. (2010). Modelling the feedbacks between rice plant microclimate and morphogenesis. First results of Ecomeristem integration into OpenAlea. In *FSPM, Functional Structural Plant Modelling*, University of California, Davis, California, USA.

Stamp, P., & Visser, R. (2012). The twenty-first century, the century of plant breeding. *Euphytica, 186*, 585–591.

Tardieu, F. (2003). Virtual plants: Modelling as a tool for the genomics of tolerance to water deficit. *Trends in Plant Science, 8*, 1360–1385.

Tardieu, F. (2012). Any trait or trait-related allele can confer drought tolerance: Just design the right drought scenario. *Journal of Experimental Botany, 63*(1), 25–31.

Thalapati, S., Batchu, A. K., Neelamraju, S., & Ramanan, R. (2012). Os11Gsk gene from a wild rice, *Oryza rufipogon,* improves yield in rice. *Functional and Integrative Genomics, 12*(2), 277–289.

Tidd, J., & Bessant, J. (2011). *Managing innovation: Integrating technological, market and organizational change*. John Wiley & Sons, West Sussex, England. 638 p.

Tomekpe, K., Jenny, C., & Escalant, J. (2004) Revue des stratégies d'amélioration conventionnelle de Musa. *Infomusa (FRA), 13*(2), 2–6.

Tsaftaris, A. S., Polidoros, A. N., Kapazoglou, A., Tani, E., & Kovačević, N. M. (2008). Epigenetics and plant breeding. *Plant Breeding Reviews, 30*, 49–177.

Vaksmann, M., Traoré, S. B., & Niangado, O. (1996). Le photopériodisme des sorghos africains. *Agriculture et développement, 9*, 13–18.

Vaksmann, M., Kouressy, M., Chantereau, J., Bazile, D., Sangnard, F., Touré, A., et al. (2008). Utilisation de la diversité génétique des sorghos locaux du Mali. *Cahiers Agricultures, 17*(2), 140–145.

Varshney, R. K., Glaszmann, J. C., Leung, H., & Ribaut, J. M. (2010). More genomic resources for less-studied crops. *Trends in Biotechnology, 28*, 452–460.

Vodouhè, S. R., & Achigan-Dako, E. G. (2006). Digitaria exilis (Kippist) Stapf. In MBaG Belay (Ed.), *Plant resources of Tropical Africa* (Vol. 1, pp. 59–63). Wageningen: PROTA Foundation, CTA, Backhuys Publishers.

Vom Brocke, K., Trouche, G., Zongo, S., Abdramane, B., Barro Kondombo, C. P., Weltzien, E., et al. (2008). Création et amélioration de populations de sorgho à base large avec les agriculteurs au Burkina Faso. *Cahiers Agricultures, 17* (2), 146–153.

Vom Brocke, K., Trouche, G., Weltzien, E., Barro Kondombo, C. P., Gozé, & E., Chantereau, J. (2010). Participatory variety development for sorghum in Burkina Faso: Farmers' selection and farmers' criteria. *Field Crops Research, 119*, 183–194.

Vom, Brocke K., Trouche, G., Hocdé, H., & Bonzi, N. (2011). Sélection variétale au Burkina Faso: un nouveau type de partenariat entre chercheurs et agriculteurs. *Grain de sel, 52–53*, 20–21.

Welcker, C., Boussuge, B., Bencivenni, C., Ribaut, J. M., & Tardieu, F. (2007). Are source and sink strengths genetically linked in maize plants subjected to water deficit? A QTL study of the responses of leaf growth and of anthesis-silking interval to water deficit. *Journal of Experimental Botany, 58*: 339–349.

Wolfe, M. S., Baresel, J. P., Desclaux, D., Goldringer, I., Hoad, S., Kovacs, G. et al. (2008). Developments in breeding cereals for organic agriculture. *Euphytica, 163*, 323–346.

Yu, J., Holland, J. B., McMullen, M. D., & Buckler, E. D. (2008). Genetic design and statistical power of nested association mapping in maize. *Genetics, 178*, 539–551.

Chapter 5
Ecological Interactions Within the Biodiversity of Cultivated Systems

Alain Ratnadass, Éric Blanchart and Philippe Lecomte

Various types of biodiversities can be found within the cultivated plot and in its surrounding environment: plant, animal and microbial biodiversities; aboveground and belowground biodiversities; productive, resource, destructive biodiversities, etc.

How much do we know about the interactions between these worlds? They have long been forgotten, relegated to black boxes, which are now being opened to shed light on the complexity of stands and interactions (Fig. 1). How can we make the most of the ecological functioning of stands? What do we lose, what do we risk and what do we stand to gain by rendering the systems more complex? Nurturing biodiversity in all its complexity is necessary but not enough; its introduction into agroecosystems must be planned and organized, and, more importantly, should be managed properly.

Chapter 3 discusses several facilitation/competition types of interaction between cultivated plants, 'pericultivated' plants (= sanitizing plants) and weeds in agronomy. Similarly, Chap. 4 partially covers direct interactions between productive and destructive biodiversities in relation to the resistance of crop varieties to pests and diseases, including the effective use of techniques such as varietal mixtures that are obtained from genetic improvement.

A. Ratnadass (✉)
Department Persyst, Hortsys Internal Research Unit (Agroecological Functioning
and Performances of Horticultural Cropping Systems)—CIRAD, Montpellier, France
e-mail: alain.ratnadass@cirad.fr

É. Blanchart
Researcher at the Eco & Sols Joint Research Unit (Functional Ecology and Biochemistry
of Soils and Agroecosystems), IRD, Montpellier, France
e-mail: eric.blanchart@ird.fr

P. Lecomte
Department ES, Director of the Selmet Joint Research Unit (Mediterranean and Tropical
Livestock Systems)—CIRAD, Montpellier, France
e-mail: philippe.lecomte@cirad.fr

É. Hainzelin (ed.), *Cultivating Biodiversity to Transform Agriculture*,
DOI: 10.1007/978-94-007-7984-6_5, © Éditions Quæ 2013

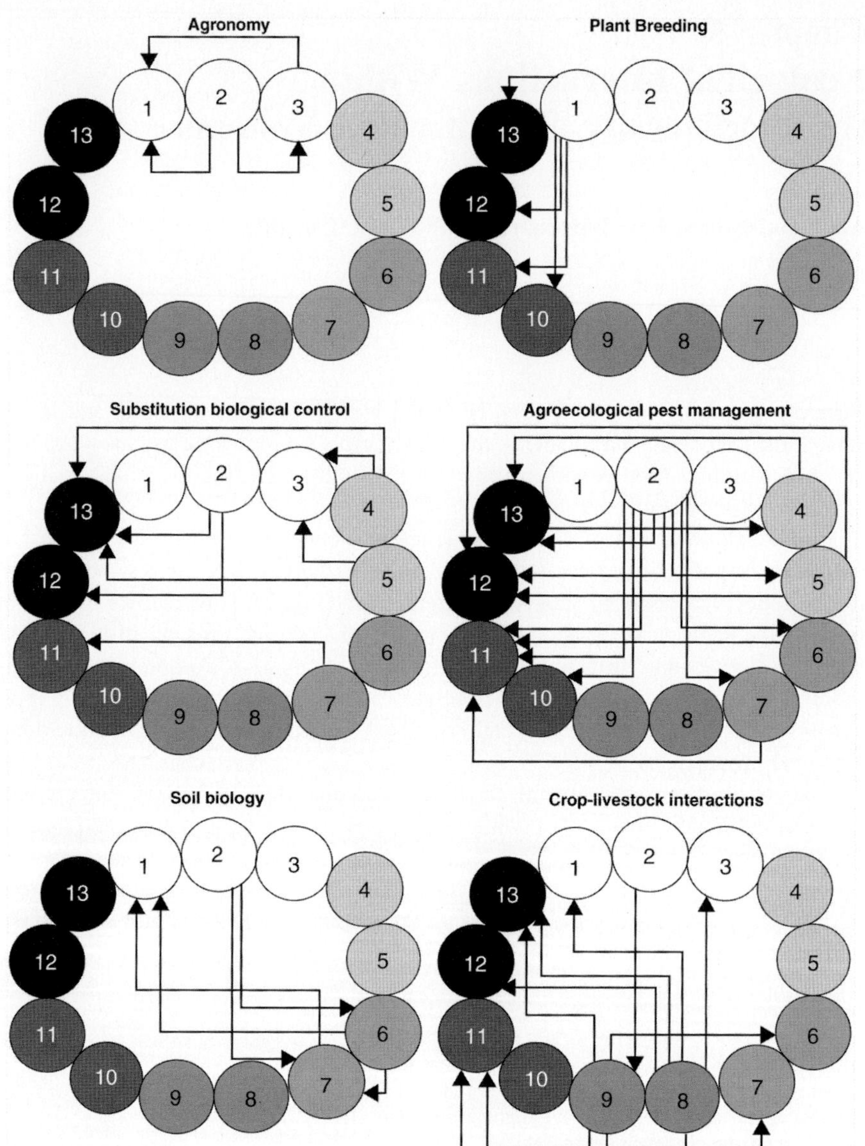

Fig. 1 Diagram of different biodiversity components in the cultivated plot and of the network of key interactions between them

This present chapter covers the use of pericultivated plant biodiversity in the form of plant extracts with biocidal properties, as also the effective use of natural animal or microbial enemies released or applied to fight pests and crop diseases. It also describes various pest control processes that result from the introduction of specific plant biodiversities in cultivated plots or their immediate surroundings in a variety of ways and at different spatio-temporal scales.

Telluric processes related to soil microflora and fauna are also presented (especially biodiversity of type 'resource'), as are their interactions with the cultivated or pericultivated plant biodiversity at the plot level.

Interactions between livestock and biodiversity, mainly plant biodiversity in the plot, either directly in the plot or outside it (via transfer of organic matter: fodder in one direction, effluents in the other), are also discussed.

1 Biodiversity and Pest Control

This section describes how ecological intensification can be applied by managing cultivated and pericultivated biodiversities to control pest populations and the damage they cause (resistant varieties, sanitizing plants and biological control agents) from the standpoint of managing a cultivated system.

1.1 Genetic Plant Biodiversity and Pest Control

1.1.1 Varietal Resistance of Plants to Pests

Research and breeding programmes have identified cultivated plant varieties resistant to hundreds of pathogens and pests (Painter 1951; Gallun 1977; Wilhoit 1992; McIntosh 1998; Mundt 2002; Thomas et al. 2002). In wheat alone, in fact, there were reported instances of resistance to at least 28 bacterial, fungal and viral pathogens, four nematode species and nine insect species (McIntosh 1998). A knowledge of the resistance mechanisms involved (non-preference, antibiosis, tolerance or compensation), as well as of the heritability and genetic determinism of resistance, is key to defining optimal breeding strategies. The main objective of a breeding programme aimed at strengthening pest resistance is to ideally select plants or genotypes that produce more than susceptible varieties when subjected to high pest or pathogen pressure, and produce at least as much in the absence of these pests or pathogens. The ideal situation is to combine, in a variety, genes that use various resistance mechanisms available against a pest in order to ensure long-term resistance and be equipped to deal with any possible changes in the pest or pathogen. Genes of resistance to different pests can sometimes be combined in the same variety, as was done by pedigree breeding for panicle pests of sorghum: midge, head-bugs and, indirectly, grain moulds (Ratnadass et al. 2002, 2006; Dakouo et al. 2005) (Box 1).

Box 1. Varietal resistance in sorghum to panicle insect pests in West Africa

Sorghum is the most important food crop in the savannas of West and Central Africa, particularly in Nigeria, Burkina Faso and Mali. Panicle-feeding mirid bugs (especially *Eurystylus oldi*) are a major constraint to the adoption of improved varieties with compact panicles (*Caudatum* race) which, despite being more productive than local *Guinea* varieties with loose panicles, are more susceptible to attack by these pests.

The problem is particularly severe in the Kolokani region, north of Bamako, where such varieties were extensively adopted over the last 25 years as their short cycle was more suited to low rainfall (attributed to climate change). The problem also affects the northern regions of Nigeria, where hybrids with compact panicles are widely cultivated, mainly to supply industrial breweries, a practice that gained currency following a government ban in 1988 on the import of cereals, including barley and malt barley.

The introduction of *Caudatum* varieties, with a reasonable resistance level to bugs (and grain mould associated with them, Ratnadass et al. (1995a, 2003)), has therefore emerged as an effective and low-cost method of reducing quantitative and qualitative losses that small and large farmers in the region are subjected to.

Unfortunately, it has always been difficult to combine the traits of the *Guinea* and *Caudatum* races in a single variety. Consequently, it is not certain that the variety that is resistant to bugs, Malisor 84–7 (even though with a compact panicle), which resulted from a recurrent breeding program based on an open-pollinated population from Mali actually has *Guinean* 'blood' (Shetty et al. 1991). Screening and breeding work conducted over several years has confirmed the strong and stable resistance to bugs in Malisor 84–7, as well as the possibility of transferring the same resistance to its progeny via pedigree breeding (Ratnadass et al. 1995b).

Penetrometry studies have also revealed the factor behind this resistance: the albumen in this cultivar hardened faster than in more susceptible cultivars, thus reducing the period during which the grain is vulnerable to bug attacks (Fliedel et al. 1996).

Heritability studies (diallel analysis) have shown that it was heritable, much like resistance to midge (the most destructive sorghum pest in the world), and since resistances to both pests were independent, it was possible to combine them in a single variety (Ratnadass et al. 2002). Studies to map *quantitative trait loci* (QTL) (Deu et al. 2005) confirmed the recessive nature of the resistance to bugs and the possibility of transgressive segregation.

This led to the breeding of a variety (CIRAD 441) which combined, on the one hand, the productivity and quality of the grain and, on the other, a resistance to both panicle pests (Dakouo et al. 2005; Ratnadass et al. 2006).

The resulting grain, however, was smaller than those of varieties that were susceptible (by having a longer grain-filling time). In other words, a cost of resistance (for consumers, at least).
For further information: Ratnadass et al. (1998, 2006).

1.1.2 Modalities of Spatio-Temporal Deployment of Resistance in the Cultivated Plot

It seems possible to use the modalities of spatio-temporal deployment of resistance to limit the spread of pests.

This type of intervention has been particularly encouraging and fruitful in the case of pathogens, such as those responsible for rust and powdery mildew on wheat or blast on rice (Castilla et al. 2003; Cox et al. 2004; Finckh et al. 2000; Mundt 2002; Zhu et al. 2000; de Vallavieille-Pope 2004; Raboin et al. 2012).

Varietal mixtures act on the specialist and polycyclic foliar pathogens mentioned above via three key mechanisms in a combined manner and over several generations (Chin and Wolfe 1984). The first is a dilution effect based on a lower probability of a spore producing a new infection due to a reduced density of susceptible plants in the mixture. The second consists of a barrier effect created by resistant plants which inhibit the dispersion of spores. Finally, we have an induced resistance effect, caused by the presence of non-virulent spores in the crop, which triggers defence mechanisms in plants (Box 2).

Box 2. Varietal mixtures

The varietal resistance of crops to diseases often decreases over time as it can be circumvented by pathogens. Growing varietal mixtures is one way to limit the spread of several diseases transmitted through the air. This happens due to the dilution effect, i.e., the number of susceptible plants in the varietal mixture is reduced; through the barrier effect to spore dispersal created by resistant plants; and finally, through induced resistance in these plants built up after contact with non-virulent spores of the pathogen. The combination of these mechanisms helps slow the emergence of pathogenic strains that can circumvent this resistance, as was demonstrated in the case of specific polycyclic leaf diseases like rust or mildew in cereals (Wolfe 1985; Finckh et al. 2000; Mundt 2002).

This strategy's effectiveness is increased if such resistance is deployed over a large area and over time, as was demonstrated in the case of mildew on spring barley in Eastern and Northern Europe, for septoria and brown rust in winter wheat in France (Mille and de Valavieille-Pope 2001), and for blast on rice in China (Zhu et al. 2000) and Madagascar (Raboin et al. 2012). This approach, it must be admitted, has been less successful and less convincing in the case of insect pests (Tooker and Frank 2012).

European legislation allows the marketing of varietal associations, but the adoption of this strategy is subject to the homogeneity of agronomic traits in the varietal mixture, especially that of cycle length. This strategy is less restrictive in countries of the South where manual harvesting is the norm. It also allows the cultivation of crops that have become susceptible to certain pathogens but possess organoleptic characteristics that are sought by the consumer. Such varieties would have been overwhelmed if they were grown in monogenotypic stands. Examples are the F-152 and F-154 upland rice varieties in Madagascar and their susceptibility to blast (Raboin et al. 2012). **For further information:** Raboin et al. (2012).

As a result, nearly 50 % of wheat fields in Europe and tens of thousands of hectares of rice in China were planted with mixtures of varieties. The United States followed a similar pattern for its winter wheat over extensive surfaces in states like Washington and Kansas (Zhu et al. 2000; Bowden et al. 2001; Mundt 2002).

Several studies have demonstrated the potential (and limitations) of this approach as far as insect pests are concerned, but it has rarely been translated into practical applications (Tooker and Frank 2012).

1.2 Animal and Microbial Biodiversity and Pest Control: Biocontrol by Introduction and Augmentation

The introduction of natural enemies of pests, which were also introduced, is another way to mobilize biodiversity (animal or microbial in this case) with natural enemies being identified in habitats where the pests originated. This constitutes the basic approach of conventional biocontrol. A well-known example is the introduction of the parasitoid *Epidinocarsis lopezi* into Africa by the International Institute of Tropical Agriculture (IITA) to fight the cassava mealy bug *Phenacoccus manihoti* which was introduced to the continent from South America (Herren et al. 1987). The island of Reunion offers another biocontrol example: for combating white grubs in sugarcane (*Hoplochelus marginalis*), originally introduced from Madagascar, the entomopathogenic fungus *Beauveria brongniartii*—also from Madagascar—was brought in to fight the grub in the island (Vercambre et al. 2008) (Box 3).

Box 3. A successful example of conventional biological control: control of white grubs in sugarcane in Reunion with an entomopathogenic fungus

Sugarcane is grown on half of the agricultural area on the island of Reunion (i.e., between 25,000 and 35,000 ha every year for the last 35 years), with an annual production of raw cane touching 2 million tonnes, equivalent to nearly 200,000 tonnes of sugar.

In June 1981, sugarcane crops in a small area in the northern part of the island were seriously affected by root damage, resulting in significant losses. The problem was caused by 'white grubs', i.e., the larvae of the *Melolonthidae* beetle (cockchafer) that was quickly identified as *Hoplochelus marginalis*. Within a span of 15 years, it had spread all over the island, leading to considerable losses and resulting in a shortfall of sugar production of between 35,000 and 45,000 tonnes in 1989–1990.

It was soon established that the pest was accidentally introduced from Madagascar 3 years before it was actually noticed in 1981. The 'harmful' biodiversity of the white grubs in Madagascar is significant (Randriamanantsoa et al. 2010).

Studies were undertaken in Madagascar within a framework of an inter-island cooperation to identify natural enemies of this white grub, since it had no effective enemies in Reunion. CIRAD organized surveys in Madagascar between 1984 and 1987 that helped identify a promising entomopathogenic fungus from among the telluric fungal biodiversity of the island. This fungus, *Beauveria brongniartii*, specific to this white grub, was tested on Reunion, and resulted in a mortality rate of greater than 50 % in the grub population.

It was then initially produced in an artisanal way in the form of spores on rice, and then at an industrial level, on clay granules (by INRA and the Calliope company), under the Betel® brand. It quickly became the mainstay of a compulsory biocontrol operation, in conjunction with the application of an insecticide (chlorpyrifos-ethyl). This led to a general decline in the white grub population.

The fungus adapted itself particularly well to the hot and humid climate and to the young volcanic soils of Reunion which are rich in organic matter but relatively poor in antagonistic organisms. The mortality rate among the white grubs consequently increased to 100 % in 1993. With a little help from adult beetles that disseminate its spores, the fungus can spread over a radius of dozens of centimetres in the soil with the help of mycelial cords whose growth is assisted by earthworm burrows. In addition, the fungus can also sustain itself in the soil in a saprophytic form, resulting in fruiting bodies called 'nuggets' that produce virulent spores (Callot et al. 1996).

Even though the pest was not completely eradicated, even 35 years after its introduction, and continues to exhibit periods of resurgence, those show a tendency towards degeneration/loss of vitality. This conventional biocontrol method was implemented through a coordination of different human functions (administration, research, development) and was based on an exploitation of 'useful' biodiversity. It can be considered a success because it helped reduce attacks to an acceptable economic level, even after the use of chemical insecticides was finally dropped.

For more information: Vercambre et al. (2008).

Biological control by augmentation, which consists of releases or application of beneficial organisms (e.g., Trichogramma or ladybugs) as a substitute for conventional phytosanitary treatment, is another way of mobilizing resource biodiversity (exogenous to the plot) to conserve the plot's productive biodiversity. This form of biocontrol is being used successfully against the spotted sugarcane borer in Reunion (Goebel et al. 2005; 2010). Unlike conservation biocontrol (see 'Protecting natural enemies and facilitating their fight against aboveground pests and pathogens'), the beneficial organisms are released in a 'hostile' environment which was not prepared in advance to encourage their activity and reproduction.

1.3 Biocides Derived from Plants Used as a Substitute for Synthetic Pesticides

The use of pericultivated plant biodiversity in the form of plant extracts with biocidal properties is another example of the use of biodiversity for pest control. The use of this method dates back to antiquity; it was used in ancient China, India, Egypt and Greece. Its use developed in Europe in the mid-nineteenth-century before it was replaced a century later by synthetic chemical pesticides.

However, biocides have become the focus of renewed interest in the last quarter century, mainly due to the increasing awareness of the harmful effects of synthetic chemical pesticides, including the new generation ones, on human health and the environment. Innocuity to human beings and auxiliary organisms is, however, not necessarily guaranteed merely because biocides originate from plant matter. Nicotine and rotenone, for example, are highly toxic and non-selective insecticides with (very broad spectrum). The most widely used products of plant origin are derived from rotenone, pyrethrum, neem, and some essential oils (Isman 2006).

In this regard, Jatropha (*Jatropha curcas*) is ideal because, unlike neem or pyrethrum, it can be planted in fields or on their periphery for other uses (such as fencing for vegetable gardens or as anti-erosion hedges in fields with annual crops) (Kumar and Sharma 2008). However, again unlike neem or pyrethrum, its extracts have rarely been used for crop protection. This is so despite several studies undertaken over the past 30 years having demonstrated its effectiveness on some 40 species of pests and a dozen crops (on the field or during storage) and despite a dramatic increase in its cultivation as agrofuel in the tropics and in spite of societal concerns regarding the excessive use of chemical pesticides (Ratnadass and Wink 2012).

1.4 'Pericultivated' Plant Species Biodiversity and Pest Control Processes

Natural biodiversity present around cultivated fields can be a source of pathogens as well as of natural enemies (predators, parasitoids, entomopathogenic microorganisms), as demonstrated by several examples where greater plant diversity has resulted in a higher incidence of pests or diseases.

However, the planned and rational integration of sanitizing or companion plants in agroecosystems can reduce the impact of pests in several ways, either individually or in combination (Ratnadass et al. 2012) (Fig. 2).

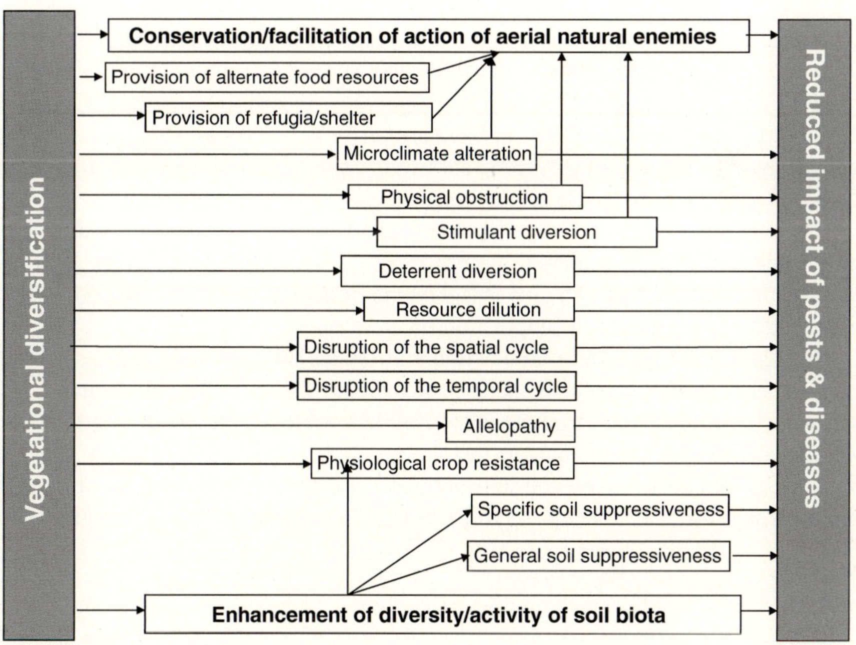

Fig. 2 The main pathways of reducing diseases and pests through the introduction of biodiversity (Ratnadass et al. 2012)

1.4.1 Resource Dilution and Breaking of the Spatial Cycle

Spatial separation of host and non-host plants limits the spread of pathogens and pests, and can be achieved through intercropping with different species or with different genotypes of the same crop, as was shown in the control of rice diseases (see 'Modalities of spatio-temporal deployment of resistance in the cultivated plot'). Cultivated plants are less visible or exposed when they are intercropped than when they are grown as a monocrop. In this way, they suffer fewer pest attacks, at least from those pests which have a reduced range of hosts and/or a reduced dispersal ability. For example, the number of insect pests found in a cruciferous crop is considerably reduced with a clover cover, since pests find it difficult to locate their host plant (Finch and Kienegger 1997; Finch and Collier 2000).

1.4.2 Stimulo-Deterrent Diversion of Pests

Repellent semio-chemicals can also be produced by a plant grown as an intercrop or as a cover plant, with the outcome being bottom-up effects (from a lower trophic level to a higher one) against pests of the main crop. In contrast, pests attracted by associated plants (traps) are less likely to wander onto the main crop, whereas natural enemies may be attracted to the crop and help regulate the pest population.

Such processes are part of the push-pull system. This system involves the combined use of trap plants and repellent plants, with a view to optimize their individual partial effects. Such a system was successfully implemented in controlling stem borers in cereals (especially maize) by ICIPE (International Centre of Insect Physiology and Ecology) and its partners in East Africa, where borers were kept away from the main maize crop and simultaneously attracted by the trap plant (Khan et al. 1997a, b, 2003).

Elephant grass (*Pennisetum purpureum*) and Sudan grass (*Sorghum sudanense*) have exhibited a good potential as trap plants, whereas molasses grass (*Melinis minutiflora*) and the Spanish clover (*Desmodium uncinatum*) have been identified as repellents with regard to oviposition by stem borers.

Repellent plants grown in association with maize not only help reduce infestation by stem borers, but also increase parasitism on the latter by their natural enemies (top-down effect: from a higher trophic level to a lower one). This is due to the production of semio-chemicals that are normally produced in response to plant damage by herbivorous insects, thus acting as a deterrent to oviposition by borer moths. They also serve as signals to parasitoids for foraging (Fig. 3 and Box 4).

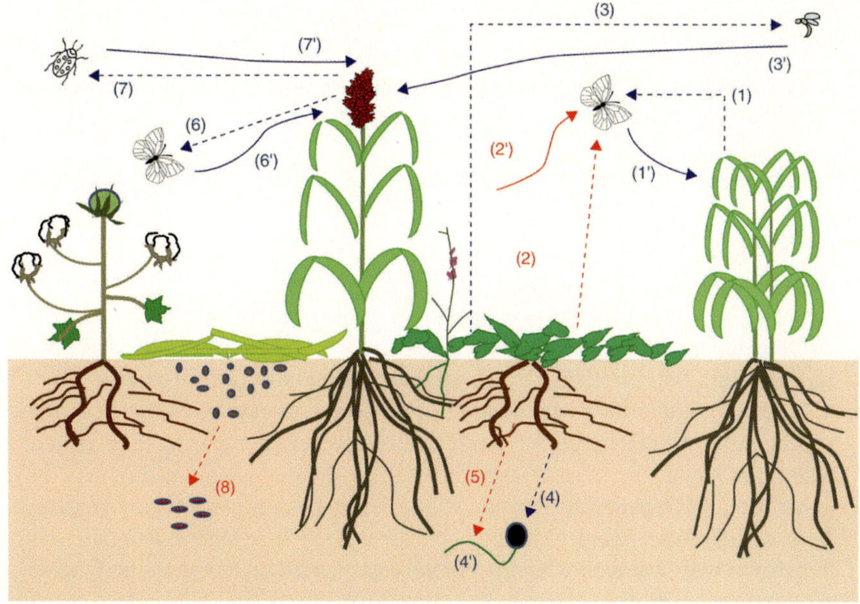

Fig. 3 Principles of pest/pathogen control by stimulo-deterrent diversion (push-pull) and conservation agriculture (DMC) approaches applied to sorghum as a main crop and sanitizing plant (trap crop or dead plant cover = mulch)

On the left. Effects of grain sorghum on fruit worm (*Helicoverpa armigera*) in cotton (*Gossypium hirsutum*) or okra (*Abelmoschus esculentus*) and its natural enemies, and of sorghum residue mulch on the pathogenic microflora or microfauna of *Malvaceae*. Dashed arrows indicate attractive or stimulating effects (blue) and repellent or biocidal/antagonistic ones (red). Solid arrows indicate reactions of the 'targeted' organisms (positive: blue, negative: red).

On the right. Effects of Sudan grass (*Sorghum sudanense*) on the periphery, and of *Desmodium* in association on stem borers (*Chilo partellus*) of grain sorghum (*Sorghum bicolor*) and their parasitoids.

(**1**) Sudan grass attracts female stem borers *Chilo partellus*. (**1'**) The females lay their eggs here rather than in the grain-sorghum. (**2**) The *Desmodium* repels females from plots where it is intercropped with sorghum. (**2'**) These females move away from the plot. (**3**) The *Desmodium* attracts parasitoids of the borer. (**3'**) These parasitoids that were attracted, at the level of the plot, find their host/prey in sorghum stalks. (**4**) The roots of *Desmodium* give out substances that stimulate the germination of *Striga* seeds. (**4'**) *Striga* seeds germinate. (**5**) The root exudates of *Desmodium* also contain allelopathic substances that inhibit the binding of the hyphae of *Striga* on sorghum roots. (**6**) The panicles of grain-sorghum planted as trap plants on the periphery of the *Malvaceae* (cotton, okra) plots attract females of the moth *Helicoverpa armigera*. (**6'**) The females lay their eggs here rather than on the cultivated *Malvaceae*. (**7**) The panicles of sorghum also attract predators of the moth. (**7'**) These predators find their prey in the sorghum panicles. (**8**) The decomposition of litter (mulch) of sorghum residues activates a telluric microflora competitive or antagonist to the pathogen/pest microflora and/or microfauna of the roots of the cultivated *Malvaceae*.

In addition, live or dead cover (*Desmodium*, sorghum residue) controls weeds using a 'barrier' effect by preventing their emergence.

Box 4. Applying the push-pull strategy to the sugarcane borer on Reunion

A team from CIRAD in Reunion recently demonstrated the potential of applying the push-pull strategy to the stem borer *Chilo sacchariphagus*, a significant sugarcane pest. This strategy was developed by ICIPE in Kenya to fight stem borers of maize and sorghum, especially *Chilo partellus*.

The technique relies on attracting (pull) female borers on the lookout for nesting sites to the leaves of a trap plant, *Erianthus arundinaceus,* which resembles sugarcane. Females thus 'lured' lay their eggs on it rather than on sugarcane. Moreover, the hatching larvae fail to complete their cycle on *Erianthus* and die 'trapped' in the plant stem. A 90 % reduction in attacks on sugarcane was observed when *Erianthus* was planted on the plot's periphery, while the average yield increased by over 20 %. These effects were observed up to a distance of 40 metres from the edge of the plot.

The push component (repellency) of the strategy remains to be implemented, which could be achieved by planting cover crops with borer-repelling properties which exist in Reunion, like those tried out in East Africa against *C. partellus*, namely the grass *Melinis minutiflora* (molasses grass) and the legume *Desmodium intortum*. In addition, developing cover crop techniques could help control weed invasions, thus reducing the cane industry's reliance on herbicides.

For further information: Nibouche et al. (2012).

Note that plants with 'attracting' traits that are grown on the periphery are also an excellent source of fodder, much like associated crops, a fact that adds to the success of the push–pull technique (see 'Biodiversity and agriculture-livestock interactions').

1.4.3 Breaking the Time Cycle

Crop rotation with non-host plants reduces the inoculum of telluric pathogens or pest populations leading to a reinfestation (e.g., rotation of strawberry with oats to control nematode infestation: La Mondia et al. (2002)).

Similarly, several types of plants can be used in rotation with banana, since they do not host the telluric nematode *Radopholus similis* (while being less effective against the more polyphagous species *Pratylenchus coffeae*). The choice of plants ranges from certain sugarcane and pineapple varieties to several species of forage plants, either grassy (*Digitaria decumbens, Brachiaria humidicola* and *Panicum maximum*) or leguminous (*Neonotonia wightii, Stylosanthes hamata* and *Macroptilium atropurpureum*) (Risède et al. 2010).

In addition, antibiotic compounds produced and released into the soil by certain plants can directly affect the ability of feeding, infection, or attachment of pests or pathogens to cultivated host plants. For example, growing maize in association with *Desmodium* results in a clear suppressive allelopathic effect on *Striga*, involving both a chemical stimulation of germination and an inhibition of the development of the root system of this parasitic weed and its attachment (by *haustoria*) to that of the host plant (Khan et al. 2002).

1.4.4 Overall and Specific Suppressiveness of Soil Against Telluric Pests and Pathogens

Different crop rotations also help encourage specific soil-dwelling enemies of pests and pathogens, or the induction of overall soil suppressiveness (characterized by a low incidence of disease, despite the presence of pathogens). This is because organic matter derived from a diversified range of crops increases the overall level of microbial activity. Moreover, the greater the number of microorganisms in the soil, the greater is the chance that at least some of them will turn out to be enemies of the pathogens.

1.4.5 Physiological Resistance of the Crop

These rotations/associations also contribute to better crop nutrition thanks to minerals from organic matter decomposition which, in turn, has a positive effect on crop resistance to pests and diseases. These and other processes are probably involved in conservation agricultural systems which use direct seeding mulch-based cropping systems (DMC).

1.4.6 Protecting Natural Enemies and Facilitating their Action Against Aboveground Pests and Pathogens

The conservation or deployment of a diverse vegetation in agroecosystems also provides essential food supplements in the form of pollen and nectar to adult parasitoids and accommodates alternative hosts/preys. This helps maintain parasitoid/predator populations in anticipation of the onset of the targeted pests. Furthermore, such vegetation can also provide shelter against hyper-predators or

nesting/egg-laying sites for natural enemies or modify the microclimate to negatively impact the growth of pests or pathogens. It may also encourage the development of natural enemies leading, in other words, to 'raising' them in the
agricultural ecosystem.

1.4.7 Direct and Indirect Architectural/Physical Effects

These effects encompass the "barrier" effects in the DMC and push-pull systems,
within the field or at its immediate surroundings (see above)

At the landscape scale, fragmentation or non-connectivity with the natural
vegetation serves as a barrier to pest movements and the spread of diseases. The
arrangement of landscape elements (biodiversity components) to this end will
require the help of landscape ecology, and pertains to a sort of 'territorial
agronomy'. Nevertheless, all these processes can also promote the development
and movement of certain pests, or even protect them from some of their natural
enemies, leading to conficting effects.

In Cameroon, mirid bug populations of traditional cacao agroforestry systems
are concentrated on cocoa trees that grow exposed to sunlight, in areas where there
is a break in the canopy cover (Babin et al. 2010). Thus, highly concentrated
Sahlbergella singularis populations can be found in 'mirid pockets' that spread
over 20–30 adjacent infested cocoa trees. In addition, these 'mirid pockets' are
usually found in areas where there is a marked insolation of cocoa trees through
the canopy shading. A possible explanation could be that cocoa trees exposed to
direct sunlight provide more food sources to mirids than those growing in shade,
thus encouraging mirid outbreaks that can cause irreversible damage to these trees.
In contrast, an infection by *Phytophthora megakarya*, the pathogen responsible for
black pod rot is favoured by shading (through its effects on the microclimate,
especially humidity).

Promising effects of landscape features were also highlighted in the case of the
coffee berry borer (CBB) *Hypothenemus hampei* (*Coleoptera: Curculionidae*) in
Costa Rica (Avelino et al. 2012). The abundance of CBB in coffee plots has a
positive correlation with the surface area covered by coffee plants within a radius
of 150 m around the plots. Negative correlations were obtained with other land
uses, especially forests, pasture lands and sugarcane plantations. As the CBB is
dependent on coffee plants, its presence in large interconnected coffee cultivations
probably increases the likelihood of flying borers finding new berries to colonize.
This is particularly marked in the period following the coffee harvest, when berries
are uncommon. Thus, it is only natural that the possibility of survival is greater in
the post-harvest period, as is the likelihood that the infestation after the subsequent
harvest will be even worse. On the other hand, fragmentation of the landscape by
land uses other than coffee cultivation reduces the chances of survival of the pests.
The most distinct effects were observed when forest patches were present, sug-
gesting a barrier effect for beetle movement.

However, the authors also found a higher incidence of coffee leaf rust, a disease caused by the fungus *Hemileia vastatrix*, in coffee plantations fragmented with pasture lands. This is apparently because such landscape arrangements favour swirling air currents that help disseminate these spores. These results demonstrate that what is considered a barrier for one species could actually encourage the growth and spread of another. It is therefore necessary to take into account the entire range of pests to ensure effective management. In this case, fragmentation of coffee landscapes with forest patches was proposed as a means to limit the abundance of berry borers without favouring leaf rust.

2 Hidden Soil Biodiversity: What Potential for Agriculture?

In addition to their effects on telluric pests, the ecological functions of soil bio-diversity can be used to improve fertility (toxicity, decontamination, remediation, etc.) and crop viability. A major issue is to open the 'black box' of soil trophic networks to better understand them, and thus better manage them.

2.1 Diversity of Soil Organisms and their Functions

Soils support an extraordinary biodiversity which is still not fully understood. Soils support the three great branches of life: Bacteria, Archaea (prokaryotes composed of cells without a nucleus) and Eukarya (made up of cells with nuclei, can be unicellular, like protozoa, or multicellular, like plants, fungi and animals). There is therefore a great species diversity in soil organisms, representing about 25 % of all the species described on the Earth, much more than the species in tropical canopies (May 1990; Decaëns et al. 2006). In addition, the diversity of soil organisms for a given site is higher than that of plants and other soil-surface organisms (Decaëns 2010). And yet, scientists know little about the diversity of soil species, and estimate that they have identified only a very tiny percentage of them. For example, it is estimated that we have identified only 0.1 % of bacterial species, 1 % of fungal species, and just half of the earthworm species, etc. (Wall et al. 2005). Molecular tools and metagenomic methods have confirmed this huge underestimation of the number of soil species. It is surprising to note that pro-karyotes, protozoa and fungi have extraordinarily high specific diversities (Lee et al. 1996). 'Barcode' type molecular tools have helped describe mysterious species that, up to now, could not be identified using conventional morphological approaches (Rougerie et al. 2009). Studies have estimated that one square metre of soil (with a depth of 20 cm) could hold several hundreds of invertebrate species,

Table 1 Relationship between size-based classification and functional classification of organisms

Size class	Functions	Functional class
Microorganisms	Decompose organic matter, recycle nutrients, fix nitrogen, control some pathogens	Chemical engineers
Microfauna	Regulate microorganisms by predation, some are plant or animal parasites	Trophic microregulators or microbivores or micropredators
Mesofauna	Break down organic matter, some are predators of microfauna organisms	Saprophagous or litter comminutors
Macrofauna	Break down organic matter, affect soil structure, some are predators or rhizophagous	Soil engineers

while one gram of soil can hold several thousands of bacterial species representing close to a billion bacterial cells, and several metres of fungal hyphae.

Scientists generally classify these organisms according to their size: microorganisms (bacteria and fungi), microfauna (protozoa and nematodes), mesofauna (mainly springtails and mites), macrofauna (earthworms, centipedes, isopods, adults and larvae of insects) and, finally, megafauna (mostly vertebrates such as mole) (Swift et al. 1979) (Table 1).

The ecological roles played by these soil organisms are relatively well known (Table 1) (Lavelle et al. 2006; Kibblewhite et al. 2008). Microorganisms are the principle decomposers of organic matter, enabling the recycling of nutrients. They are sometimes referred to as soil chemical engineers (Turbé et al. 2010). Bacteria also have the ability to fix atmospheric nitrogen, either on their own or with the help of certain plants. Similarly, some fungi, by associating themselves with plant roots, form mycorrhizae that influence nutrient cycling. Microfauna organisms (protozoa and nematodes) are known primarily as the regulators of microbial

Fig. 4 From *left to right*, and *top to bottom*: springtail, termite, earthworm, millipede, ant, earwig

communities, and are sometimes known as microbivores. Mesofauna organisms are considered to be primarily scavengers, consuming and breaking down organic matter on the ground. Finally, the principle function of macrofauna organisms is to modify the soil structure, incorporate organic matter into the soil and regulate the availability of resources for other organisms through non-trophic interactions. They are thus called soil engineers, with reference to ecosystem engineers defined by Jones et al. (1994) Fig. 4.

2.2 Diversity of Interactions Between Organisms

Research undertaken to understand the biological functioning of soil are increasingly focusing on interactions between soil organisms. These complex, multiple interactions are usually divided between trophic and non-trophic interactions.

2.2.1 Trophic Interactions

The immense diversity of soil organisms is partly explained by the abundant availability of the diversified resource that is decaying organic matter. Decomposer chains established from this resource determine its decomposition and the mineralization of soil nutrients. Although nutrient mineralization is mainly carried out by microorganisms (bacteria and fungi), their activities are highly regulated by soil organisms at higher trophic levels. These organisms, known as trophic microregulators, primarily include nematodes, protozoa, mites and springtails. Numerous studies have shown that they can affect the biomass and microorganism activity, either directly by predation, or indirectly by breaking down organic matter, disseminating microbial propagules or altering nutrient availability (Cragg and Bardgett 2001). Predation of microorganisms by invertebrates normally releases soil nutrients either as a result of increased microbial activity or by the excretion of excess nutrients by predators. A recent microcosm study showed that predation by nematodes (*Rhabditis* sp.) of rhizosphere bacteria (*Bacillus subtilis*) increases the availability of nitrogen and phosphorus for the plant *Pinus pinaster* (Irshad et al. 2011). Soil engineers, like earthworms, are also known to directly modify communities of bacteria, fungi or soil protozoa through digestion (Bonkowski and Schaefer 1997; Bonkowski et al. 2000; Bernard et al. 2012). A number of studies even suggest that these organisms are an important component of earthworm nutrition. Similarly, springtails and mites feed selectively on soil fungi, which can alter the structure of fungal communities in a given environment (Bonkowski et al. 2000).

2.2.2 Non-Trophic Interactions

Non-trophic interactions are mainly related to changes in the distribution of resources through the action of ecosystem engineers. Their physical actions lead to a greater availability of trophic resources or nutrients for other soil organisms, as well as the creation of new habitats with specific physicochemical features (Blanchart et al. 2009). As a result, both biological assemblages and functions are modified by bioturbation. For example, Loranger et al. (1998) have shown that, in the Vertisols of Martinique, the microarthropod species richness was higher in patches with elevated densities of earthworm (*Polypheretima elongata*) than in areas with low densities. Similarly, several studies show that casts (excrements) of earthworms support microbial communities different from that of the surrounding soil (Tiwari and Mishra 1993, for fungi; Chapuis-Lardy et al. 2010; Bernard et al. 2012, for bacteria). Also, soils of termite mounds or anthills are characterized by different microbial communities (Brauman et al. 2000; Dauber and Wolters 2000). These changes in microbial communities, and possibly in micro-trophic chains, are brought about by ecosystem engineers and lead to changes in the cycles of carbon and other major nutrients. They also alter nutrient availability and the emission of greenhouse gases (mainly N_2O and CO_2), as was shown, for example, by (Postma-Blaauw et al. 2006; Coq et al. 2007; Mariani et al. 2007 and Chapuis-Lardy et al. 2009, 2010).

2.3 Impacts on Ecosystem Services

The activities of soil organisms in the assemblages and complex interactions seen above influence primary soil functions at the source of ecosystem goods and services, i.e., goods and services provided by ecosystems to humanity, as defined in the Millennium Ecosystem Assessment (MEA 2005). Soil organisms are related to ecosystem services through ecological functions. Kibblewhite et al. (2008) suggest that four ecological functions form the basis of all ecosystem services provided by the soil: (1) transformation of carbon molecules (decomposition of residues and soil organic matter, as also the synthesis of new molecules), (2) nutrient recycling, (3) maintenance of soil structure (aggregation, transport of particles, formation of poral networks), and (4) biological control of pest populations. The latter ecological function was covered in the previous section. Agricultural food production, for example, relies on these four functions. Erosion control depends almost exclusively on the maintenance of soil structure. The quality and supply of water resources is dependent on the soil structure (which controls run-off, infiltration, and water retention) as well as the recycling of nutrients (which releases varying quantities of elements which could leach, including some—such as nitrates—that could cause pollution). Soil use actually determines the diversity of organisms that grow in it, and in turn, these latter provide a number of ecosystem services. On the other hand, crop production is

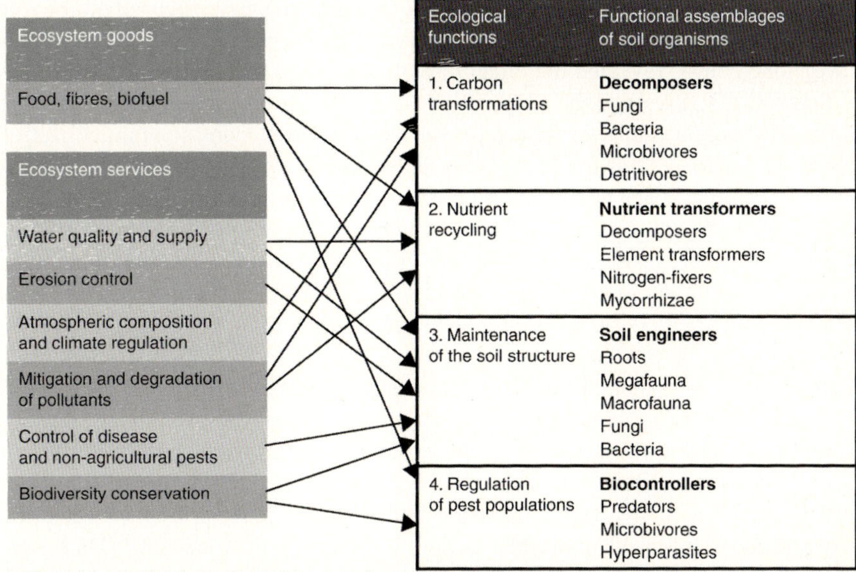

Fig. 5 Relationships between functional assemblages of soil organisms and ecosystem services through ecological functions (according to Kibblewhite et al. (2008))

based on the four ecological functions and consequently, on all functional groups that help achieve it, as shown in Fig. 5. Therefore, it is functional diversity that is of prime importance (Altieri 1999). The question is how to implement this functional diversity in order to best fulfil ecological functions that sustain ecosystem services.

2.4 Relationship Between Belowground Diversity and Aboveground Diversity

2.4.1 Biological Functioning of Soil and Plant Growth

Plants absorb nutrients through complex interactions in the rhizosphere between roots, symbiotic or non-symbiotic microorganisms and soil fauna. Competition for nutrients and energy in this functional area known as the rhizosphere is stiff. Several studies have described the role of nitrogen-fixing bacteria or mycorrhizal fungi in plant growth, or even of free microorganisms and their involvement in the mineralization and release of nutrients available to plants. We also begin to perceive the highly complex mechanisms by which plants defend themselves against pathogens and control organisms present in the rhizosphere (Bonkowski et al.

2009, and previous section). In contrast, far fewer studies have focused on the role of the soil fauna, especially soil engineers, in crop production. Several studies have, nonetheless, revealed the effect of earthworms on plant growth (see reviews of Scheu 2003; Brown et al. 2004; Wurst 2010), with most observing a positive effect. This action can be explained by various processes: increased mineralization of nutrients; changes in the availability of water and oxygen in the rhizosphere; hormonal effect; dispersion of microorganisms that aid plant growth; and control of pathogens by the dispersal of microorganisms that are enemies of root pests. Recent studies suggest that soil fauna activities are probably much more complex and significant than previously thought for the resistance of plants to stresses or in relation to plant diversity, and that the strength and trend of the effects on the plant depend on the invertebrate species involved (Laossi et al. 2010). Blouin et al. (2005) have shown that earthworms can change plant physiology and increase its tolerance to parasitic nematodes. Similarly, studies have recently shown that the presence of earthworms alters resource allocation in plants and the ratio of belowground/aboveground biomasses. This can be explained by an improved availability of nutrients, especially phosphorus, for the plant or by the expression of plant genes involved in cellular division and response to stresses (Jana et al. 2010). In Guadeloupe, it was shown that earthworm activity could reduce damage caused by plant-parasitic nematodes on banana (Loranger et al. 2012). According to these authors, this result is indirect and linked with changes in phosphorus mineralization by earthworms and a subsequent improvement in plant nutrition, making it more tolerant to nematodes. Interactions between crops (competition versus facilitation) can also be modified by the presence of earthworms. A recent study conducted in laboratory conditions has shown that the presence of earthworms cancelled the competition between durum wheat and chickpea when these plants were grown in association (unpublished).

2.4.2 Interaction Between Belowground and Aboveground Diversities

Terrestrial ecosystems can be divided in two distinct compartments: aboveground and belowground. Ecologists have only recently started taking interest in linkages between the two compartments, leading to major advances in understanding the functioning of ecosystems and agrosystems (Kardol and Wardle 2010). Studies have shown that changes in aboveground trophic networks could influence those belowground, and vice versa, and that these relationships affect the composition of the plant community (De Deyn et al. 2007). Studies have also shown that the diversity of mycorrhizal fungi determines plant diversity and productivity (van der Heijden et al. 1998). On the other hand, it has also been shown that plant diversity strongly influences the soil microbial community, and that each plant species contributes to the functioning of the belowground system (Eisenhauser et al. 2010). It also appears that reducing plant biodiversity affects earthworms more than it does microorganisms (Spehn et al. 2000). Other authors suggest that the reduction of plant diversity has little effect on soil organisms, depending on functional plant groups: the presence of leguminous plants is significantly correlated with belowground earthworm density (Gastine et al. 2003).

2.5 Utilizing this Biodiversity in Agriculture

2.5.1 Direct Manipulation

Inoculations of soil organisms (in the form of biofertilizers) to improve crop production and plant health were mainly undertaken for the group of bacteria known as PGPB (plant growth-promoting bacteria) or PGPR (plant growth-promoting *rhizobacteria*) and mycorrhizal fungi. The PGPB (mainly strains of the genera *Rhizobium*, *Azotobacter*, *Azospirillium*, *Pseudomonas*) can stimulate plant growth through various mechanisms: nitrogen fixation, phosphate solubilization and mineralization, siderophore production and synthesis of hormones or plant vitamins (Vessey 2003). These microbial technologies have been used to address several agricultural and environmental issues, with results being generally positive (see Whipps 2001; Glick et al. 2007; Thuita et al. 2012), even though the resulting benefits are difficult to replicate in a consistent manner (Singh et al. 2011). To be effective, inoculated organisms must have soil conditions that are suitable for development, which is not always the case. Increasingly, the inocula include assemblages of microorganisms that combine PGPR, nitrogen-fixing bacteria, mycorrhizae, actinomycetes and other useful microbes such as those that produce fungicides to protect plants against diseases (Singh et al. 2011). Research is being carried out to understand PGPR diversity, their ability to colonize, their interaction mechanisms and formulations for inoculations.

As far as invertebrates are concerned, few studies have been conducted on the inoculation of earthworms on a large scale. The main limitation lies in the difficulty of raising earthworms in large numbers. A second difficulty consists of maintaining relatively large densities of earthworms over a period of time. In Martinique, Blanchart et al. (2004) raised and introduced 4,500 earthworms (*Polypheretima elongata*) in an experimental field of 50 m^2 (equivalent to a density of 90 ind.m^{-2}) in cultivated soil which supported an initial earthworm density of about 2 ind.m^{-2}. The density dropped to 25 ind.m^{-2} within 6 months, before increasing steadily during the following 4 years of the experiment. Earthworms can be grown in large numbers in the humid tropics using a soil substrate mixed with sawdust or a mix of organic matter of different qualities (Senapati et al. 1999). As suggested by Senapati et al. (1999), it is cheaper to raise earthworms than harvest them in nature: it costs 3.6 Euros to raise 1 kg of them, whereas it takes as much as 6.125 Euros to gather an equivalent biomass. In southern India, earthworms were introduced in large quantities in over 200 ha of tea plantations. This was achieved by digging trenches along contours, incorporating organic materials of various qualities (compost from household waste and prunings from tea bushes) and earthworms. This technique of bioorganic fertilization increased soil quality and, subsequently, boosted tea production. Unfortunately, the difficulty of raising earthworms in large numbers did not allow this technique to be adopted widely.

2.5.2 Indirect Manipulation

A large number of studies have shown that the diversity, density, biomass and community structure of soil organisms are dependent on farming practices: depth, intensity, frequency of tillage, type and frequency of fertilization, management of organic matter, plant rotations and associations. For several years now, researchers have demonstrated the alarming effects of intensive conventional farming on soil organisms, biotic interactions and the availability of resources (Kennedy and Smith 1995; Matson et al. 1997). In this kind of agriculture, biological functions were systematically replaced by chemical fertilisers and tillage. In recent years, the introduction of agroecological practices based on minimizing chemical inputs and tillage has helped intensify ecological processes in agricultural soils while significantly increasing the diversity and density of microorganisms and soil fauna (for instance, in the tropics: Blanchart et al. (2006, 2007); Rabary et al. (2008); Villenave et al. (2009)) As noted above, the vegetation composition, or more precisely, the choice of functional groups of crops may also affect the composition of soil-organism communities and, ultimately, the ecological functions of soil. Yet few studies have addressed this issue in agricultural environments.

3 Biodiversity and Agriculture-Livestock Interactions

Since Neolithic times, agriculture has most often been associated with livestock farming, with obvious mutual benefits in terms of animal feed and replenishment of soil nutrients. With the advent of synthetic fertilizers, however, industrial agriculture has dissociated the two activities, choosing to optimize them independently, i.e. by intensely using synthetic fertilisers and creating an industrial sector to produce animal feed. The introduction, or reintroduction, of animal husbandry in the agricultural system could lead to multiple impacts on biodiversity. They could be significant, multiscale and contribute positively—or sometimes, admittedly, negatively—to biodiversity.

An extreme case of negative impact is the naturalization of exotic animal species in the indigenous environment they are introduced in. The problem begins when they become invasive enough to threaten the survival of the local ecosystems. An example is the case of camels (*Camelus dromedarius*) which were introduced to the Australian continent in the nineteenth-century as draught animals, and later released into the wild in the early twentieth-century. They multiplied freely in the absence of natural predators. The intense grazing pressure they exerted led to the erosion of the endemic biodiversity, which in turn encouraged the spread of weeds and invasive plants. They are also thought to participate indirectly to the erosion of global biodiversity by virtue of their belching which contributes to the greenhouse effect.

Without seeking to fuel the major controversies between livestock rearing and 'cultivated' biodiversity, we provide a few examples of the successful integration of raised animal biodiversities with other types of biodiversity. They are expressed as interactions between cycles at the level of the soil, plant, animal, plot, herd, or farm. Indeed, an integrated management of agricultural and livestock activities can enhance complementarities between cropping systems (production of fodder, symbiotic nitrogen fixation and nutrient cycling) and livestock systems (inputs of organic manure and energy). This allows a general intensification while reducing the consumption of non-renewable energy, chemical fertilizers and feed concentrates. Livestock that transforms biomass on non-arable marginal land of the farm should also be considered for its vital role of 'valorising' agricultural by-products such as crop residues and for adding value to primary farm products (Dugue et al. 2004).

In farms and landscapes, livestock rearing is at the origin of a great variety of domestic breeds and typical products, and has a significant capacity to maintain large extents of natural areas. It can thus contribute to the maintenance of dry steppes, humid grasslands or high altitudes prairies. It can also be useful in preventing forest fires (MAB 2012[1]). Extensive traditional grazing has a positive impact on biodiversity, in a broad sense, through the creation and maintenance of heterogeneous landscapes, as well as through its role in the dispersal of propagules (zoochory).

3.1 Interactions Between Livestock Rearing and Telluric and Plant Biodiversity

Despite stereotypes of the mechanical effects of agricultural soil compaction as a result of trampling by herds—studies show this only occurs in situations of excessive livestock pressure (Bell et al. 2011)—livestock rearing practices (restoration of pastures, manure spreading, etc.) contribute to soil enrichment and the diversification of macrofauna and microflora. Compared to the use of inorganic fertilizers, the application of organic manure in maize or cotton fields (Peacock et al. 2001; Acosta-Martínez et al. 2010) results in significant effects on microbial biomass, on the profile of existing species and, consequently, on the enzymes they circulate in the soil and its pool of organic matter. In this way, it interacts with the overall fertility of the environment. Such changes in the soil ecosystem influence the primary production capacity and floristic biodiversity of the vegetation cover

---

[1] http://mab-france.org/fr/concilier-activites-et-environnement/elevage-et-biodiversite/ (retrieved: 25 May 2013).

that colonizes the soil, whether in agricultural fields or in pastures. Interactions also take place between soil microorganisms, phyllosphere microflora, the microflora used for producing cheese, or the bio-contaminating microflora. Innovative research conducted by INRA on milk from farms in mountain pastures demonstrated the usefulness of microbial diversity in reducing the pathogenicity of *Listeria monocytogenes* in raw milk cheese (Retureau et al. 2010). Similarly, the dominant factor in the natural environment with regard to the floristic composition of pastures appears to be related to the richness of the land. Heavily fertilized and rich in nutrients, these pastures have a low florisitic diversity dominated by nitrophilous species. In contrast, pastures subject to less intensive practices are rich in flora, whose diversity determines the constitution of secondary composites including terpenes, a key factor for the organoleptic diversity of dairy products (Cornu et al. 2005).

Numerous interactions also exist between aboveground and belowground animal biodiversities in the cycle of returning organic matter back to the soil. A good example is the optimization of the decomposition of dung of ruminants. Several studies devoted to this have examined the introduction of dung beetles (*Scarabaeus laticollis*) in Australia to facilitate cattle dung decomposition and ensure the return of nutrients to the soil (Edwards 2007). Similarly, studies in the United States have analyzed the effects of their activities in controlling eggs and nematode cycles in the dung, the latter being parasites of ruminants (Fincher 1975).

3.2 Use of the Local Plant Biodiversity in Livestock Rearing

Local natural or pericultivated plant biodiversity can, in lieu of synthetic molecules, contribute to the sustainability of livestock rearing by helping improve animal health. Several examples illustrate this, including the use of essential oils as alternatives to antibiotics in aquaculture in Madagascar (Sarter et al. 2011; Randrianarivelo et al. 2009, 2010) or on terrestrial livestock to repel biting insect (e.g., geranium oil used on cattle in Reunion against *Stomoxys calcitrans*), the use of *Jatropha* extracts as an anthelmintic (Ratnadass and Wink 2012) or of the extract of cassava leaf (*Manihot esculenta*) against *Haemonchus contortus*, a major helminth parasite of small ruminants (Marie-Magdeleine et al. 2010). In each case, it was a matter of identifying alternatives to counter the harmful effects of the massive use of anthelmintics on natural soil macrofauna (see 'Hidden soil biodiversity: what potential for agriculture?').

The microflora and microfauna of the rumen ecosystem inside the animal, and the balance between their populations contribute to the functioning of a particularly efficient reactor to enhance the value of the diverse resources from agricultural fields and areas of natural vegetation. Here too, the bacterial biodiversity interacts with that of plants consumed by the animals. In addition to variations observed of major constituents (proteins, cellulose, starch, etc.), plants are also rich in various secondary compounds (tannins, saponins, alkaloids, etc.). Despite the dearth of studies on them, these constituents can be of great use, through a modification of the species profile and microbial activity, in controlling methane emissions, ammoniacal degradation of proteins, hydrogenation of fatty acids, stimulation of low-quality fodder intake, etc. (Durmic et al. 2008).

3.3 Interaction Between Organic Matter from Livestock Rearing and Plant Pests

Not only can livestock animals be integrated into certain systems for managing weeds and insect pests (see above) but, in ways similar to those presented above (see 'Specific "pericultivated" plant biodiversity and pest control processes'), they can also directly interact with plant pests. For example, organic fertilization affects pest infections and infestations: dung that carries Striga seeds or white grubs larvae (*Scarabaeidae* larvae) which infest 'rainfed' rice as observed in Madagascar. These effects can be positive or negative. For example, the smell of manure increases the appetite of white grubs and, as a result, increases the damage they inflict on the rice. Excessive fertilization with organic nitrogen (e.g., with livestock manure) increases the vulnerability of plants to diseases such as rice blast, or their attractiveness to phloem-feeding insects. On the other hand, better nutrition (i.e., balanced nutrition) leads to greater tolerance of plants to pests.

3.4 Examples of Integration at the Plot Level

There exist 'ancient' examples of fighting pests without the use of herbicides, e.g., rice-fish farming systems, especially in Vietnam, or the use of ducks in rice fields to combat weeds and other pests (Box 5).

Box 5. The enhancement of agro-biodiversity in rice systems

Rural populations in Asia have perfected the agrobiodiversity of rice systems over thousands of years by using cultivated plants, domesticated animals and aquaculture to ensure food security and a regular income.

The potential for diversification in a rice ecosystem is high because of a continuous presence of fresh water. Fish, frogs, snails, insects, etc. can constitute the primary source of animal proteins and fatty acids for rural populations. Such species could be part of the intrinsic natural biodiversity of rice fields, or could be introduced on purpose, e.g., species of tilapia, barbel and carp in rice-fish farming systems.

Rice systems can also host several livestock species. Ducks live on small fish, aquatic organisms and weeds (including the Common barnyard grass or

Cockspur Grass, *Echinochloa crus-galli*) in rice fields. They also eat insect pests that cause devastating outbreaks, such as brown planthopper (*Nilaparvata lugens*), and the sclerotia of *Rhizoctonia solani*, a pathogen responsible for the sheath blight disease in rice grain, whose occurrence is thus significantly reduced occurrence (Su et al. 2012). This ancient practice from China was first adopted by Japan (Furuno 2001) and more recently by other Asian countries like Vietnam (Men et al. 2002) and Bangladesh (Hossain et al. 2005) and even in Camargue, France (Falconnier et al. 2012). Ducklings are introduced after rice seedlings are transplanted. In the event that rice is planted directly, ducklings are introduced when the seedlings are sufficiently developed. In either case, care is taken to ensure the ducklings do not damage the seedlings. Ducks stimulate the growth of rice as they waddle in between the rows of rice, even as they eliminate weeds, either by consuming them or trampling them. Their droppings too help improve the fertility of the rice fields and increase rice yield. A welcome bonus is the supply of duck meat.

Larger livestock, such as buffalo, cattle, sheep and goats consume rice straw and bran, a by-product of rice milling, as well as lower quality rice grains or those from surplus crops. In addition, cattle can be used in transportation and in preparing the fields, and their dung can be used as organic fertilizer.

Rice fields also play host to a large number of natural enemies or predators that help control harmful insects and pests, thus reducing the need for pesticides. Fish feed on harmful plants and help the fight against weeds. Some plant species share a symbiotic relationship with the rice. For example, azolla, a nitrogen-fixing fern, can be grown in paddy fields to increase the availability of nutrients, reduce the number of weeds and facilitate the integration of fish with livestock. These associated plant species are also used by farmers as a source of food and medicine, and as feed for fish and livestock. The poor, especially the landless, can increase their side revenues through fishing and by gathering medicinal plants.

One of the most promising alternatives to help reduce the use of herbicides or mechanical mowing in the horticultural sector in the Caribbean is to associate a live plant cover to orchard plants. In banana plantations, not only does this help manage weeds, but it also helps in soil remediation. Such associated plants can be legumes, e.g., *Arachis pintoi*, or grasses. The associated plant cover could be kept under control by mowing or by grazing, using free-range animals that consume and maintain the biomass (Archimède et al. 2012).

In this context, we see systems that associate orchards and sheep in the Caribbean, particularly in Cuba, where the association between citrus and sheep has been studied for several years (Mazorra 2006). This is similar to the Normandy 'meadow-orchard' system—the first important agroforestry system in France, and a good example of integration between agriculture and livestock rearing, associating apple orchards and dairy cattle. Sheep in orchard-sheep systems are trained to graze without damaging the trees using different learning techniques. One such technique involves the use of tethers to prevent animals from eating the bark and leaves. A more sophisticated technique is to spray a repellent on citrus leaves (lithium chloride or syrup of ipecacuanha) that induces distaste for the leaves of this species right from when the animals are introduced. These sheep are thus conditioned not to eat the leaves, and can even transmit this behaviour to their young (Lavigne et al. 2011).

More 'simply', in Martinique, sheep can help manage plant cover in orchards that is either natural (weeds) or grown (service plants). However, to prevent the sheep from harming the main crop, it is possible to leverage their natural aversion to certain foods by 'judiciously' associating the crop with *Annonaceae* species such as soursop, custard apple and apple-cinnamon. Similarly, grazing by poultry has been an effective method of weed control in guava orchards; geese are the preferred fowl as they are more herbivorous than the largely granivorous chicken (Lavigne et al. 2012). This can be compared to regular mechanical mowing. However, the pressure exerted by the poultry on the grass cover is much greater on some grasses and broadleaf weeds than on sedges, thus bringing about a gradual change in the flora of the orchard. Such selective grazing must be compensated for while managing the system, e.g., by regular reseeding (Fig. 6).

Fig. 6 Goose grazing in a guava orchard

3.5 Production of Fodder in the Agricultural System

As far as fodder production is concerned, its integration into the agricultural system should be considered at the farm level, with the biomass produced contributing to building up a stock of feed for harsh periods (winter, dry seasons). However, this can also be done at the level of the plot, by using the crop itself (crop residues or dual-purpose cereal varieties) or through recourse to peripheral or associated plants.

3.5.1 Dual-Purpose Cereals and Legumes

This method concerns several types of legumes, such as cowpea and groundnut, and grains like rice, maize and sorghum.

Sorghum is a plant adapted to semi-arid areas where maize cannot be grown. Several studies have been carried out on the dual-purpose use of the sorghum plant, i.e., for agriculture and for livestock rearing (Fig. 7). Efforts that are currently underway (Chantereau et al. 2004) indicate a real genetic variability that could potentially be exploited to increase yield and improve the 'digestibility' of straw (Box 6).

Fig. 7 Sorghum straw: a considerable amount of biomass usable as animal feed in dry seasons

> **Box 6. Genetic biodiversity and the 'dual-use' trait of sorghum**
>
> The stems and leaves of cereals contribute to the plant structure; they form the base that determines the photosynthetic yield in terms of grain. During harvesting, the 'straw' component constitutes 40–60 % of the total biomass produced during plant growth.
>
> Straw is considered a low-value by-product in developed countries. It is either discarded, incorporated in the soil or sold for use as animal bedding in livestock stalls. However, in family farming systems in warmer regions, it is considered a forage resource of prime importance for the maintenance—and even survival—of farm animals during difficult periods.
>
> Its nutritional value, however, is limited due to the high content of carbohydrates and lignocellulose, combined with low protein levels. Efforts are underway for some time now to develop and demonstrate physical processing techniques (fine chopping, molasses treatments) or chemical ones (ammonia, urea) to improve the digestibility of straw. Agro-pastoralists have yet to adopt these techniques due to problems of availability of labour, high cost of inputs and efficiency concerns.
>
> In recent years, the implementation of participatory breeding practices, the interest in multiple objectives and a knowledge of small-scale agriculture have led to a renewed look at the biodiversity of cereal crops and the 'dual-use' model for plants.
>
> This concept helps address demands from farmers for plant varieties that are adapted to local conditions and have an increased grain yield potential along with an improved forage value.
>
> **For further information:** de Alencar Figueiredo et al. (2006).

3.5.2 Associations/Rotations of Fodder Crops with Other Crops

These methods can be applied to push-pull systems (see 'Stimulo-deterrent diversion of pests'), some agroforestry systems with woody forage legumes and direct-seeding mulch-based cropping systems (DMC) (Séguy et al. 2009; see Chap. 3).

Therefore, even though these conservation agriculture systems in the highlands of Madagascar contribute to the development of sustainable agriculture, mainly by helping control erosion, and were initially focused primarily on food, systems which include fodder-production as an additional goal are also potential vehicles for disseminating the use of DMC (Anonymous 2008).

Naudin et al. (2012) have shown that the production and conservation of biomass in DMC systems at Lake Alaotra are not always sufficient to fulfil the agroecological functions expected from mulch. However, some species, like *Vicia villosa*, do perform these functions and ensure 90 % ground cover, while producing 3 tonnes of biomass per hectare as fodder (Box 7).

Box 7. Fodder production in a direct seeding mulch-based system

The emergence of innovative cultivation techniques called 'conservation' agriculture, especially direct sowing on existing plant cover without ploughing of the soil, represents an attractive alternative. In these systems, the soil cover is provided by intercropping different species in the same plot and by a crop rotation that ensures the field is never left bare.

The use of such systems in family farms is complex. It requires the development of specific operational itineraries that are adapted to local needs in terms of the choice of crop associations, management of plant cover and planning of crop rotations. We are also confronted with a basic issue of an intense competition based on two types of biomass uses: on the one hand, a practice that generates optimal results only over a long period of time, provided large amounts of biomass are produced and are, ideally, left on the field and, on the other hand, a short-term requirement of farmers, pastoralists, agro-pastoralists to use almost all the biomass within the cropping year as livestock feed. The stakes are high as this impacts the agrarian system (absence of individual property rights), the rights to traditional uses (grazing of crop residues) and livestock rearing (localized or transhumant). The latter is deprived of a resource, despite animals being, at the economic and social level, a major factor in income generation and capital building at the regional scale. Impediments to the relationship between livestock rearing and DMC lead to discussions and issues that have to be addressed through research and by local stakeholders in order to reach solutions.

A possible and interesting hypothesis is that partial and especially rational use of the plant cover for livestock rearing can result in a gradual appropriation of the plant cover system. The originality of the partial use of the plant cover is to fully exploit it for its nutritional value to help derive an additional resource high in energy and nitrogen, rather than just as fodder (some successive and very early mowing of the regrowth of *Brachiaria spp.*, *Stylosanthes*, *Vicia spp.* or other species produces large quantities of very high protein content fodder), and then allow growth in order to get plant cover. Conservation strategies allow postponing the utilization to the time when the economic value is at its highest (off-season milk or fattening). In Burkina Faso and Madagascar, the managed grazing of crop cover and/or the making of silage or hay from a part of the biomass cover (Naudin et al. 2012; Kueneman et al. 2002) are some examples that suggest that the combination of a technical innovation providing high value addition to the 'no-till cover crop' innovation can be highly synergistic in the adoption of conservation agriculture itineraries Fig. 8.

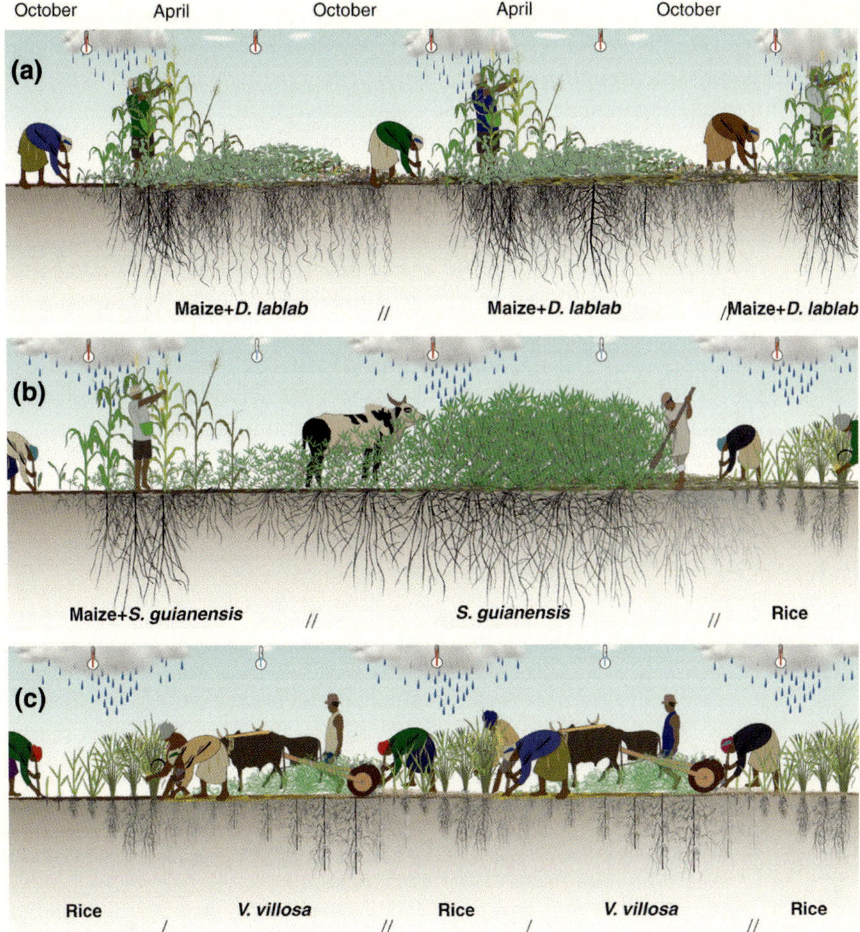

Fig. 8 Examples of the succession of cover crops in conservation cropping systems in the Lake Alaotra region, Madagascar. (**a**) A two-year cycle on the slopes of maize + *Dolichos lablab* in the nth year and 'rainfed' rice in the year n + 1. (**b**) A multi-year rotation on the slopes with a maize crop + *Stylosanthes guianensis* in the nth year, only *S. guianensis* used as forage in the year n + 1/2/3, upland rice in the last year. (**c**) In fields on the plains, a double crop rotation in the year with *Vicia villosa* in the off season and rice as the main crop (according to Séguy et al. (2009))

4 Conclusion

There are numerous and varied interactions between different biodiversity components within a cultivated plot which lead to synergistic or conflicting effects. Several examples were used to illustrate the effects of the use of certain modalities of deploying plant biodiversity in the cultivated plot against numerous types of pests, as well as the mainly synergistic effects between 'raised' animal biodiversity and weed control and pest regulation.

Taking soil organisms in a sustainable productive agriculture into account remains a real challenge for research. Despite the accumulated knowledge on soil organism groups, their biology and the functions they individually fulfil in the soil, we are still struggling to understand their interactions and hierarchize the processes involved in accomplishing a particular function. Similarly, very few studies have focused so far on the relationships between aboveground and belowground diversities. Nevertheless, recent research shows that ecological functions and ecosystem services are based on the presence of a great diversity of functional groups. Any move towards sustainable agriculture must therefore encourage the presence of this functional diversity of soil, through direct or indirect actions.

While the need exists to understand basic processes underlying these effects, it is also equally important to contextualize the cultivation of biodiversity because of the extreme diversity of situations. In this respect, the diversity of situations in which CIRAD works with its partners is both a constraint and an opportunity to help open up new fields of research.

References

Acosta-Martínez, V., Bell, C. W., Morris, B. E. L., Zak, J., & Allen, V. G. (2010). Long-term soil microbial community and enzyme activity responses to an integrated cropping-livestock system in a semi-arid region. *Agriculture, Ecosystems and Environment,137*, 231–240.

Altieri, M. A. (1999). The ecological role of biodiversity in agroecosystems. *Agriculture, Ecosystems and Environment,74*, 19–31.

Anonymous, (2008). Conduite des systèmes de culture sur couverts végétaux et affouragement des vaches laitières: guide pour les Hautes Terres de Madagascar, pp. 90.

Archimède, H., Gourdine, J. L., Fanchone, A., Tournebize, R., Bassien-Capsa, M., & González-García, E. (2012). Integrating banana and ruminant production in the French West Indies. *Tropical Animal Health and Production,44*, 1289–1296.

Avelino, J., Romero-Gurdián, A., Cruz-Cuellar, H. F., & Declerck, F. A. J. (2012). Landscape context and scale differentially impact coffee leaf rust, coffee berry borer, and coffee root-knot nematodes. *Ecological Applications,22*, 584–596.

Babin, R., ten Hoopen, G. M., Cilas, C., Enjalric, F., Yede, P., Gendre, Lumaret, J. P. (2010). Impact of shade on the spatial distribution of *Sahlbergella singularis* in traditional cocoa agroforests. *Agricultural and Forest Entomology, 12*, 69–79.

Bell, L. W., Kirkegaard, J. A., Swan, A., Hunt, J. R., Huth, N. I., & Fettell, N. A. (2011). Impacts of soil damage by grazing livestock on crop productivity. *Soil and Tillage Research,113*, 19–29.

Bernard, L., Chapuis-Lardy, L., Razafimbelo, T., Razafindrakoto, M., Pablo, A. L., Legname, E., et al. (2012). Endogeic earthworms shape bacterial functional communities and affect organic matter mineralization in a tropical soil. *The ISME Journal,6*, 213–222.

Blanchart, E., Albrecht, A., Chevallier, T., & Hartmann, C. (2004). The respective roles of biota (roots and earthworms) in the restoration of physical properties in vertisol under a *Digitaria decumbens* pasture (Martinique). *Agriculture, Ecosystems and Environment,103*, 343–355.

Blanchart, E., Villenave, C., Viallatoux, A., Barthès, B., Girardin, C., Azontonde, A., et al. (2006). Long-term effect of a legume cover crop (*Mucuna pruriens* var. *utilis*) on the communities of soil macrofauna and nematofauna, under maize cultivation, in southern Benin. *European Journal of Soil Biology,42*, 136–144.

Blanchart, E., Bernoux, M., Sarda, X., Siqueira Neto, M., Cerri, C. C., Piccolo, M., et al. (2007). Effect of direct seeding mulch-based systems on soil carbon storage and macrofauna in Central Brazil. *Agriculturae Conspectus Scientificus, 72*, 81–87.

Blanchart, E., Marilleau, N., Chotte, J. L., Drogoul, A., Perrier, E., & Cambier, C. (2009). SWORM: an agent-based model to simulate the effect of earthworms on soil structure. *European Journal of Soil Science,60*, 13–21.

Blouin, M., Zuily-Fodil, Y., Pham-Ti, A. T., Laffray, D., Reversat, G., Pando, A., et al. (2005). Belowground organism activities affect plant aboveground phenotype, inducing plant tolerance to parasites. *Ecology Letters,8*, 202–208.

Bonkowski, M., & Schaefer, M. (1997). Interactions between earthworms and soil protozoa—a trophic component in the soil food web. *Soil Biology and Biochemistry,29*, 499–502.

Bonkowski, M., Cheng, W., Griffiths, B. S., Alphei, J., & Scheu, S. (2000). Microbial-faunal interactions in the rhizosphere and effects on plant growth. *European Journal of Soil Biology,36*, 135–147.

Bonkowski, M., Villenave, C., & Griffiths, B. (2009). *Rhizospher fauna*: the functional and structural diversity of intimate interactions of soil fauna with plant roots. *Plant and Soil,321*, 213–233.

Bowden, R., Shroyer, J., Roozeboom, K., Claassen, M., Evans, P., Gordon, B., et al. (2001). Performance of wheat variety blends in Kansas. Keeping up with Research 128, Kansas State University Agricultural Experiment Station and Cooperative Extension Service Manhattan, Kansas.

Brauman, A., Doré, J., Eggleton, P., Bignell, D., & Kane, M. D. (2000). Molecular phylogenetic profiling of microbial communities in guts of termites with different feeding habits. *FEMS Microbiology Ecology,35*(1), 27–36.

Brown, G. G., Edwards, C. A., & Brussaard, L. (2004). How earthworms affect plant growth: burrowing into the mechanisms? In C. A. Edwards (Ed.), *Earthworm Ecology* (pp. 13–49). CRC: Press.

Callot, G., Vercambre, B., Neuveglise, C., & Riba, G. (1996). Hyphasmata and conidial pellets: an original morphological aspect of soil colonization by *Beauveria brongniartii*. *Journal of Invertebrate Pathololology,68*, 173–176.

Castilla, N. P., Vera Cruz, C. M., Mew, T. W. (2003). Using rice cultivar mixtures: a sustainable approach for managing diseases and increasing yield. *International Rice Research Notes, 28*, 5–11.

Chantereau, J., Trouche, G., Rami, J. F., Deu, M., Barro, C., & Grivet, L. (2004). RFLP mapping of QTLs for photoperiod response in tropical sorghum. *Euphytica,120*, 183–194.

Chapuis-Lardy, L., Ramiandrisoa, R. S., Randriamanantsoa, L., Morel, C., Rabeharisoa, L., & Blanchart, E. (2009). Modification of P availability by endogeic earthworms (*Glossoscolecidae*) in ferralsols of the Malagasy Highlands. *Biology and Fertility of Soils,45*, 415–422.

Chapuis-Lardy, L., Brauman, A., Bernard, L., Pablo, A. L., Toucet, J., Mano, M. J., et al. (2010). Effect of the endogeic earthworm *Pontoscolex corethrurus* on the microbial structure and activity related to CO_2 and N_2O fluxes from a tropical soil (Madagascar). *Applied Soil Ecology,45*(3), 201–208.

Chin, K. M., & Wolfe, M. S. (1984). The spread of *Erysiphe graminis* f. sp. *hordei* in mixtures of barley cultivars. *Plant Pathology,33*, 89–100.

Coq, S., Barthès, B. G., Oliver, R., Rabary, B., & Blanchart, E. (2007). Earthworm activity affects soil aggregation and soil organic matter dynamics according to the quality and localization of crop residues: an experimental study (Madagascar). *Soil Biology and Biochemistry,39*, 2119–2128.

Cornu, A., Kondjoyan, N., Martin, B., Verdier-Metz, I., Pradel, P., Berdague, J.-L., et al. (2005). Terpene profiles in Cantal and Saint-Nectaire type cheese made from raw or pasteurized milk. *Journal Science Food Agriculture,85*, 2040–2046.

Cox, C. M., Garrett, K. A., Bowden, R. L., Fritz, A. K., Dendy, S. P., & Heer, W. F. (2004). Cultivar mixtures for the simultaneous management of multiple diseases: tan spot and leaf rust of wheat. *Phytopathology,94*, 961–969.

Cragg, R. G., & Bardgett, R. D. (2001). How changes in soil fauna diversity and composition within a trophic group influence decomposition processes. *Soil Biology and Biochemistry,33*, 2073–2081.

Dakouo, D., Trouche, G., Bâ Malick, N., Neya, A., Kaboré, K. B. (2005). Lutte génétique contre la cécidomyie du sorgho, *Stenodiplosis sorghicola*, une contrainte majeure à la production du sorgho au Burkina Faso. *Cahiers Agricultures*, *14*, 201–208.

Dauber, J., & Wolters, V. (2000). Microbial activity and functional diversity in the mounds of three different ant species. *Soil Biology and Biochemistry,32*, 93–99.

de Alencar Figueiredo, L. F., Davrieux, F., Fliedel, G., Rami, J. F., Chantereau, J., Deu, M., et al. (2006). Development of NIRS equations for food grain quality traits through exploitation of a core collection of cultivated sorghum. *Journal of Agriculture and Food Chemistry,54*(22), 8501–8509.

Decaëns, T. (2010). Macroecological patterns in soil communities. *Global Ecology and Biogeography,19*, 287–302.

Decaëns, T., Jiménez, J. J., Gioia, C., Measey, G. J., & Lavelle, P. (2006). The values of soil animals for conservation biology. *European Journal of Soil Biology,42*, S23–S38.

De Deyn, G. B., van Ruijven, J., Raaijmakers, C. E., de Ruiter, P. C., & van der Putten, W. H. (2007). Above- and belowground insect herbivores differentially affect soil nematode communities in species-rich plant communities. *Oikos,116*, 923–930.

Deu, M., Ratnadass, A., Ag Hamada, M., Diabaté, M., Noyer, J. L., Chantereau, J. (2005). Quantitative trait loci for head-bug resistance in sorghum. *African Journal of Biotechnology*, *4*, 247–250.

Dugué, P., Vall, E., Lecomte, P., Klein, H. D., & Rollin, D. (2004). Évolution des relations entre l'agriculture et l'élevage dans les savanes d'Afrique de l'Ouest et du Centre. Un nouveau cadre d'analyse pour améliorer les modes d'intervention et favoriser les processus d'innovation. *OCL,11*, 268–276.

Durmic, Z., McSweeney, C. S., Kemp, G. W., Hutton, P., Wallace, R. J., & Vercoe, P. E. (2008). Australian plants with potential to inhibit bacteria and processes involved in ruminal biohydrogenation of fatty acids. *Animal Feed Science and Technology,145*(1–4), 271–284.

Edwards, P. (2007). *Introduced Dung Beetles in Australia 1967–2007 Current Status and Future Directions*. Australia: Dung Beetles for Landcare Farming Committee. 66 p.

Eisenhauser, N., Hessler, H., Engels, C., Gleixner, G., Habekost, M., Milcu, A., et al. (2010). Plant diversity effects on soil microorganisms support the singular hypothesis. *Ecology,91*, 485–496.

Falconnier, G., Mouret, J. C., Hammond, R. (2012). Des canards pour désherber les rizières : une intégration agriculture-élevage prometteuse pour les riziculteurs biologiques camarguais. *ORP Conférence 2012*, 27–30 August 2012, 124. Montpellier.

Finch, S., & Collier, R. H. (2000). Host-plant selection by insects: a theory based on 'appropriate/ inappropriate landings' by pest insects of cruciferous plants. *Entomologia Experimentalis et Applicata,96*, 91–102.

Finch, S., & Kienegger, M. (1997). A behavioural study to help clarify how undersowing with clover affects host plant selection by pest insects of brassica crops. *Entomologia Experimentalis et Applicata,84*, 165–172.

Fincher, G. T. (1975). Effects of dung beetle activity on the number of nematode parasites acquired by grazing cattle. *The Journal of Parasitology,61*(4), 759–762.

Finckh, M. R., Gacek, E. S., Goyeau, H., Lannou, C., Merz, U., Mundt, C. C., et al. (2000). Cereal variety and species mixtures in practice, with emphasis on disease resistance. *Agronomie,20*, 813–837.

Fliedel, G., Ratnadass, A., Yajid, M. (1996). Study of some physico-chemical characteristics of developing sorghum grains in relation with head bug resistance. In: D. A. V. Dendy (Ed.), *Proceedings of the ICC International Symposium on Cereals Science and Technology: Impact on a Changing Africa*. (pp. 46–63, 9–13 May 1993, ICC, Vienna, Austria). Pretoria: South Africa.

Furuno, T. (2001). *The power of duck: integrated rice and duck farming*. Tasmania: Tagari Publications, Australia.

Gallun, R. L. (1977). Genetic basis of Hessian fly epidemics. *Annals of the New York Academy of Science,287*, 223–229.

Gastine, A., Schere-Lorenzen, M., & Leadly, P. M. (2003). No consistent effects of plant diversity on root biomass, soil biota and soil abiotic conditions in temperate grassland communities. *Applied Soil Ecology,24*, 101–111.

Glick, B. R., Cheng, Z., Czarny, J., & Duan, J. (2007). Promotion of plant growth by ACC deaminase-containing soil bacteria. *European Journal of Plant Pathology,119*, 329–339.

Goebel, R., Tabone, E., Karimjee, H., Caplong, P. (2005). Mise au point réussie d'une lutte biologique contre le foreur de la canne à sucre Chilo sacchariphagus (Lepidoptera, Crambidae), à la Réunion. *7e Conférence internationale sur les ravageurs en agriculture*, Montpellier.

Goebel, F. R., Roux, E., Marquier, M., Frandon, J., Do Thi Khanh, H., Tabone, E. (2010). Biocontrol of Chilo sacchariphagus (Lepidoptera: Crambidae) a key pest of sugarcane: lessons from the past and future prospects. *Sugar Cane International, 28*, 128–132.

Herren, H. R., Neuenschwander, P., Hennessey, R. D., & Hammond, W. N. O. (1987). Introduction and dispersal of *Epidinocarsis lopezi* (*Hym., Encyrtidae*), an exotic parasitoid of the cassava mealybug, *Phenacoccus manihoti* (*Hom., Pseudococcidae*), in Africa. *Agriculture Ecosystems and Environment,19*, 131–144.

Hossain, S. T., Sugimoto, H., Ahmed, G. J. U., & Islam, M. R. (2005). Effect of integrated rice-duck farming on rice yield, farm productivity, and rice-provisioning ability of farmers. *Asian Journal of Agriculture and Development,2*, 79–86.

Irshad, U., Villenave, C., Brauman, A., & Plassard, C. (2011). Grazing by nematodes on rhizosphere bacteria enhances nitrate and phosphorus availability to *Pinus pinaster* seedlings. *Soil Biology and Biochemistry,43*, 2121–2126.

Isman, M. B. (2006). Botanical insecticides, deterrents, and repellents in modern agriculture and an increasingly regulated world. *Annual Review of Entomology,51*, 45–66.

Jana, U., Barot, S., Blouin, M., Lavelle, P., Laffray, D., & Repellin, A. (2010). Earthworms influence the production of above- and belowground biomass and the expression of genes involved in cellular proliferation and stress responses in *Arabidopsis thaliana*. *Soil Biology and Biochemistry,42*, 244–252.

Jones, C. G., Lawton, J. H., & Shachak, M. (1994). Organisms as ecosystem engineers. *Oikos,69*, 373–386.

Kardol, P., & Wardle, D. A. (2010). How understanding aboveground-belowground linkages can assist restoration ecology? *Trends in Ecology and Evolution,25*(11), 670–679.

Kennedy, A. C., & Smith, K. L. (1995). Soil microbial diversity and the sustainability of agricultural soils. *Plant and Soil,170*, 75–86.

Khan, Z. R., Overholt, W. A., & Ng'eny-Mengech, A. (2003). Integrated pest management case studies from ICIPE. In K. Maredia, D. Dakouo, & D. Mota-Sanchez (Eds.), *Integrated pest management in the Global Arena*. MI: Michigan State University, East Leasing.

Khan, Z. R., Chiliswa, P., Ampong-Nyarko, K., Smart, L. E., Polaszek, A., Wandera, J., et al. (1997a). Utilization of wild gramineous plants for management of cereal stemborers in Africa. *Insect Science and its Application,17*, 143–150.

Khan, Z. R., Hassanali, A., Overholt, W., Khamis, T. M., Hooper, A. M., Pickett, J. A., et al. (2002). Control of witchweed *Striga hermonthica* by intercropping with *Desmodium* spp., and the mechanism defined as allelopathic. *Journal of Chemical Ecology,28*, 1871–1885.

Khan, Z. R., Ampong-Nyarko, K., Chiliswa, P., Hassanali, A., Kimani, S., Lwande, W., et al. (1997b). Intercropping increases parasitism of pests. *Nature,388*, 631–632.

Kibblewhite, M. G., Ritz, K., & Swift, M. J. (2008). Soil health in agricultural systems. *Transactions of the Royal Society B,363*, 685–701.

Kueneman, E., Hoffmann, I., Kebe, B., Belem, C., Lecomte, P.H., Rollin, D., Carsky, R. (2002). FAO Joint mission FAO Cirad Crop livestock integration in the PRODS Program in Burkina Faso. Travel summary report Reg. file code PL6/1 Kueneman, p. 20.

Kumar, A., & Sharma, S. (2008). An evaluation of multipurpose oil seed crop for industrial uses (*Jatropha curcas* L.): a review. *Industrial Crops and Products,28*, 1–10.

LaMondia, J., Elmer, W. H., Mervosh, T. L., & Cowles, R. S. (2002). Integrated management of strawberry pests by rotation and intercropping. *Crop Protection,21*, 837–846.

Laossi, K. R., Decaëns, T., Jouquet, P., & Barot, S. (2010). Can we predict how earthworm effects on plant growth vary with soil propoerties? *Applied and Environmental Soil Science,*. doi:10.1155/2010/784342.

Lavelle, P., Decaëns, T., Aubert, M., Barot, S., Blouin, M., Bureau, F., et al. (2006). Soil invertebrates and ecosystem services. *European Journal of Soil Biology,42*, S3–S15.

Lavigne, A., Dumbardon-Martial, E., & Lavigne, C. (2012). Les volailles pour un contrôle biologique des adventices dans les vergers. *Fruits,67*, 341–351.

Lavigne, C., Lesueur-Jannoyer, M., de Lacroix, S., Chauvet, G., Lavigne, A., & Dufeal, D. (2011). De la production fruitière intégrée à la gestion écologique des vergers aux Antilles. *Innovations agronomiques,16*, 53–62.

Lee, S. Y., Bollinger, J., Bezdicek, D., & Ogram, A. (1996). Estimation of the abundance of an uncultured soil bacterial strain by a competitive quantitative PCR method. *Applied Environmental Microbiology,62*, 3787–3793.

Loranger, G., Ponge, J. F., Blanchart, E., & Lavelle, P. (1998). Impact of earthworms on the diversity of microarthropods in a vertisol (Martinique). *Biology and Fertility of Soils,27*, 21–26.

Loranger, G., Cabidoche, Y. M., Delone, B., Quénéhervé, P., & Ozier-Lafontaine, H. (2012). How earthworm activities affect banana plant response to nematodes parasitism. *Applied Soil Ecology,52*, 1–8.

Marie-Magdeleine, C., Udino, L., Philibert, L., Bocage, B., & Archimede, H. (2010). *In vitro* effects of Cassava (*Manihot esculenta*) leaf extracts on four development stages of *Haemonchus contortus*. *Veterinary Parasitology,173*(1–2), 85–92.

Matson, P. A., Parton, W. J., Power, A. G., & Swift, M. J. (1997). Agricultural intensification and ecosystem properties. *Science,277*, 504–509.

May, R. M. (1990). How many species? *Philosophical Transactions of the Royal Society B: Biological Sciences,330*, 293–304.

Mazorra, C. (2006). *Manejo de la selección del alimento para reducir el ramoneo de ovinos integrados a plantaciones de cítricos*. CIBA-UNICA-ICA, La Habana: Tesis Doctor en Ciencias Veterinarias. 121 p.

McIntosh, R. A. (1998). Breeding wheat for resistance to biotic stresses. *Euphytica,100*, 19–34.

Men, B. X., Ogle, R. B., & Lindberg, J. E. (2002). Studies on integrated duck-rice systems in the Mekong delta of Vietnam. *Journal of Sustainable Agriculture,20*, 27–40. doi:10.1300/J064v20n01_05.

Mille, B., de Valavieille-Pope, C. (2001). Associations variétales et interventions fongicides contre les septorioses et la rouille brune du blé d'hiver. *Cahiers Agricultures, 10*, 125–129.

MEA (Millennium Ecosystem Assessment) (2005). *Ecosystem and Human Well-Being: Synthesis*. pp. 137, Washington: Island Press.

Mundt, C. C. (2002). Use of multiline cultivars and cultivar mixtures for disease management. *Annual Review of Phytopathology,40*, 381–410.

Naudin, K., Scopel, E., Andriamandroso, A. L. H., Rakotosolofo, M., Andriamarosoa, Ratsimbazafy, N. R. S., Rakotozandriny, J. N., Salgado, P., Giller, K. E. (2012). Trade-offs between biomass use and soil cover. The case of rice-based cropping systems in the Lake Alaotra region of Madagascar. *Experimental Agriculture, 48*, 194–209.

Nibouche, S., Tibère, R., & Costet, L. (2012). The use of *Erianthus arundinaceus* as a trap crop for the stem borer *Chilo sacchariphagus* reduces yield losses in sugarcane: preliminary results. *Crop Protection,42*, 10–15.

Painter, R. H. (1951). *Insect Resistance in Crop Plants*. New York: Macmillan.

Peacock, A. D., Mullen, M. D., Ringelberg, D. B., Tyler, D. D., Hedrick, D. B., Gale, P. M., et al. (2001). Soil microbial community responses to dairy manure or ammonium nitrate applications. *Soil Biology and Biochemistry,33*(7–8), 1011–1019.

Postma-Blaauw, M. B., Bloem, J., Faber, J. H., van Groeningen, J. W., de Goede, R. G. M., & Brussaard, L. (2006). Earthworm species composition affects the soil bacterial community and net nitrogen mineralization. *Pedobiologia,50*, 243–256.

Rabary, B., Sall, S., Letourmy, P., Husson, O., Ralambofetra, E., Moussa, N., et al. (2008). Effects of living mulches or residue amendments on soil microbial properties in direct seeded cropping systems of Madagascar. *Applied Soil Ecology,39*, 236–243.

Raboin, L. M., Ramanantsoanirina, A., Dusserre, J., Razasolofonanahary, F., Tharreau, D., Lannou, C., et al. (2012). Two-component cultivar mixtures reduce rice blast epidemics in an upland agrosystem. *Plant Pathology,*. doi:10.1111/j.1365-3059.2012.02602.x.

Randriamanantsoa, R., Aberlenc, H. P., Ralisoa, O. B., Ratnadass, A., Vercambre, B., (2010). Leslarves des Scarabaeoidea (Insecta, Coleoptera) en riziculture pluviale des régions de haute etmoyenne altitudes du centre de Madagascar. *Zoosystema, 32*, 19–72

Randrianarivelo, R., Danthu, P., Benoit, C., Ruez, P., Raherimandimby, M., & Sarter, S. (2010). Novel alternative to antibiotics in shrimp hatchery: effects of the essential oil of *Cinnamosma fragrans* on survival and bacterial concentration of *Penaeus monodon* larvae. *Journal of Applied Microbiology,109*, 642–650.

Randrianarivelo, R., Sarter, S., Odoux, E., Brat, P., Lebrun, M., Menut, C., et al. (2009). Composition and antimicrobial activity of essential oils of *Cinnamosma fragrans*. *Food Chemistry,114*, 680–684.

Ratnadass, A., & Wink, M. (2012). The phorbolester fraction from *Jatropha curcas* seed oil: potential and limits for crop protection against insect pests. *International Journal of Molecular Sciences,13*, 16157–16171.

Ratnadass, A., Chantereau, J., Gigou, J. (eds.) (1998). Amélioration du sorgho et de sa culture en Afrique de l'Ouest et du Centre. In: *Actes de l'Atelier de restitution du programme conjoint sur le* sorgho Icrisat-Cirad. (pp. 17–20 March 1997, Bamako, Mali, coll. Colloques, Montepellier). CIRAD–CA: France.

Ratnadass, A., Doumbia, Y. O., Ajayi, O. (1995a). Bioecology of sorghum head bug Eurystylus immaculatus and crop losses in West Africa. In: K. F. Nwanze, O. Youm (Eds.), *Panicle Insect Pests of Sorghum and Pearl Millet: Proceedings of an International Consultative Workshop* India, (pp. 91–102, 4–7 October 1993, ICRISAT Sahelian Center, Niamey, Niger, 502 324), Patancheru: Andhra Pradesh.

Ratnadass A., Ajayi O., Fliedel, G., Ramaiah, K. V. (1995b). Host plant resistance in sorghum to Eurystylus immaculatus in West Africa. In: K. F. Nwanze, O. Youm (Eds.), *Panicle Insect Pests of Sorghum and Pearl Millet: Proceedings of an International Consultative Workshop*, India, (pp. 191–199, 4–7 October 1993, ICRISAT Sahelian Center, Niamey, Niger, 502 324), Patancheru: Andhra Pradesh.

Ratnadass, A., Chantereau, J., Coulibaly, M. F., & Cilas, C. (2002). Inheritance of resistance to the panicle-feeding bug *Eurystylus oldi* and the sorghum midge *Stenodiplosis sorghicola* in sorghum. *Euphytica,123*, 131–138.

Ratnadass, A., Fernandes, P., Avelino, J., & Habib, R. (2012). Plant species diversity for sustainable management of crop pests and diseases in agroecosystems: a review. *Agronomy for Sustainable Development,32*, 273–303.

Ratnadass, A., Chantereau, J., Cissé, B., Ag Hamada, M., Fliedel, G., Grabulos, J., Luce, C. (2006). Selection of a sorghum line CIRAD 441 combining productivity and resistance to midge and head bugs. *International Sorghum and Millets Newsletter, 47*, 30–32.

Ratnadass, A., Marley, P. S., Hamada, M. A., Ajayi, O., Cissé, B., Assamoi, F., et al. (2003). Sorghum head-bugs and grain molds in west and central Africa. 1. Host plant resistance and bug-mold interactions on sorghum grains. *Crop Protection,22*, 837–851.

Retureau, É., Callon, C., Didienne, R., & Montel, M.-C. (2010). Is microbial diversity an asset for inhibiting *Listeria monocytogenes* in raw milk cheeses? *Dairy Science and Technology,90*(4), 375–398.

Risède, J. M., Lescot, T., Cabrera Cabrera, J., Guillon, M., Tomekpe, K., Kema, G. H. J., Cote, F. (2010). Challenging short and mid-term strategies to reduce the use of pesticides in banana production. *Banana Field Study—Guide Number 1.* Retrieved May 25, 2013) from http://www.endure-network.eu/endure_publications/endure_publications2.

Rougerie, R., Decaëns, T., Deharveng, L., Porco, D., James, S. W., Chang, C.-H., et al. (2009). DNA barcodes for soil animal taxonomy: transcending the final frontier. *Pesquisa Agropecuaria Brasileira,44,* 789–801.

Sarter, S., Randrianarivelo, R., Ruez, P., Raherimandimby, M., & Danthu, P. (2011). Antimicrobial effects of essential oils of *Cinnamosma fragrans* on the bacterial communities of the water rearing of *Penaeus monodon* larvae. *Vector Borne and Zoonotic Diseases,11*(4), 433–437.

Scheu, S. (2003). Effects of earthworms on plant growth: patterns and perspectives. *Pedobiologia,47,* 846–856.

Séguy, L., Husson, O., Charpentier, H., Bouzinac, S., Michellon, R., Chabanne, A., Boulakia, S., Tivet, F., Naudin, K., Enjalric, F., Chabierski, S., Rakotondralambo, P. Rakotondramanana (2009). La gestion des écosystèmes cultivés en semis direct sur couverture végétale permanente. In: *Manuel pratique du semis direct à Madagascar. 1,* Chap. 2, p. 32.

Senapati, B. K., Lavelle, P., Giri, S., Pashanasi, B., Alegre, J., Decaëns, T., Jimenez, J. J., Albrecht, A., Blanchart, E., Mahieu, M., Rousseaux, L., Thomas, R., Panigrahi, P. K., Venkatachalam, M. (1999). In-soil earthworm technologies for tropical agroecosystems. In: (P. Lavelle, L. Brussaard, P. Hendrix (Eds.), *Earthworm Management in Tropical Agroecosystems.* CABI Publishing, 199–238.

Shetty, S. V. R., Beninati, N. F., Beckerman, S. R. (1991). Strengthening sorghum and pearl millet research in mali. (ICRISAT, p. 85, Patancheru 502 324), Andhra Pradesh: India.

Singh, J. S., Pandey, V. C., & Singh, D. P. (2011). Efficient soil microorganisms: a new dimension for sustainable agriculture and environmental development. *Agriculture, Ecosystems and Environment,140,* 339–353.

Spehn, E. M., Joshi, J., Schmid, B., Alphei, J., & Körner, C. (2000). Plant diversity effects on soil heterotrophic activity in experimental grassland ecosystem. *Plant and Soil,224,* 217–230.

Swift, M. J., Heal, O. W., & Anderson, J. M. (1979). *Decomposition in terrestrial ecosystems.* Oxford: Blackwell Scientific.

Su, P., Liao, X. I., Zhang, Y., & Huang, H. (2012). Influencing factors on rice sheath blight epidemics in integrated rice-duck system. *Journal of Integrative Agriculture,11,* 1462–1473.

Tao, J., Chen, X., Liu, M., Hu, F., Griffiths, B., & Li, H. (2009). Earthworms change the abundance and community structure of nematodes and protozoa in a maize residue amended rice-wheat rotation agro-ecosystem. *Soil Biology and Biochemistry,41,* 898–904.

Thomas, J., Hein, G., Baltensperger, D., Nelson, L., Haley, S. (2002). Managing the Russian wheat aphid with resistant wheat varieties. NebFact, Nebraska Cooperative Extension, NF96–307.

Thuita, M., Pypers, P., Herrmann, L., Okalebo, R. J., Othieno, C., Muema, E., et al. (2012). Commercial rhizobial inoculants significantly enhance growth and nitrogen fixation of a promiscuous soybean variety in Kenyan soils. *Biology and Fertility of Soils,48,* 87–96.

Tiwari, S. C., & Mishra, R. R. (1993). Fungal abundance and diversity on earthworm casts and undigested soil. *Biology and Fertility of Soils,16,* 131–134.

Tooker, J. F., & Frank, S. D. (2012). Genotypically diverse cultivar mixtures for insect pest management and increased crop yields. *Journal of Applied Ecology,49,* 974–985. doi:10.1111/j.1365-2664.2012.02173.x.

Turbé, A., De Toni, A., Benito, P., Lavelle, P., Lavelle, P., Ruiz, N., Van der Putten, W. H., Labouze, E., Mudgal, S. (2010). Soil biodiversity: functions, threats and tools for policy makers. Bio intelligence service, IRD, and NIOO, report for European commission (DG environment).

de Vallavieille-Pope, C. (2004). Management of disease resistance diversity of cultivars of a species in single fields: controlling epidemics. *Comptes Rendus Biologies. 327*(611–620).

van der Heijden, M. G. A., Klironomos, J. N., Ursic, M., Moutoglis, P., Streitwolf-Engel, R., Boller, T., et al. (1998). Mycorrhizal fungal diversity determines plant biodiversity, ecosystem variability and productivity. *Nature,396,* 69–72.

Vercambre, B., Charbonnier, G., Launois, M., Laveissière, G. (2008). Le ver blanc au paradis vert, ou l'histoire d'un bioagresseur de la canne à sucre en milieu insulaire. Enquête scientifique, coll. Les savoirs partagés, CIRAD, p. 75.

Vessey, J. K. (2003). Plant growth promoting rhizobacteria as biofertilizers. *Plant and Soil,255*, 571–586.

Villenave, C., Rabary, B., Chotte, J. L., Blanchart, E., & Djigal, D. (2009). Impact of direct seeding mulch-based cropping systems on soil nematodes in a long-term experiment in Madagascar. *Pesquisa Agropecuaria Brasileira,44*, 949–953.

Wall, D. H., Fitter, A. H., Paul E. A. 2(005). Developing new perspectives from advances in soil biodiversity research. In R. D. Bardgett, M. B. Usher, D. W. Hopkins (Eds), *Biological Diversity and Function in Soils*. Cambridge: Cambridge University Press, 3–27.

Whipps, J. M. (2001). Microbial interactions and biocontrol in the rhizosphere. *Journal of Experimental Botany,52*, 487–511.

Wilhoit, L. R. (1992). Evolution of herbivore virulence to plant resistance: influence of variety mixtures. In R. S. Fritz & E. L. Simms (Eds.), *Plant resistance to herbivores and pathogens: ecology, evolution and genetics* (pp. 91–119). Chicago: University of Chicago Press.

Wolfe, M. S. (1985). The current status and prospects of multiline cultivars and variety mixtures for disease resistance. *Annual Review of Phytopathology,23*, 251–273.

Wurst, S. (2010). Effects of earthworms on above- and belowground herbivores. *Applied Soil Ecology,45*, 123–130.

Zhu, Y., Chen, H., Fan, J. H., Wang, Y., Li, Y., Fan, J. X., et al. (2000). Genetic diversity and disease control in rice. *Nature,406*, 718–722.

Chapter 6
Conserving and Cultivating Agricultural Genetic Diversity: Transcending Established Divides

Sélim Louafi, Didier Bazile and Jean-Louis Noyer

Did we wait for a discussion on conserving biodiversity before actually doing so? Clearly no: various categories of farmers were conserving biodiversity well before scientists formulated the concept. Indeed, while the very first practices of agriculture included a logic of seed conservation, man was already conserving biodiversity even before the advent of agriculture. For example, studies on forests (Guillaumet 1996) clearly show that the development of a proto-agriculture included the management of biological resources that can be likened to a process of in situ conservation/protection that persists even today among populations of mobile hunter-gatherers in a constrained environment. The genus *Dioscorea* (yam) is an example of a crop for which the concept of conservation was established even before the process of its domestication had begun. It continues so even today, given its predominant mode of vegetative reproduction (Hladik et al. 1984; Hamon et al. 1992; Dounias 1996; Tostain et al. 2005; Chaïr et al. 2010). Conversely, for cereals—or more generally, for annual seed plants—farmers are mainly interested in the conservation of seed necessary to maintain the crop from year to year. This is an ad hoc short-term conservation strategy, one that is not even thought of as

S. Louafi (✉)
Department Bios, Agap Joint Research Unit (Genetic Improvement and Adaptation of Mediterranean and Tropical Plants)—Cirad, Montpellier, France
e-mail: selim.louafi@cirad.fr

D. Bazile
Department ES, Green Internal Research Unit (Renewable Resources Management and Environment)—Cirad, Montpellier, France
e-mail: didier.bazile@cirad.fr

J.-L. Noyer
Department Bios, Cirad, Montpellier, France
e-mail: jean-louis.noyer@cirad.fr

É. Hainzelin (ed.), *Cultivating Biodiversity to Transform Agriculture*,
DOI: 10.1007/978-94-007-7984-6_6, © Éditions Quæ 2013

such. In doing so, farmers have gradually let disappear—or more frequently, let become scarce—the wild or ancestral forms at the origin of the varieties and populations they grow. The work of Vavilov in the 1930s and later that of Harlan in 1951 were seminal in their discussion on centres of origin and non-centres of domestication.[1] They raised awareness in the scientific community and also amongst influential agricultural-sector groups of the existence and importance of feral or domesticated wild genetic resources. The fear of genetic erosion continues to prevail in debates on different and sometimes competing intellectual approaches for conserving agricultural genetic diversity,[2] i.e., to prevent its loss and to maintain its availability. Private initiatives, governmental measures, non-governmental actions, regional, national or international actions for conservation have gradually appeared since World War II. They have encompassed both ex situ conservation in different forms—botanical gardens or 'acclimatization' gardens (the precursors), chilled seeds, living collections in fields, etc.—and in situ conservation—nature reserves, contracts with the human populations who are the guardians of resources, etc.

These conservation mechanisms are the result of mankind's commitment to knowledge advancement, awareness of the importance of the environment, humanitarian visions, and commercial or national interests. However, reproductive biology has constrained these mechanisms to a great extent. Cross-pollination, self-pollination, the dominance of the vegetative reproduction regime, the length of biological cycles, recalcitrance, annual plants, perennial plants, photoperiodism, etc. have led conservation structures to a certain specialization. The cultivated plants that are easiest to conserve and which are dominant in the 'industrial' agrosystems benefit from huge and expensive, and sometimes redundant, conservation mechanisms. Civil society is right to ask questions about these mechanisms' cost-effectiveness and their impact on the maintenance of diversity. In contrast, for underutilized plants (sometimes called neglected or orphan plants) or those more difficult to conserve from a biological point of view, the conservation efforts are more haphazard and almost always poorly funded and inadequately identified. The recent combination of biological and social sciences sheds new light on issues of conservation and of the use of agricultural genetic diversity.

In this chapter, we propose to go beyond an approach that oscillates between a traditional and rigid vision of conservation, on the one hand, and a vision predicated on sustained human intervention, on the other. The first is a utilitarian, almost commercial, vision, one that is found in major international ex situ mechanisms. The second is more secure because it is based on a constant human intervention, albeit of variable quality and uncertain durability. Neither approach is fully satisfactory. We want to highlight the complexity and structural inequalities between

---

[1] Harlan (1971) subsequently proposed a more elaborate classification, which distinguishes three centres where agriculture could have first appeared, and three non-centres (due to their large size) where it would later spread. Centres and non-centres could have exchanged ideas, techniques and varieties.

[2] Genetic diversity of crop species and their wild relatives.

mechanisms to conserve agricultural genetic diversity. The idea of cultivating biodiversity to transform agriculture appears already well advanced given the numerous debates and socio-economic discussions on ensuring access to and control of agricultural genetic diversity. This shows *a contrario* that the part of biodiversity judged to be without economic value risks being inadequately conserved.

We start with a historical analysis of the implementation of conservation mechanisms. We then take up the international political strategies that govern conservation and mobilization of agricultural genetic diversity, their shortcomings and possible solutions. Finally, we describe the advances in the knowledge of this genetic diversity and its conservation, and the reasons for transcending the in situ–ex situ divide.

1 History of the Conservation of Genetic Resources in Agriculture

1.1 Early Developments

A gradual domestication and selection of plant species has always formed part of the development of agriculture. Seed exchanges between continents began as soon as man first started to travel and experienced a significant increase with expeditions to the New World in the fifteenth century. It was also at this time that the first botanical gardens were set up. They had the responsibility of receiving and classifying plants collected during these expeditions (Brockway 1988). Subsequently, the policies accompanying the development of colonies accelerated the international movement of genetic resources. It was not until the early twentieth century, however, that the first systematic work to set up collections was undertaken, at the instigation of plant breeders who had begun to appear and to organize themselves professionally (Garrison Wilkes 1988). But these efforts were narrowly targeted and were mainly conducted on an ad hoc basis, with no clear long-term strategy. When some resources became useless to breeding programmes, collections were no longer properly maintained or simply destroyed.

At the same time, the famous Russian scientist Nikolai Vavilov (1887–1943) defined and identified the origin of plants which were of major agricultural interest and undertook extensive botanical-agronomic expeditions to collect material for research purposes.

After World War II, advances in genetics, the appearance of resistance in improved varieties, and growing international exchanges between scientists made possible by advances in transportation led to a growth in the global exchange of genetic resources (Kloppenburg 1988). Given this spurt, the need soon arose to coordinate documentation and collection efforts for a more effective exchange of materials. This is precisely what FAO did in 1948 by seeking to establish a catalogue of collected and globally available genetic resources. Despite these efforts, however, exchanges remain limited to developed countries (including the USSR) in what were then called the 'plant introduction stations'. This was essentially resources that were not intended to be preserved over the long term (Kloppenburg 1988).

The turning point came in the 1960s with the introduction of the Green Revolution under the aegis of major American foundations, Ford and Rockefeller in particular. The need for access to a broad-based genetic diversity increased along with the growing awareness of the high risk of genetic erosion by the development of an industrialized agricultural production. These two facts pushed the FAO, in association with the International Biological Program, to hold a conference in order to come up with a coordinated response to the problems of conservation and the availability of genetic diversity. Held in 1967, this conference was intended to increase awareness on the effects of the erosion of Plant Genetic Resources for Food and Agriculture (PGRFA) and on the need to establish an international network dedicated to this and related issues. But while genetic erosion was accepted as a fact among scientists at the conference, there were marked differences of opinion on how to respond to it (Pistorius 1997). The debate finally focused on the question of what PGRFA must be collected and how. An initial, utilitarian opinion advocated an ex situ conservation of genetic resources of major crops and their wild relatives. A second train of thought, originating with population ecology, judged in situ conservation to be paramount, even for species having only a very local interest or of no immediately known interest. Without actually taking a final position, the conference report put forward practical reasons (time and money), rather than scientific ones, to justify favouring the first option at the expense of the second. An expert panel was set up within the FAO to move forward on the issue, but the solution that emerged originated from outside the panel, from a group of donor countries led by the United States and American foundations under the aegis of the World Bank. Relying on the existence of collections already set up in important breeding programmes funded by the Ford and Rockefeller foundations, the Consultative Group on International Agricultural Research (CGIAR) was created in 1971. It brought together existing international agricultural research institutes and those newly created around the world in a unique institutional framework hosted by the World Bank.

This new global governance of PGRFA has three specific characteristics. The first is that the CGIAR, supported by a donor consortium, is outside the framework of the United Nations. The second is its exclusive choice of ex situ conservation. The third concerns the decision to focus almost exclusively on major agricultural crops of interest, each centre being in charge of a limited number of crops important for food security.

The establishment of this network, however, did constitute the first effort at systemizing and formalizing conservation strategies and at sharing plant breeding resources at the international level. Effective coordination of these efforts became the responsibility of the International Board on Plant Genetic Resources (IBPGR), an international centre created primarily for this purpose. Responsible for coordinating the tasks of collecting, conservation and development of standard descriptors, it quickly became an international institution of reference for those involved in PGRFA. IBPGR—which later became IPGRI (International Plant Genetic Resources Institute) and, finally, Bioversity International—took the 'genetic resource' object into a new dimension, making it, for better or for worse, an object of international policies and politics.

1.2 The Functioning of the CGIAR System

The activities carried out by international agricultural research centres (IARCs) in the field of PGRFA consist of the collection, storage, characterization, multiplication and evaluation, pre-breeding, data management and supply of information, provision of PGRFA, and research and training (Fig. 1).

Five types of genetic material are considered by the CGIAR: wild relatives of crop plants; local varieties and primitive cultivars; obsolete ancient cultivars; advanced breeding lines, mutations and other products of plant breeding programmes; and high-yield elite modern cultivars. Collection and conservation activities concern mainly the first three categories.

IBPGR's coordinating role has three components: promotion and dissemination of information on PGRFA; support for theoretical and practical research activities (concerning sampling, collection, conservation, evaluation, training and creation of databases[3]); and the establishment of an international network of national and international PGRFA centres.

In more concrete terms, IBPGR establishes regional priorities with regard to the collection and conservation of PGRFA, funds scientific expeditions for collection (including the plants included in IARC mandates), and provides the technological means of conservation to IARCs and national centres. In 1978, there were 25 gene banks, including 13 in industrialized countries. The majority of the banks in developing countries were in the IARCs.

It should be noted that evaluation is required for plant genetic resources to be usable by a plant breeder. Valorising the breeding work is only possible after an evaluation of the PGRFA's agronomic qualities. This evaluation consists of four distinct stages: establishment of 'passport' data (geographical and botanical data); description of phenotypic traits; preliminary assessment (pre-breeding) to identify (in the field) traits likely to interest breeders; and in-depth assessment. The last of these mainly consists of laboratory work for screening traits of resistance to specific pathogens or environments, followed by work in the field under appropriate conditions. IBPGR has never really focused, through research or funding, on this last stage (Hawkes 1985). Consequently, IARCs have assumed the

---

[3] Providing support for breeding work (preselection), testing and distribution is not part of IBPGR's mandate.

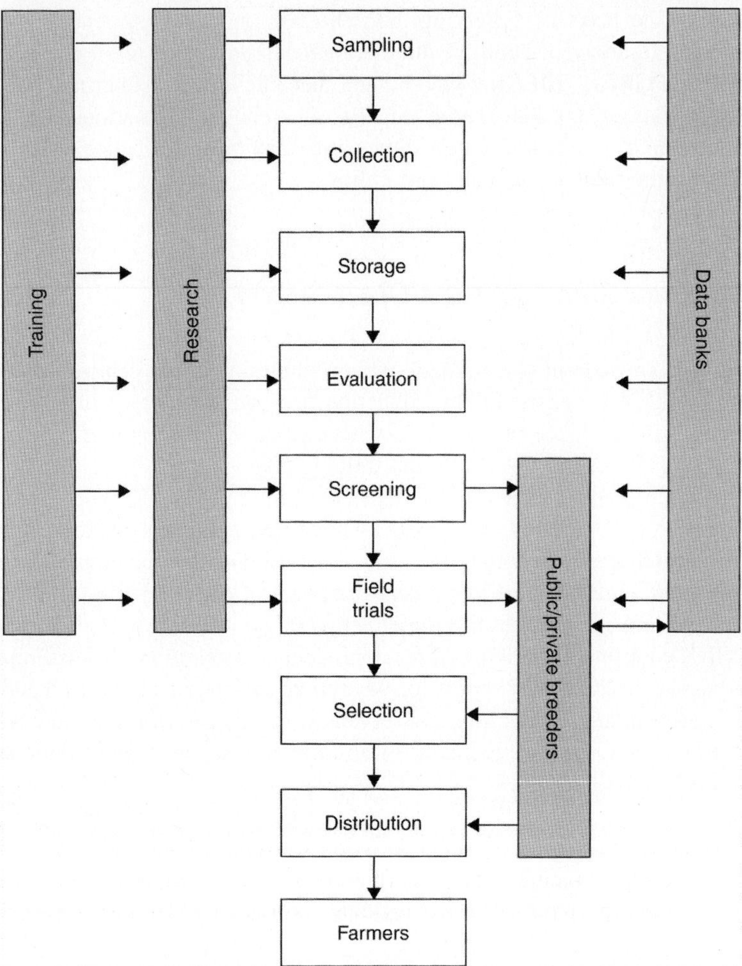

Fig. 1 The process of R&D on PGRFA in IARCs

responsibility for this stage but they have only concentrated on plants for which they were commissioned (major plants). For other species, only private entities or national research organizations with substantial human and technological resources can undertake this work.

The same problem exists for PGRFA improvement (Hawkes 1985). This activity, called 'pre-breeding' or 'selection of parental lines', is essential in the use of PGRFA of primitive cultivars or wild relatives. The genes of wild relatives can be useful to breeders in developing cultivars suitable for agriculture which still relies on traditional species. But even IARCs which collected this material have little incentive in using it for their own breeding activity for advanced lines for fear of losing years of work because of the 'backward' step it would entail. Pre-breeding helps sidestep this danger in the use of the collected material. By neglecting this activity, IBPGR has helped increase the divide between potential breeders in the South and IARCs. Thus, even though species such as millet, sorghum, cassava, yam and cowpea were collected by IARCs or national centres, they have been neglected by most breeders.

Finally, networks help promote the establishment of national and regional committees on genetic resources. Their operating mechanism relies on collaboration between national research organizations and private breeders, most notably for the selection and distribution of PGRFA. The network is used to centralize ex situ gene banks. This consists of bringing together the genetic material from a variety of regions in one place, thus facilitating the work of breeders.

IARCs have focused their work on collecting, conserving and breeding their mandated crops. In addition, since IBPGR does not undertake the work of comprehensive evaluation or improvment, the IARCs' specific mandate or donor interests determine the orientation of collection and evaluation programmes.

This manner of working continued throughout the 1970s and 1980s. It was not until the mid-1990s that the first participatory breeding programmes struck up within the CGIAR and the term in situ or 'on-farm' conservation retook centre stage. These changes resulted not only from criticisms of the governance model adopted by the CGIAR to manage PGRFA but also from technical limitations of the ex situ conservation model.

1.3 Criticisms of Ex Situ Conservation Strategies

CGIAR opened itself to criticism over its policies by opting for this model of networked gene banks managed by Northern donors which emphasized the agro-economic value of PGRFA. Criticism became even stronger subsequent to the rise of intellectual property issues in the 1980s.

The criticism was directed both at CGIAR's mode of governance and at its conservation model. As far as governance was concerned, criticism focused mainly on two areas.

First, there was a transition from a network of ex situ regional gene banks (an approach that represented a consensual scientific view—although oriented mainly towards an ex situ model) to a network of banks which were more specifically oriented. In other words, the idea of preserving large pools of agricultural genetic diversity, which had been partly motivated by the risk of genetic erosion, fell by the wayside. The focus shifted to conserving the major agricultural crops, justified by the needs of international agricultural research in charge of propagating the Green Revolution and coordinated exclusively by funding entities. By pushing the ex situ approach to this extreme, this mode of regulation of agricultural genetic resources confirmed the scientific option which had surfaced in 1967 during the technical conference.

By the end of the 1970s, the legal status of the collections set up by the CGIAR had begun to be questioned, as had the modes of mobilizing this diversity and of its distribution to users worldwide. By making available the material stored in international gene banks without the consent of farmers or States, this mode of conservation attracted criticism, mainly from developing countries. A dispute arose over the sharing of the benefits arising from such use. It crystallized the conflict between farmers' rights and breeders' rights. On one side, breeder-exporters of improved varieties defended the existence of a mechanism for protection of intellectual property—at that time UPOV, Union for the Protection of New Varieties of Plants, was their reference—as compensation for the investments they had made to develop these new varieties. On the other side, developing countries stressed that this work of genetic improvement would not have been possible if they themselves had not provided the raw material, i.e., the genetic resources. They wanted compensation for these contributions (Brahy and Louafi 2004). As a result of these tensions, these collections were put under the ambit of the FAO (through an International Undertaking at first and later via the International Treaty), and the farmers' rights were formally recognized in resolutions 4/89 and 5/89 under the International Undertaking (and later in Article 9 of the International Treaty).

But apart from these political and legal aspects, the model of ex situ conservation itself also poses technical challenges that have weakened and still weaken its implementation. Ex situ conservation follows one of several methods. In most cases, it consists of seed banks conserved for the short, medium or long term based upon temperature (-4 to -20 °C), or sometimes the pollen is frozen. Less used alternative methods exist for specific biological materials which respond poorly to conservation. For example, some species cannot be conserved through seeds and require living ex situ collections. Such is the case of trees or plants using vegetative reproduction, for example, banana, cassava or yam. For these plants, in vitro propagation and/or cryopreservation of somatic tissues (apex or embryo) is used.

These ex situ seed bank collections offer the advantage of preserving a large amount of genetic resources at a single, secure location. However, this arrangement comes with its own attendant risks, especially concerning long-term durability and permanence. Technical problems, mainly power outages and failures of refrigeration systems, have often resulted in permanent loss of important collections in smaller such establishments. Continuous funding, often difficult to ensure even in the richer countries, is essential to maintaining the quality of preserved material and to guarantee the proper management of samples. Often the collections exist, but only in name as the material is not really usable. To this can be added the possibility of the stored material reacting badly to the storage conditions, resulting in an early loss of material; possible changes of the samples, especially through an uncontrolled modification of their genetic structure during regeneration phases; and a cessation of the plant-environment co-evolution (in the broadest sense). Finally, by isolating and separating the biological material from the sociocultural pressures of which it is part, ex situ conservation causes a disconnect between the preserved material and the knowledge and practices associated with it, even if these latter are recorded in databases.

Even though under fire from critics, the ex situ conservation model continues to remain the standard. Political debates continue to be centred around it and strategies across the globe espouse its adoption in one way or another. These international frameworks only partially reflect the various existing conceptions of genetic resources and of methods to valorise them. However, these frameworks are being subjected to various pressures all the time and hence evolving constantly.

2 International Strategies and Policies in Favour of Mobilizing Genetic Diversity

The issue of the conservation of agricultural genetic diversity is inseparable from its use. The erosion of this diversity is as much a consequence of its under-utilization as of its over-exploitation. So irrespective of the model chosen, conservation only makes sense if it is put to use by a wide variety of users with differing capacities and needs. However, since the early 1980s, the issue of supply of—and access to—genetic diversity remains the topic of intense and contentious debate. Viewpoints vary widely, originating as they do from different international conceptions of the legal and economic statuses of these resources.

2.1 Different Conceptions of the Status of Genetic Resources

The Convention on Biological Diversity (CBD) established the States' sovereignty over these resources. While this issue is no longer challenged, the same cannot be said of the question of the legal status of these resources, which remains open. Within this general principle, at least four concepts coexist and are expressed through discourses and practices of actors and communities using genetic resources.

2.1.1 Global Public Good

There are many definitions of the concept of global public good, some of them conflicting. Without going into a semantic debate about this concept, we compare it here with the concept of the common heritage of mankind, which existed—without ever being formalized—before national sovereignty was even recognized. Unlike the common heritage of humanity, which presupposes that these resources belong to no one, the notion of global public good refers to the fact that these resources actually belong to everyone. The direct consequence is that in the absence of an international government, it is incumbent on the international community to define the rules of access and use of these resources and, if necessary, to conserve them. It may seem that this concept goes somewhat against the idea of national sovereignty, but that is not true. One has only to think of the current discussions on genetic resources beyond national jurisdictions (i.e., genetic resources of the oceans), or of a case of particular interest here, the resources that come under the aegis of the International Treaty on Plant Genetic Resources for Food and Agriculture (ITPGRFA). States which have signed this treaty exercise their sovereignty over these resources to create a common pool of genetic resources that comply with common rules agreed to by all the member States. A form of transfer of sovereignty to the international level takes place through this decision. The treaty enshrines the common and shared responsibility of States to manage such resources[4] grouped within the multilateral system, mainly because of the strong interdependence that binds them to these resources.

Practices that ensure open access to plant material remain, for various reasons, linked to the specificities of PGRFA, deeply rooted in the practices of the community of breeders and agronomists. An increasing number of claims of sovereignty over PGRFA are being put forward, mainly by delegates in CBD negotiating forums (and often originating from their Ministry of the Environment or that of Foreign Affairs). We note that, faced with these claims, the international agricultural research centres (IARCs) of the GCRAI succeeded *in extremis* in inserting a non-retroactivity clause to the CBD so that it does not apply to collections established before 1992. The consequence of this clause is that pre-1992 collections do not have a clearly defined status. This was not a problem earlier but given the post-CBD context, the issue had to be addressed. The solution arrived at was to place these collections under the auspices of FAO on behalf of the international community. This decision was embodied in the trust agreement of 1994, which was not only confirmed by the ITPGRFA, but also extended to all national collections under the control of the member countries of the treaty. Signatory member States will commit to the multilateral system established by the International Treaty and will put the material they have sovereignty over (i.e., under their control and in the public domain) at the disposal of the international community.

This multilateral system can thus be seen as the most successful expression of a collective management of common resources through a system of sharing of genetic resources at the international level (Louafi 2012). It offers two immense advantages: first, in the context of a legal vacuum on the status of the collections held by IARCs, it accords them a legal status, one that, most importantly, does not block their movement. And, secondly, it expands the logic of easy access to all national collections of signatory countries in a context of tensions and strong sovereignist claims. Without this system, the international research network which, *de facto,* freely exchanges material internationally from these collections

---

[4] The ITPGRFA multilateral system covers a list of 64 species considered to be of major agronomic importance and essential to global food security. These resources are listed in Annex 1 of the ITPGRFA. See http://www.planttreaty.org/content/crops-and-forages-annex-1 (retrieved: 9 May 2013).

(consisting of samples acquired from around the world) could not have existed. It would have been impossible to regulate the conditions of access according to specific rules of each country of origin.

2.1.2 Private Good

The ability to individually appropriate genetic material is a recent development. It emerged at the beginning of the twentieth century when a new professional category—the plant breeders—began investing increasing amounts of money for breeding and looked for ways to recoup their investments (Brahy and Louafi 2004). Variations within a variety and their self-replicating character make this a complex issue. While technology provided some solutions—in particular through hybrid varieties—it was the creation in 1961 of a Plant Variety Protection Certificate (PVPC) by UPOV which allows breeders to enjoy a monopoly of 15 years over a plant variety. Four conditions must be met to benefit from this protection: novelty, distinctness, uniformity and stability. This protection, however, applies only to the variety (and not to the genes that compose it) and only for direct use for agricultural production. The material developed is, in fact, freely available for further improvement (research exemption). It can also be used from 1 year to another—at least its 1961 version can—by farmers for their own seed requirements (farmer's privilege).

The concept of protecting plant material through intellectual property rights got a fresh boost starting in the early 1980s in the United States with the recognition of the patentability of living organisms. This concept quickly spread to other industrialized countries (Europe and Japan) and subsequently became universal through the Agreement on Trade-related Aspects of Intellectual Property Rights (TRIPS Agreement) included in the Marrakesh Agreement of 1994. Here too, the concept of genetic resources as private property may seem to run counter to national sovereignty. But in reality, each States is free to define the scope of its sovereignty as it sees fit, through its national legislation on intellectual property.

2.1.3 National Public Good

National sovereignty applies quite 'naturally' to the material directly under the control and management of governmental organizations. This view is widely shared within the community of conservators since they are strongly conscious of the public nature of their efforts which often depend on resources provided by governments.

2.1.4 Common Good

Beyond the dichotomy between private good and public good (and the particular case intentionally restricted to 'global public good' as described above) exists a fourth category, that of the common good, which is often overlooked even though it is the most prevalent. This category includes genetic resources which are communally managed by human groups governed by their own rules and procedures. These groups may be composed of a combination of non-State collective actors, State actors and private actors. Their coordination takes a hybrid form of private and public regulation.[5] Often associated with locally managed genetic resources, these hybrid forms of regulation are now coming up increasingly at a global level. The development of information-technology communication tools and progress in conservation biology and in genomics have indeed made possible a distributed management of genetic resources and their associated information at the global level (Parry 2004).

2.2 An International Framework Unrepresentative of these Different Conceptions

Despite this diversity of conceptions, there are only two main approaches that currently dominate in contemporary debates on the governance of cultivated biodiversity and that have a direct influence on the solutions developed at the international level.

The first approach—which can be described as an incentive-based approach—is based on rational choice theory. It relies on monetary incentives and a contractual approach to regulate exchanges. The second, at the other extreme, can be best described as a hierarchical or command-and-control model. It aims to use the law to regulate the behaviour of different actors and to ensure compliance with their obligations.

We find combinations and juxtapositions of these two approaches in mechanisms for regulating access to genetic diversity and sharing of benefits arising from their use within the Convention on Biological Diversity.

As for the framework for intellectual property rights, it is directly inspired by the incentive-based approach, with the aim of fostering biodiversity-based innovation.

---

[5] This concept of genetic resources as a common good is particularly relevant in the case of genetic resources for agriculture and food because of several reasons described in detail by Schloen et al. (2011).

2.2.1 The Shortcomings of the Rules for Access and Benefit-Sharing, and Improvements Envisaged in the Agricultural Sector

The Shortcomings

The access to and exchange of genetic resources are regulated through benefit-sharing agreements between a provider and a recipient in the form of bilateral contracts under private law. This approach is grounded in Coase's theory of externalities (Coase 1974). The market does not take the value of diversity for individuals and society into account and, at the same time, no one can easily be denied its use (so there is no incentive for an individual to make it available at his own expense). Consequently, the allocation of ownership rights between private parties through a negotiated contract is seen as an effective way to better reflect the value of genetic diversity, since these contracts formalize direct or indirect monetary incentives for sharing the benefits.

But the high degree of uncertainty of this value at the time of accessing genetic resources, coupled with the lack of legal security in case one party deviates from the contract, has led to a movement to go beyond a strictly contractual approach. Thus, these contracts are embedded in a wider set of agreements or mechanisms to limit opportunistic behaviour (Dedeurwaerdere 2004). They are made subject to national legislation to regulate by law these contractual practices. These regulatory mechanisms can take many forms: for example, mandating the use of standardized contracts or terms; the establishment of mechanisms for monitoring and compliance (such as disclosure of origins in patents); or the need to obtain consent of actors who are not parties to the exchange in order to exercise sovereign rights of States over these resources or to protect the rights of local and indigenous people.

However, it has been shown that the contractual approach to access and benefit-sharing, even when framed in this way, is inadequate to achieve social (equitable) and environmental objectives (Dedeurwaerdere 2004; Goëschl and Swanson 2002). Even when subject to legal (hierarchical) public regulation, monetary incentives defined in these contracts do not take into account the diversity and complexity of motivations at work in the exchange of genetic resources, which, in most cases, go beyond sole monetary considerations. They do not meet the needs of all user communities involved and remain, therefore, limited to a certain category of users and to uses amenable to such incentives.

The exchange of genetic resources is, in fact, subject to a more complex set of motivations. They include, for example, societal motivations (global public goals such as increasing knowledge, conservation of biodiversity and the fight against hunger) and, more prosaically, non-monetary social motivations (reputation, reciprocity). Indeed, it has been shown that reputation (as judged by the quality of material and information exchanged, or by publications) and reciprocity (exchange of information expected in return) figure prominently in the motivations of conservation and exchange of genetic resources (Dedeurwaerdere et al. 2012).

Moreover, even assuming that monetary incentives alone work as expected, they will never generate sufficient investment to conserve and exchange genetic resources because a large proportion of these resources are and will long remain of unknown value.

Finally, in some cases, it might even be counterproductive to rely on monetary incentives for all types of exchanges. Recourse to market value may lead to a reduced desire to contribute to the collective effort in the communities concerned, mainly due to mistrust and suspicion generated by a monetary angle where one did not previously exist (crowding-out effect, as described by Frey and Jegen (2001)). In other words, a contractual approach may undermine the cooperative practices necessary for conservation.

Some Solutions

Given the inadequacies of this Access and Benefit Sharing framework, the PGRFA sector has endeavoured to develop alternative strategies. These have consisted of amending existing solutions or developing completely new ones, more tailored to these resources' specific nature and to their forms of use in R&D processes.

The new approaches have tried to move away from the bilateral contractual approach for regulating the exchange of genetic resources. The specific attributes of these resources (man-made diversity, the importance of intraspecific diversity for breeding, difficulty in determining their origin, high interdependence between countries, constant need for new variations, importance for food security) argue for the establishment of a more collective way of managing access to these resources and for sharing their benefits. The International Treaty, with its multilateral system of access and benefit-sharing, is the most successful example of this approach of pooling resources. Such an approach reduces transaction costs for access to genetic diversity in ex situ banks and also reduces redistribution costs by detaching the sharing of benefits from individual providers. And, finally, it emphasizes the non-monetary aspects of the benefits generated that could take place irrespective of whether a product is put on the market or not.

The mechanism which allows this pooling of resources remains contractual in nature—the standard material-transfer agreement—but rather than being bilateral, it works at a common international level, oscillating between a global public good and a global commons (see Halewood et al. (2012), for a detailed discussion of the distinction between them). This collective logic remains, however, compatible with the vision of genetic resources as a private good. Genetic resources held privately are free to be made part of the multilateral system. Moreover, private ownership of genetic resources originating with the multilateral system (via a patent) is made possible on payment of a tax on revenues generated by this appropriation. This tax is a sanctioned break with the collectively approved logic of access. It goes into a collective fund also managed by all contracting parties to the treaty on the basis of allocations they have agreed upon. This mechanism thus forms the monetary aspect of the sharing of benefits.

But the treaty does not stop at this multilateral system, which remains essentially a tool designed for and dedicated to ex situ genetic resources. Its Article 9 deals with farmers' rights and, as restrictive as this article is, it does recognize the legitimacy of an appropriation of resources that is neither a private good nor a public good (national or international) but which is a common good shared by a group of local actors, the farmers themselves.

This right is difficult to implement and, except for some local initiatives, it is not often supported by States (Andersen 2008). Nevertheless, the fact remains that, as of now, the treaty remains the only instrument to offer this pluralistic framework that recognizes the legitimacy (albeit with large differences in effective implementation) of competing concepts on the status of genetic resources.

However, the equilibrium brought about by the treaty remains imperfect and fragile. The treaty's various elements are implemented at different rates, and sensitivities remain high among stakeholders about their fair implementation. While the facilitated access to genetic resources that the treaty makes possible is crucial to the agricultural and food sector, one of the main perceived inequities is the fact that all countries and stakeholders cannot benefit from it in the same way. The poorest stakeholders, and more generally developing countries, lack research capabilities (particularly in the domain of plant breeding). They usually do not have the means and capacity to benefit from the pooled genetic resources. They perceive, rightly or wrongly, that the sustained and exclusive importance accorded to ex situ conservation primarily serves the interests of developed countries and the richer stakeholders.

Apart from the International Treaty, we note that discussions are currently on under the aegis of the Commission on PGRFA of FAO concerning the specific features of all genetic resources for food and agriculture in all three kingdoms (animal, plant and microbial). These discussions are calling for the implementation of solutions different from the conventional ones considered in the Nagoya Protocol to the Convention on Biological Diversity. Efforts to systematize these features as well as to identify current PGRFA exchange practices are underway (Schloen et al. 2011; Chiarolla et al. 2012).

2.3 The Shortcomings of the Intellectual Property Framework for Genetic Resources and Envisaged Improvements

The Shortcomings

The framework of intellectual property rights over living organisms is also based on monetary incentives designed to promote biological innovations. By providing protection for inventions based on genetic diversity, intellectual property rights are supposed to encourage its use and therefore its conservation. As mentioned in the section on the different conceptions of the status of genetic resources (see above), the agricultural sector is characterized by the coexistence of at least two systems of intellectual property: the patent system and the system of Plant Variety Protection (PVP). Both systems are regulated internationally by two agreements respectively: Trade Related Aspects of Intellectual Property Rights

(TRIPS) of the World Trade Organization (WTO) and UPOV, affiliated with the World Intellectual Property Organization (WIPO).

The UPOV system, *sui generis,* is the one that applies more to the self-replicating and evolving nature of plant material. Since the product that originates from innovation (variety) is itself a genetic resource, a balance must be struck between protecting the innovation and allowing access to genetic resources. This balance is achieved within UPOV through a specific exemption that allows the subsequent use of a PVP-protected innovation for research purposes.

The UPOV system also provides greater legal security than the patent system. Whereas the same product can be the subject of several patents, a variety is protected by a single PVP (Dutfield 2011). Situations of 'patent thickets' where patents depend on other patents (Shapiro 2000; Heller and Eisenberg 1998) or of 'patent hold-ups' involving involuntary patent violations can result in a higher number of cases covered in the patent system than with a PVP.

In practice, however, except in a limited number of countries (including the United States), the patent system is not used to protect varieties themselves but is used for biotechnological inventions such as processes or genetic sequences.

But, beyond the relative technical merits of these two systems, the intellectual property system itself raises fundamental issues which have been discussed extensively in the literature. One of the main limitations identified for conservation and the use of genetic diversity is the fact that this mechanism only comes into play at the end of the value chain. It therefore works effectively only for improved material or for material whose value is already known, even partially (through characterization data or available evaluation). It is therefore far from providing sufficient incentives to share the vast majority of genetic diversity found ex situ—even less for in situ—especially if the value is unknown at the time of access (Swanson and Goëschl 2000; Goëschl and Swanson 2002).

In addition, the incentive system works poorly for innovating in orphan domains (for which effective demand does not exist). It also does not serve the interests of countries that do not undertake innovation activities and therefore are not eligible for benefits accruing from the protection of intellectual property. Finally, in the same way as for the effects of exclusion (crowding-out) previously described for mechanisms for access and benefit-sharing, the introduction of economic incentives may affect exchanges of material or information previously considered common, with holders of this material or information wanting to leverage it themselves. These so-called anti-commons situations lead to an intrusion of intellectual property in areas where it has no reason to be by modifying existing cooperative and altruistic behaviour (Heller and Eisenberg 1998; Cassier 2002).

All these problems are even more pronounced in the agricultural sector. It seems that innovation in this sector is more a question of research coordination between many actors, with different statuses, than a matter of individual incentives alone.

While the private seed sector is able to work well on the basis of these incentives, it also depends on the exploration of genetic diversity by public research.

The monetary incentives in this sector do not constitute all the motivations at work for the exchange and use of genetic diversity (see Sect. 3 below: 'Need for in situ conservation and complementarities with ex situ conservation'). Similarly, those who defend community rights and rights over traditional knowledge highlight the existence of collective rights for access, exchange and use of seeds and genetic resources that are not reducible to individual rights as defined by intellectual property mechanisms.

Some Solutions

As in the case of Access and Benefit Sharing regulations, various strategies have been attempted to overcome the limitations of the existing global intellectual property frameworks.

A first category of solutions aims to amend the existing framework of intellectual property rights by using existing flexibilities. These solutions attempt to implement collective approaches better suited to certain uses and innovation of plant material. This is the case, for example, of the establishment of information clearing houses in order to make information more transparent and available from applications for intellectual property rights, especially patents. A good example is Patent Lens, an initiative of the Australian organization, Cambia, to help make intellectual property rights more transparent. This set of tools has already proven itself in 'patent landscape' analyses of rice gene promoters. Using genetic sequencing data, the data contained in patent applications are linked to data on the appropriation of genetic resources, to data on the state of the art and to data on the varieties. In promoting free access to this newly generated information, Cambia's initiative bolsters the ability of users to navigate the patent system. It also helps make better use of the available (but dispersed) information with the aim of disseminating existing technologies and of identifying neglected areas of research with greater accuracy.[6]

Patent pools are another type of mechanism in this category. This term refers to the collective management of a patent portfolio by a group of actors (often from public research) with the aim of increasing their negotiating skills against private monopolies and to identify complementarities within their respective portfolios. The Epipagri initiative,[7] supported by the European Commission, is based on this principle. It helps member institutions manage their intellectual property capital more cooperatively.

Other cooperative initiatives include collaborative research in pre-competitive research phases where common rules are established for the sharing of information and research materials, sharing of new databases and immediate sharing of research outcomes. The Apomixis consortium serves as a good example of this type of initiative. It was established as a public/private partnership between two

---

[6] See http://www.cambia.org/daisy/cambia/2458.html (retrieved: 10 May 2013).

[7] The report of the Epipagri initiative, supported by the EU through its Sixth Framework Programme, is available at http://cordis.europa.eu/search/index.cfm?fuseaction=result.document&RS_LANG=EN&RS_RCN=12437527&q (retrieved: 10 May 2013).

public centres (CIMMYT and IRD) and three private companies (Pioneer Hi Bred, Limagrain and Syngenta) to identify and characterize the components required for apomixis in maize. A system of non-exclusive licensing for this research's products allows a segmented and differentiated management that meets the needs of return on investment of private companies in profitable markets as well as the need of distributing these products to small farmers in developing countries.

Other systems currently under development offer themselves as alternatives to existing intellectual property systems. They operate through the concept of open (but codified) access to material and associated knowledge.

This is the case with Creative Commons[8] or Science Commons[9] copyright systems which aim to restrict use to only non-commercial purposes (and therefore not patentable by anyone) through the use of standardized contracts for access to published material.

As an extension of this idea, attempts to license free and open access to seeds have been discussed (Aoki 2009; Beck 2010). The idea is to reverse the logic behind monopoly rights granted to the inventor in the system of intellectual property rights in order to promote free and open access while maintaining the legal restrictions that prevent others from obtaining monopoly rights over this free material. Unlike the concept of the public domain, the open access licensing system originates from private ownership (Beck 2010). It is thus a defensive mechanism to guard against misappropriation. Applied to seeds, the system consists of contracts for exchanges between members of this open access system to increase the pool of material available under these conditions. The most practical contractual system for open access licensing is the one called the shrink-wrap system. In this system, opening of a packet of seeds indicates acceptance of the conditions of the open access agreement. Implementations of such a system are still in their infancy and face significant practical difficulties.

Finally, it is worth noting the development of solutions that are complementary to the logic of appropriation, called stewardship approaches (as opposed to ownership approaches discussed above). Described more in detail in Sect. 3, stewardship approaches are intended to recognize the role of the farmers as managers of genetic diversity and to create the appropriate legal mechanisms—in addition and 'compensatory' to intellectual property—so that farmers can continue to play this role. Mechanisms of empowerment such as participatory plant breeding which associates farmers upstream of research processes with breeding activities, those for providing support to local conservation community programmes or those for protection and valorisation of farm seeds form part of these approaches.

---

[8] See http://creativecommons.org/ (retrieved: 10 May 2013).

[9] See http://sciencecommons.org/about/ (retrieved: 10 May 2013).

3 Need for In Situ Conservation and Complementarities with Ex Situ Conservation

Even though the issue of the role of farmers in the conservation of genetic resources is not new (Brush 1989), it is not taken into account in conservation programmes to any great degree. Farmers have long evaluated and appropriated genetic diversity and the heritability of traits in particular. They have always experimented, crossbred, selected and made use of observed differences for the next generation in terms of the variability they have seen within and between cultivated plots of the previous generation. The diversity of farming systems and management methods provides a wide scope for experiments for the creation of variability and the use of new cultivated biodiversities.

Agroecosystems cover 30 % of the Earth's surface (Altieri 1992). They encompass a wide range of situations involved in the creation of biodiversity in its broadest sense, agrobiodiversity and cultivated biodiversity in particular. The most visible portion of the latter is meant mainly for our food, whether animal or vegetable. It hides a whole section of agrobiodiversity consisting of soil micro-organisms, pollinators, pests and diseases, and many other elements of biodiversity essential for their contributions to regulating ecosystem services and for supporting agriculture.

Altieri's (1987) studies show how the diversity of family farming contributes to the conservation not only of crops but also of their wild relatives. The maintenance of traditional farming systems, inserted into adjacent ecosystems, enhances the flows between systems to enable continuous evolution of cultivated varieties (Collins and Qualset 1999). These studies show that a successful in situ conservation cannot be considered in isolation from the society whose practices have made this diversity possible (Jarvis et al. 2007). It is important to be able to link local knowledge, requirements of agricultural production and the need to conserve genetic resources in a holistic way. If the dimension of cultural diversity is ignored, it will not be possible to grasp the dynamics of genetic diversity in their entirety.

Even if we attempt to manage agrobiodiversity by limiting ourselves to in situ genetic diversity, we will find it necessary to consider both its conservation and its uses, since the latter promotes the former. Characterization of in situ agrobiodiversity actually raises the question of the type of information associated with the direct management of genetic diversity and also of the actions derived from biodiversity that will shape future progress. That is why it is customary to classify species into three categories according to their functions in the agroecosystem (Wale et al. 2011):

- plant species sowed or planted intentionally to harvest food, fibres or wood, or for decorative purposes, etc.;
- the crop wild relatives with which the can interbred. These constitute the gene pool associated with crops and can evolve independently, exchange pests and

diseases with the cultivated space, and sometimes even be sources of food during famines;
- wildlife in the agricultural environment which interacts with the system of agricultural production by providing various regulation services.

We will restrict our focus to the first category by concentrating primarily on annual crop varieties maintained by farmers.

The enormous genetic diversity contained in traditional varieties (landraces) is the most accessible part of global biodiversity and one that can be economically leveraged directly. The use of this diversity of traditional varieties in livelihood strategies is key to adapting farming systems to socio-economic and/or environmental changes for many farmers around the world (Jackson et al. 2007). It is difficult to estimate its magnitude with any precision given the lack of statistics on informal and subsistence agriculture systems. However, according to Francis (1986), 'these farmers use about 60 % of agricultural land and provide 15–20 % of the world's food supply.'

Moreover, agrobiodiverstiy provides benefits other than through its direct use in production systems of family farms: traditional varieties constitute the basic material used by all plant breeders to develop new improved varieties. Indeed, ex situ seed bank collections represent only part of what existed in situ at some point in farmers' fields, i.e., the part that could be collected, conserved and is still alive. Unless major changes are made in the global organization of agriculture, the evolution of agricultural production and the future food supply of a majority of the planet's inhabitants will depend mainly on this ex situ resource. Improved seeds are provided by a decreasing number of multinational seed companies (now fewer than ten major ones worldwide) who work from their own collections. This situation has led and is leading to a reduction in the genetic diversity of cultivated varieties. Their adaptive capacities are thus becoming constrained and this in a time when major environmental changes are expected. This genetic base can only be enriched with a return to the farmers' fields or to places where wild crop relatives are found.

Beyond the technical issues related specifically to the conservation of collections is the question of the sampling required to create them in the first place. Indeed, even with a good representation of the diversity of crops and associated varieties in a region of interest, it is extremely difficult to get hold of so-called minor varieties (grown by few farmers and over small areas) without first having established a close relationship with farmers in the area and the human societies they belong to. Therefore, in order to obtain samples of these minor varieties, one would have to first build relationships of trust in all villages to be surveyed. Needless to say, this is extremely difficult and impractical to do. Moreover, it is not a move that can be defended ethically unless a real advantage can accrue to these populations. Since the objective of the sampling is the conservation of varieties in order to avoid genetic erosion, it is important to include farmers' varieties which are at the greatest risk of disappearance, i.e., the minor varieties, in the material collected. Their disappearance would automatically entail the loss of an unknown part of the genetic diversity of the species.

The immense size of the in situ diversity which farmers manage is therefore almost impossible to characterize for various reasons. On the one hand, this traditional varietal diversity is based on a high diversity of individus considered as populations in ecology, and described as farmers' varieties through our representation; it is therefore not so much a farmer's variety that is of interest but rather the gene pool consisting of all varieties of the village and its structuring at different scales, from the plot to the country or the biome (Sagnard et al. 2008). On the other hand, depending on whether we are dealing with cross-pollinated or self-pollinated plants, the structure of genetic diversity will be of interest at very different scales. For example in the case of sorghum, predominantly self-pollinated (>70 %), the intra-village diversity between varieties will be very high. Expanding the scale to that of the region will only add very limited genetic diversity (see Box 1). In the case of pearl-millet, a predominantly cross-pollinated species, the genetic structure at the two scales mentioned above will be exactly the opposite. We will find the varieties in the village to be very close whereas there will be big differences across the region.

Box 1. Dynamics of the Evolution of Varieties of Cultivated Sorghum in Mali over the Medium Term. How to Fight Entrenched Ideas About the Genetic Erosion of Farmer Varieties

The change in environmental conditions is one of the most common reasons put forward for varietal changes. That is why it is not uncommon to hear that the drought in West Africa has led farmers to adopt "modern" sorghum varieties that can be harvested earlier than traditional cultivars.

To analyze this assumption as a hypothesis, varieties from two surveys of local cultivars conducted in Mali 20 years apart were studied. The first batch of 472 varieties was from 188 villages surveyed in 1978 by the Institute of Research for Development (IRD, France). The second batch of 275 varieties was collected during a survey carried out between 1999 and 2000 by the Institute of Rural Economy (IER, Mali) in 46 villages in two north–south transects. Sampling conditions for both surveys were comparable, including the choice of the sampling date, selected so that both early and late cultivars could be collected in a single pass.

In order to compare recent developments in Malian sorghum cycles, the phenology of varieties from these two collections of local varieties surveyed 20 years apart was studied. The rainy season was characterized through its start and end dates for each village sampled. Then the photoperiodic sensitivity of the cultivars was measured using a test at an agricultural station with two sowing date. A model was then used to study the adaptation of the varieties to the climate taking into account the latitude of and rainfall pattern in their areas of origin.

The results show that the observed rainfall deficit has not resulted in a significant shortening of growing seasons. In 20 years, the average cycle of local cultivars has shortened by just 5 days. For latitudes lower than 14°N, the vast majority of cultivars are photoperiodic and varieties bloom in the 20-day period leading up to the average date of the end of the rainy season. This characteristic helps optimize watering of crops and avoids many biotic limitations. At latitudes of 14°N or further north, the average blooming coincides with the end of the season. We note the simultaneous presence of both early and late varieties. In these regions, sorghum farming is less dependent on rainfall as traditional systems leverage diversified situations and make efficient use of water on different toposequences. This diversity of cycles helps to strengthen agricultural production in arid zones.

If correct, the hypothesis of climatic deterioration should have shown greater erosion of sorghum varieties in northern Mali (Sahel) than in the south (Sudano-Guinean zone). On the contrary, the results show that the disappearance of varieties is much lower in the north (−25 %) than in the south (−60 %). Sorghum's varietal diversity is a factor for increasing its hardiness in the face of difficult and uncertain climatic conditions. Archival data on Malian agriculture provide information on agricultural production systems over the past 20 years. It then becomes possible to analyze how changes in variables of the production system correlate to the loss of bio-diversity of sorghum. The results obtained for the expansion of cultivated areas in Mali show a progressive dominance of maize in crop rotations in the south, despite the surface area devoted to sorghum remaining constant. In conclusion, changes in cropping systems are primarily related to the expansion of cotton cultivation, followed by maize in southern Mali with the development of animal traction and easier access to inputs. Improved technical skills of farmers has helped boost their farms' performance and therefore to change their production goals. Currently, demand is increasing for intensification of cotton and maize farming, which should further con-tribute to the disappearance of other sorghum varieties, for which climate change may again be blamed.

For further information: Kouressy et al. (2003, 2008); Soumaré et al. (2008).

Cultivated biodiversity in the farmer's field results from the differences observed by the farmer between plots or between individuals within a plot. These differences are based on plant morphology, productivity, grain quality, phenology of plant development throughout the vegetative cycle, resistance to diseases and pests, as well as other more subtle criteria which are not apparent to a novice. Genetic studies of sorghum' populations conducted over the past few decades and the advent of molecular biology and high-throughput genotyping has shaken up this way of classifying living organisms in the field. New tools are now available to understand genetic diversity as a fundamental element in the evolution of biodiversity. And yet, these genetic studies only offer us states of genetic diversity from which we must reconstruct history, i.e., explain the dynamics of biodiversity. This is why cultural aspects that are reflected in the diversity of farming practices will always remain relevant. Indeed, these types of agricultural systems highlight the genotype × environment relationship on the field, where the farming practices should be considered part of the environment. This diversity of practices then brings to light distinct phenotypic expressions which are permitted by the genetic variability of open-pollinated varieties maintained by farmers.

The key to knowing the diversity cultivated by farmers is based on the vernacular denomination of their varieties. While this in itself is useful for understanding the farmers' management practices, it also constitutes a limitation in grasping their genetic diversity. The varieties are identified and classified using phenotypic criteria which allow the farmer to identify and separate batches of seeds according to the expression of these traits in the field. These criteria help the farmer characterize a traditional variety and distinguish it from others by naming it uniquely. Used in this way, these criteria are essentially agromorphological. The farmer uses the criteria visible on the field to, on the one hand, select his batches of seed to reproduce the variety year after year while maintaining its traits. On the other, he selects individuals within a population because they exhibit traits that differ from the ideotype of the reference variety and which may be potentially interesting to evolve.

Names of farmer varieties are not fixed and it is common to see the name of the same variety change as it is shared between farmers. Traceability and distinction become difficult when only batches of seeds are at hand. That is why the description of varieties by farmers should be based on distinguishing criteria to facilitate interaction with researchers. Even more importantly, this will make it possible to physically contrast pairs of varieties thus described to avoid naming two varieties by the same vernacular name or assigning the same name to two distinct varieties (Table 1).

Given that each variety has a life (exchanges between farmers, incorporation of new elements from other varieties, regression or introgression) in the village or small natural region, its characterization by the farmers themselves informs us about the key stages of its dissemination and exchange between farmers. This

Table 1 Distribution of varietal diversity in Malian farms. Case of sorghum, 34 villages, 1,474 farms

Number of varieties	Number of farms	Percentage
0	150	10.18
1	1032	70.01
2	247	16.76
3	34	2.31
4	6	0.41
5	4	0.27
6	1	0.07

characterization often incorporates the social aspects of the organization and structure of the village's population or the cultural aspects of the time and occasion of the exchange which often go beyond the mere biological characterization of the farmer variety (Box 2). The entirety of this cultural diversity associated with the cultivated diversity must be taken into account if we want to understand the evolutionary dynamics at work.

> **Box 2. Social Organization of Farmers, Diversity and Adaptability of Cultivated Plants**
>
> Christian Leclerc
>
> It is necessary to understand the factors behind the structuring of the diversity of in situ genetic resources in order to define sampling strategies or the most appropriate conservation approaches. Of these factors, anthropological ones are still the least understood. The results from the 'Reproducing plants, reproducing a society' project, funded and carried out by CIRAD on the slopes of Mount Kenya, show that the social organization of Meru farmers, with its rules of marriage, residence and inheritance, contributes to the organization of the genetic diversity of sorghum and to its adaptability to contrasting environments.
>
> In Meru society, as in many others, farmers belong to lineages within which marriage is prohibited: marriage is only allowed with members of other lineages. After her wedding, the new wife moves to her husband's village. To start agriculture in this new environment, she inherit seeds from her mother-in-law, who herself had moved to the village when she got married. The diversity of varieties is thus transmitted from generation to generation along matrilineal lines (Fig. 2). This practice is particularly well suited to the steep slopes of Mount Kenya, where growing conditions depend on the altitude. Any movement of seeds along the slope would lead to adverse consequences: varieties adapted to high altitude would suffer at low altitudes, and vice versa.
>
> Instead, the rule of patrilocal residence and inheritance of seeds along matrilineal lines tends to conserve genetic resources at the same altitude by favouring its specialization in a given environment. In addition, by limiting the flow of seeds, it increases the genetic diversity of varieties in the region. This is the result of a triple interaction between genetic (G), environmental (E) and social (S) factors, according to a $G \times E \times S$ model where the social component is explicated.
>
> **For further information:** Leclerc (2009); Leclerc and Coppens d'Eeckenbrugge (2012); Project Arcad (Crop Biodiversity Research and Resource Center) http://www.arcad-project.org (retrieved: 12 May 2013).

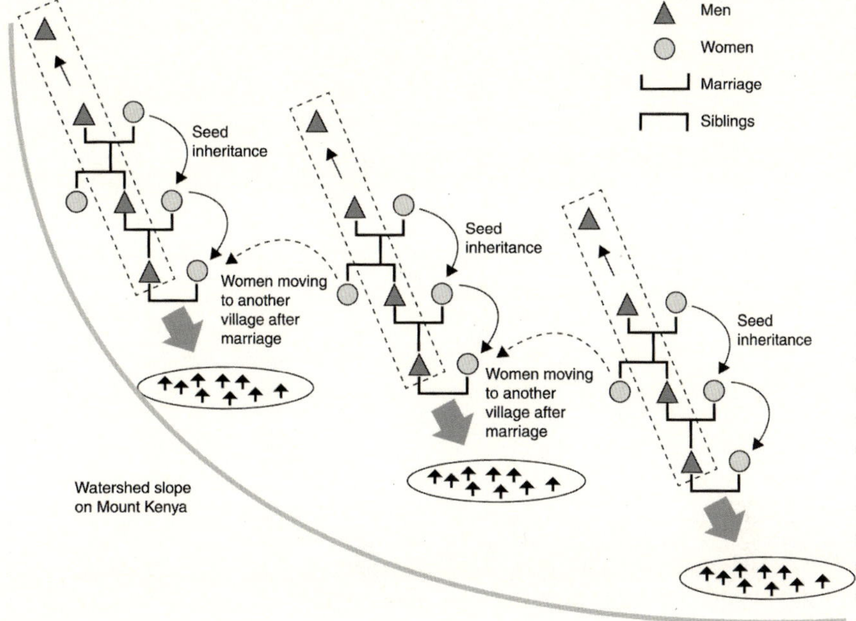

Fig. 2 Groups of patrilocal residences on the slope of Mount Kenya

After marriage, women join the villages of their husbands and start farming there. They inherit seeds from their mothers-in-law. The seeds are thus transmitted from mothers-in-law to daughters-in-law over the generations, and varieties are therefore kept at the same altitude, favouring their adaptation.

The way society is structured affects the genetic diversity of cultivated species because of exchanges which form part of specific rituals or worship. These crop diversity dynamics are also based to a great extent on the organization of work in the community and within family groups, facilitating or hindering the movement of varieties (Fig. 3).

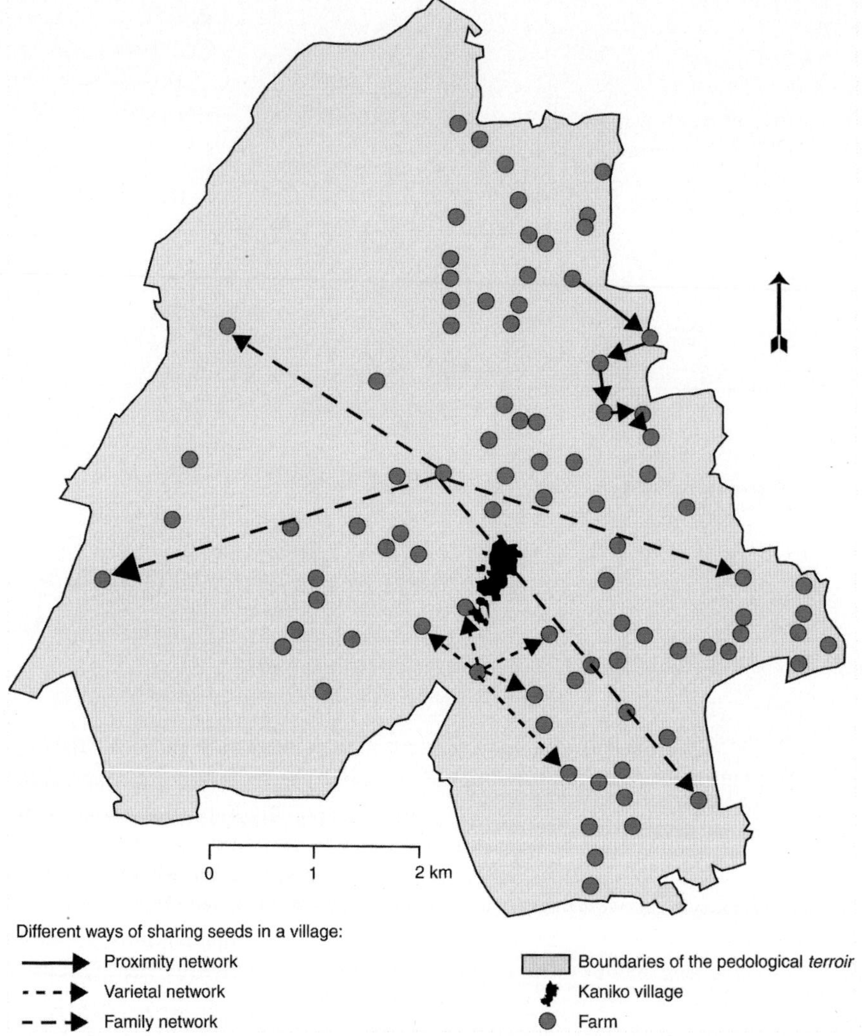

Fig. 3 Social seed-exchange networks in farmer environments. Case of sorghum in Mali (according to Bazile (2006))

Georeferenced surveys conducted over several years helped identify social networks based on the exchange of seeds. We can distinguish between:

- proximity networks (1). The farmer sows what he may find in his immediate proximity; 90 % of introductions of varieties come from neighbours' fields. Varieties spread in a snowball effect;
- varietal networks (2). The variety is defined by its adaptation to a particular ecological context: adaptation to a soil type, drought tolerance, etc. To obtain

the seed of this variety, the farmer who produces the best seed in the *terroir* (in the pedological sense of the word) will be approached;
- family networks (3). The allocation of labour between family members for certain agricultural tasks and phases when the family comes together for other agricultural activities contribute to the movement of varieties between *terroirs* within members of the same family. This provides an opportunity for farmers to try out the same variety under different ecological conditions and agricultural practices.

These types of flows of varieties highlight the importance of the concept of free sharing and of the dependence on family solidarity in traditional agriculture.

To be able to fully describe the varietal diversity which farmers hold, it is essential to link various scales to one another. At the farm level, the farmer manages a palette of varieties of one or more species within his cropping system in order to best leverage the diversity of the soils of his plots (Bazile et al. 2008) and to respond to his various objectives: food security, monetary income, etc. Nevertheless, in managing the farm, the farmer is constrained by economic viability. Therefore, few farmers grow a very large number of varieties of the same species since this generates additional labour costs. (This refers to the classical notion of economies of scale for labour productivity). A farmer always tries to optimize this ratio through the number of varieties sowed. This constraint at the individual level ensures that the average number of varieties grown for a given species by a farmer often remains low. Only a few farmers grow a large number of varieties and have knowledge of all the varieties existing in the village or in the surrounding small natural region. For example, in Mali, a very high number of farmer varieties of sorghum were identified in surveys undertaken by IER-CIRAD. Yet, a comprehensive study of 34 villages (1,474 farms) shows that 70 % of farmers sow only one variety annually, while the total number of sorghum varieties in a village ranges from 6 to 12, after accounting for synonyms[10] (Table 1). The fact that the majority of the farmers surveyed grow only one or two varieties in a year shows the importance of the community, for it is at this level that the genetic diversity of cultivated species actually exists—through a shared access to genetic resources that are subject to traditional social rules.

This echoes the point made above of farmer exchanges taking place within geographically or socially defined networks. Thus, farmers do not individually have the village's varietal diversity in their own fields because of this diversity is rather shared between all the farmers in a village. In addition, each farmer has limited knowledge, estimated at about 30 %, of the existence and the availability of all the varieties at the village level. Free sharing and village solidarity ensures that they have full access, at least in theory, to all the plant material that they know about or

---

[10] Field verification with complete plot mapping was conducted in seven villages. It showed that the methodology of the survey was valid.

are able to ask for by specifying the desired characteristics. This characteristic validates the fact that farmers tend to obtain new varieties from their neighboors.

The farmer seed system thus contributes, through non-market exchanges and solidarity, to the exchanges of genetic resources between farmers. These fluxes have a positive effect on in situ conservation of crop biodiversity. Since the 1930s, agronomists highlighted the risks associated with the expansion of monoculture areas sown with uniform cultivars (Marshall 1977). A homogeneous plant stand is more vulnerable to epidemics and outbreaks of pests and diseases. The vulnerability of the genetic composition of the plant population is then estimated only from an ecological point of view by the probability of damage, which increases in such cases. The example of the great famines of the nineteenth century in Ireland (Shumann 1991) following the destruction of its entire potato crop by a pathogen *(Phytophthora infestans)* is a classic case study. This destruction occurred due to the genetic uniformity of the cultivated species. The solutions advanced by agronomists for controlling diseases and pests can be divided into two types:

- the first relies on the contribution of biotechnology to find genes for resistance or tolerance, without changing the model of genetic 'poverty' of the plant stand. In this case, if another pathogen starts creating a problem, the same solution is repeated;
- the second takes biodiversity into account in the management of epidemic risk. The work of Zhu et al. (2000) shows how a combination of different varieties of rice facilitates disease control and offers increased resistance to pathogens.

Analyses of the contribution of biodiversity in improving agricultural production or its maintenance and stability, irrespective of the environmental conditions, remain poorly documented at the scientific level. Farmers in the tropics, though, have been using biodiversity since time immemorial on empirical bases. Indeed, using agrobiodiversity to reduce vulnerability and improve stability and risk management requires permanent changes of scales, from plots to landscapes, as illustrated in Box 3 (Jackson et al. 2007).

Box 3. Heterogeneity of Landscapes and In Situ Conservation of Biodiversity
Emmanuel Torquebiau

Agricultural activity is not limited by the boundaries of fields. It is, in fact, the result of a network of relationships with the environment that actually works at the landscape scale (Dale and Polasky 2007; Swinton et al. 2007). By taking this network of relationships into account, we are led to believe the principle that the satisfaction of human needs by agricultural production, on the one hand, and the conservation of biodiversity, on the other, are not necessarily antagonistic propositions (Robson 2007). Maintaining a landscape is tantamount to cultivating biodiversity thanks to the heterogeneity associated with landscape structures. Hedges, field margins, riverain forest strips, and drainage ditches are all part of the ecological infrastructure where elements of natural biodiversity and agrobiodiversity can be found. The juxtaposition of different land uses—fields, forests, pastures, protected or natural areas, etc.—helps limit certain deleterious effects nomally caused by the standardization of land use, such as erosion or spread of pathogens.

The science of landscape ecology has formalized the study of heterogeneous landscapes (Burel and Baudry 1999). A heterogeneous landscape is a mosaic of spatial units in contact with each other. The basic elements of the landscape structure are the matrix, corridor, network, patch, edge and central area. Such a landscape is analyzed in terms of diversity and spatial organization, complexity, adjacency and connectivity between units of the mosaic. It is possible to design an agricultural mosaic landscape with different roles assigned to neighbouring units, for example, production in a field, windbreak effect of a hedge, protection of rare species in a wooded area, area set aside for recreational or ornamental purposes. A natural corridor can provide connectivity between two protected areas separated by a cultivated area. In this way, this heterogeneous landscape becomes a multipurpose (or multifunctional) landscape, simultaneously capable of meeting production targets and protection objectives, most notably the conservation of biodiversity. The concept of 'ecoagriculture' was proposed to describe and manage these landscapes (Scherr and McNeely 2008), still present in the South, but threatened by the industrialization of agriculture. These landscapes can also provide some ecosystem services such as carbon sequestration, pollination or water circulation.

To encompass the landscape scale, agricultural research must reinvent some of its methods. What public policies will encourage collective action of stakeholders sharing a landscape? What ideotypes should be selected to make the most of the landscape structure? How do pathogens circulate in a landscape? Can irrigation practices be changed? What are the appropriate sizes of plots and farms? These are just some of the questions that an 'agronomy of heterogeneity' must answer.

After the ecological analysis of damage caused by a reduction of biodiversity, the next step is to undertake an economic analysis of any loss of production in terms of lost income for farmers. A multi-year assessment of the cropping system is necessary to judge the extent of damage to crops in the tropics. This risk is then incorporated into the farmer strategies and the practices subsequently adopted to manage biodiversity. Farm production can no longer be evaluated over an annual timeframe since the farmer has to assess the stability of his production in its full rotation (crop succession over several years) and he no longer seeks to optimize his production for any one given year. Instead he adopts a risk-reduction strategy to meet the recurring goal of food security for his family. He strives for the avoidance of crop failures and overall multi-year production stability. He can leverage a portfolio of varieties for each cultivated species and associations of cultivated species to reduce the pressure of pathogens on his plots by incorporating a gradient of sensitivity to various risks into his cropping system. Some studies also show that market integration can help increase biodiversity in production systems (Box 4).

> **Box 4. Changes in Biodiversity and Increased Market Exchanges: the Case of Sorghum and Millet in West Africa**
> Sandrine Dury, Maryon Vallaud and Harouna Coulibaly
>
> According to several authors, there is a loss in varieties of cultivated cereals in the world and in Africa in particular. This erosion of varietal diversity is related, in part, to the development of markets. Other authors, working in Niger, observe that the biodiversity of millet and sorghum has not experienced major erosion over the past 30 years. These latter authors advocate a passive management—a hands-off attitude—letting producers continue to conserve agricultural biodiversity as they have managed to do so far. The first case, however, obliges stakeholders to devise more proactive mechanisms to conserve and manage biodiversity in order to halt varietal erosion.
>
> A comparison was undertaken of the biodiversity of millet and sorghum in the villages and farms using surveys of 120 farmers in the two Malian villages selected to represent two contrasting situations in terms of market access.
>
> At the overall village level, the village that is more integrated into the market has a greater richness of variety. In this village, the farms generally undertake more varied activities and sell a larger share of their productions. In addition, we observe a change in the nature of species and cultivars cultivated. On the one hand, short-cycle sorghums, more flexible in terms of the work schedule, are more popular and better represented in crop rotations. On the other, millet is grown over larger areas because it is preferred by urban consumers, which allows it to be sold at slightly higher prices. Finally, a multi-factor regression highlighted that, all other things being equal, selling more cereals is accompanied by a lower biodiversity.
>
> Thus, changes that have occurred in other parts of the world can also take place in Africa. The increase in cereal sales may very well be accompanied by a reduction in the biodiversity of cultivated cereals, even if today we do not observe it at the village level where the diversity of farms helps maintain crop diversity.
>
> **For further information:** Vallaud (2011); Vallaud et al. (2011).

It is essential for the farmer to adapt to change and such a strategy to manage crop biodiversity makes this adaptation dynamic (Jackson et al. 2010). Unlike the initial responses from agronomists, who proposed varieties with resistance genes, the management of populations (or heterogeneous plant communities) by farmers allows different individuals in an annually grown population (or farmer variety) to be exposed to risks. The self-production of seeds by selecting individuals for the next generation (planted in the following year) identifies individuals which are the new type of variety desirable in terms of socio-economic-political and environmental changes.

This recent trend towards adaptation to change through traditional practices however shows its limitations. Reconciling scientific knowledge and farmer knowhow is often difficult, which is why the latter is not yet really considered reliable enough to be used by agronomists. Theoretical studies underpinning the concept of resilience today offer a new analytical framework to understand and analyze the process of adaptation and transformation of agriculture. Indeed, the recent analysis of Elfstrand et al. (2011) focuses on how 'farmer knowledge of cultivated biodiversity can contribute positively to the transformation of agricultural systems.' Through a theoretical analysis of the concept of resilience of socio-ecosystems, the authors offer a grid to better understand the role of agricultural biodiversity for food security, risk management and improvement of living conditions. Such an approach should lead to a better assessment of the constraints and of the possibilities of adapting to changes brought about by the different ways of managing agrobiodiversity. Indeed, this thinking opens up a new avenue of research on the extent to which this is recognized in decision making and policy formulation. Chevassus-au-Louis and Bazile (2008) speak of the human, social and cultural capital that must necessarily be integrated into knowledge exchange networks to build tomorrow's plant-breeding innovation. Participatory selection consists of working directly with farmers or farmer organizations (Box 3) to incorporate the criteria preserved in local varieties into the 'modern' varieties being improved through research. Farmer varieties are open-pollinated varieties (heterogeneity of individuals) which means they evolve continuously over time. Individuals resistant to changing environmental conditions are selected by the farmers to be the generation that will be planted the following year. Given the cropping conditions, co-evolution may also include some results of crosses with wild relatives growing in the vicinity of the cultivated plot. These dynamics of cultivated diversity lead to a reconsideration of plant breeding for high-stress environments subject to climatic hazards. A reflection on what constitutes an open-pollinated variety, conserved by the farmer because of its qualities of heterogeneity, is now necessary in order to meet the seed requirements of farmers in the South who use cropping systems with low levels of input use. The characteristics of varieties that permit better adaptation and stability in the face of environmental hazards must be better understood and integrated into participatory breeding methods (Box 5). These new approaches to selection and plant breeding form part of the movement for the in situ conservation of genetic diversity. They are therefore complementary to those for ex situ conservation. The advantage of these approaches is such that they can be immediately promoted by governments and can bring about regulation-induced adaptations (Box 6).

Box 5. Involving Farmers in Creating, Improving and Conserving Local Biodiversity of Sorghum in Burkina Faso
Kirsten vom Brocke

The work presented here describes a strategy of conserving and enhancing the genetic diversity of local sorghum in Burkina Faso. This strategy is based on the inclusion of a large number of interesting and important traits in populations which were improved using participatory recurrent selection in order to boost adaptation to different agroecological zones. Producer preferences and needs at all stages of the creation of these populations are respected by this process, from the choice of parents for crossbreeding to the management of the population by farmers in their fields. Four populations were thus created for three agro-climatic zones, each including eight to 15 local varieties and three to four elite varieties. Each population was then improved over two to three successive generations in its target region. The key elements of this adaptation phase were the identification, by producers, of sterile male plants during blooming, the harvest, evaluation and classification of sterile male plants by preference and maturity.

A first crossbreeding without selection was conducted at the agricultural station. Farmers undertook subsequent crossbreeding and the management of generations for adaptation in their fields. They identified sterile male plants during blooming, harvested them and classified the panicles harvested at maturity in terms of preference categories. To do so, farmers used different-coloured labels to differentiate early-flowering plants from late-flowering ones.

On maturity of these plants, farmers subdivided each sample into three classes: high, medium and low quality. This grouping of panicles was done either immediately at harvest time by groups of up to 15 farmers at each site, or, if time was limited, sterile male panicles were harvested in bulk and classified later when farmers and breeders had more time. In each case, breeders and technicians noted down the selection criteria in evaluation forms. This allowed breeders to get feedback on the properties of the crossbred population.

The final choice of panicles for the establishment of new populations is the result of a division of roles between producers, farmer organizations and breeders.

This exercise allowed breeders to exert just a little selective pressure on sterile male plants to eliminate those with the worst characteristics and to take greater account of farmer preferences over the course of crossbreeding generations. The objective was to process in bulk at least 500 panicles annually in the early years of crossbreeding. Since the total number of panicles retained rarely exceeded 300, some panicles from the moderate-quality group were also included in the bulk population. For growth cycles,

the proportion of panicles from early and late plants included in the selection depended on the needs of the farmers as dictated by their production system. Whenever necessary, breeders analyzed the selected panicles for traits that farmers could not easily evaluate, for example, to reduce the frequency of plants with brown coat, a recessive trait which reduces grain quality and which is difficult to evaluate.
For further information: Rattunde et al. (2009); Vom Brocke et al. (2008).

Box 6. The 'Seeds of Passion' in Semi-Arid Brazilian Zones: An Experiment to Valorise Local Agrobiodiversity Through Public Policies
Marc Piraux, Éric Sabourin, Luciano Silveira and Ghislaine Duque

In the Brazilian semi-arid region, in the 1980s and 1990s, the hybrid seeds designed for irrigated agricultural systems which were distributed by the State did not meet the needs of family farmers both in food quality and adaptation to recurring drought. Many rural communities in the region were already reporting success in their experiments to set up the first community seed banks in the region, with support from the Church and, later, from NGOs. It was a matter of making available local varieties that were adapted to the environment, met the demands of the regional market and conserved their gene pool from 1 year to the next. From the mid-1990s, organizations of family farmers of the semi-arid region set up a network of these community seed banks (maize and beans) and asked the government for support. In 2002, the State of Paraíba passed a seed law which recognized the significance of local seeds for biodiversity and for farmers' self-sufficiency. It authorized the absorption of these community banks—called 'Seeds of Passion' by the farmers—into the public seed distribution programme. This shows the symbolic value accorded to these seed banks by the farmers of the region. These approaches were integrated into a broader reflection on 'co-habitating with drought', promoted by a network of local associations and producer organizations (through trade unions active in the semi-arid region). From 2003, the Brazilian Ministry of Agrarian Development and various Brazilian states supported this process through several legal measures and new public standards. It was a matter of 'legalizing' local seeds, given the competition from certified seeds being sold by seed companies but which were often unsuitable and also much more expensive. The aim of the research was to support and accompany these social and institutional innovations. Farmer participation was crucial because it was they who, through local experiences, organized together to valorise local seeds and create new specific public-policy instruments. Lessons were learnt about accompanying endogenous local innovations and co-construction of knowledge. This approach has shown the importance for agriculture, on the one hand, of

incorporating a strong social component and, on the other, of developing institutional frameworks for the sustainability of local innovations.
For further information: Almeida and Cordeiro (2001); Sabourin et al. (2004, 2005); Piraux et al. (2012).

The current institutional landscape in the world is marked by the superposition of different approaches that have been attempted over time and are now sedimented. We have shown that the multiplicity of pursued objectives (innovation, conservation, equity) and social motivations at work in the exchange and use of genetic resources can hardly be encompassed in a single regulatory framework. The open model of common heritage of the International Undertaking has shown its limitations as it is unable to deal with all the diversity of expectations at once. The CBD model based on monetary incentives with the exercise of national sovereignty (public intervention) fares no better since it suffers from the same limitations. And yet, while these models are not completely effective, they cannot be discarded out of hand because each of them partially addresses the objectives. The solution will therefore have to be found in a coexistence of these different concepts in a pluralistic framework.

4 Conclusion: Hybridization or Co-Evolution of Conservation Models

Contrasting in situ conservation with ex situ conservation, seeking to compare these two concepts and to bring them together (or make one complementary to the other) is an idealistic approach, though a narrow one, to the global issue of conserving agrobiodiversity. While this comparison and exploration of complementarity would have made sense before the establishment of conservation mechanisms in the late 1960s, it no longer does so. The context has changed because these mechanisms now actually exist. It is unrealistic to think that we could return to an exclusively in situ conservation because of the needs and constraints of global agriculture. Huge amounts of information have been and continue to be acquired ex situ, and approaches of setting up core-collections (Deu et al. 2006) facilitate, for example, access to the target material in the mass of what is conserved. However, it is dogmatic to think that ex situ conservation and its use will, on their own, meet the needs for adapting agriculture to a rapidly changing environment.

It is often considered a given that each of the two approaches is a homogeneous block. This is emphatically not the case. On both sides, the situations are numerous and complex:

- At the biological level alone, the very nature of the objects conserved influences the ability to conserve. Seeds, tubers, individuals in the field, etc. do not present the same constraints, as some are easier to conserve than others. We also know that different models of reproductive biology generate more or less gamete drift in each in situ generation or in each ex situ regeneration cycle. We know that selective pressures are variable and influence with more or less effect the degree of genetic erosion, and that effective reproductive time steps are variable and accelerate or slow down the impact of genetic variability on adaptation to change. Fortunately, biological complexity presents major advantages: it is both the basis for all responses to changes in environmental stresses and an obstacle to the hegemonic or centralizing intentions of any particular public or private institution. It invalidates any attempt to consider the vegetable world as an easily leveraged business proposition.

- In terms of knowledge and practices that are now unquestionably associated with the conservation of biodiversity, it is the human skills that are found to be unequal. All farmers cannot be equally effective in conserving their biological resources and in favourably influencing the conservation of resources adjacent to the ones they use. Scientists, breeders, conservators, enthusiasts and gardeners—all involved in one way or another in the conservation of agricultural genetic diversity and of biodiversity in the wider sense—do not have the same skills nor the same resilience or responsiveness to the constraints that may impact their activities. The result is a chronic lack of sampling for ex situ collections. This has the advantage of forcing contacts and continuous back-and-forth interactions between ex situ and in situ;
- While biological and human limitations have a long history or even predate history altogether, economic considerations are new and ever changing. Since awareness dawned of the importance of biodiversity and up to the 1990s, the only important debate outside activist circles revolved around the major imbalance between the financial resources allocated to ex situ conservation and the almost nonexistent ones to in situ conservation. This imbalance persists, but again, the situation is more complex than it seems. The FAO reports on the state of the world's genetic resources show that there is an imbalance internal to ex situ conservation between the so-called major crops (maize, wheat, barley, rice, etc.) and those classified as 'others'. The latter are either under-funded or actually orphans. Their diversity is only conserved in situ by farmers in their fields or by a few rare breeders in their working collections. The intrusion of considerations of intellectual property rights and of access and benefit-sharing agreements since Rio has complicated the landscape. Much of the debate increasingly revolves around the benefits that one can derive from the use of agrobiodiversity.

Changes in this overall context are relegating conservation approaches themselves to second place. National—and nationalist—sentiments have led many States to close their borders to the flow of biological resources (including sometimes for material that had been brought into the country for the purpose of ex situ conservation for the benefit of all). All this with the often hypothetical goal of monetizing these resources. This posture has frequently led the conservators in these countries—and even the countries themselves—to lose the benefits of improvements or new knowledge generated elsewhere. Moreover, the embargo can be official (a lesser evil actually), or unofficial, through administrative barriers that are raised only for the highest bidder. This latter situation opens the door to all the excesses and risks associated with misappropriation of valuable genetic material. At the same time, pressures of economic competition have led large private entities to advocate mechanisms for patenting living organisms, not only so that they can appropriate the material but also, and especially, to keep it away from their competitors.[11] Competition now also exists between large public and/or international organizations. It finds expression through constant initiatives to consolidate, centralize, and streamline ex situ (very rarely in situ) mechanisms in order to maintain governance over these resources, even though the consolidation they desire runs up against biological constraints or those of local knowledge. In this landscape, the role of private foundations remain ambiguous because of the large amounts they could leverage and their susceptibility to lobbying, thus muddying already troubled waters.

In this chapter, we have taken stock of mechanisms for conserving agrobiodiversity by focusing on agricultural genetic diversity. It makes emphatically clear that the time has come for a paradigm shift: we have to stop thinking of in situ or ex situ in a binary way and try to contrast or combine the two approaches (Santonieri et al. 2011). We must rather define objectives of conserving/protecting agricultural biodiversity in terms of geographic levels (local, regional, North/ South, global), social-management levels (individuals, human societies, mankind) and socio-economic levels (individual income, local market, global trade). These objectives must lead to a real transformation of agriculture. An appropriate mix of conservation tools should be chosen in consultation with the actors involved in the maintenance of comparable biological objects. The approaches advocating a priori homogenization of conservation mechanisms must be rejected and economic valorisation must be returned to its proper place. Are these objectives within our reach or just wishful thinking?

References

Almeida, P., & Cordeiro, A. (2001). *Sistema de seguridade da semente da paixão* (p. 120). Rio de Janeiro: Estratégias comunitárias de conservação de variedades locais no semi-árido.

Altieri, M. A. (1987). Peasant agriculture and the conservation of crop and wild plant resources. *Conservation Biology, 1*(1), 49–58.

Altieri, M. A. (1992). Agroecological foundations of alternative agriculture in California. *Agriculture, Ecosystems and Environment, 39*, 23–53.

---

[11] Some plant-breeders who defend the UPOV system would not recognize this logic, since a research exemption applies to a PVPC and their competitors can therefore always have access to the protected material. Once again we see how complex the situation is, with hetereogeneity extending even into the industrial seed sector.

Andersen, R. (2008). *Governing agrobiodiversity: Plant genetics and developing countries* (p. 420). Aldershot: Ashgate.

Aoki, K. (2009). Free seeds, not free beer: Participatory plant breeding, open source seeds, and acknowledging user innovation and agriculture. *Fordham Law Review, 77*(5), 2275–2300.

Bazile, D. (2006). State-farmer partnerships for seed diversity in Mali. *Gatekeeper series 127* (p. 22). London: IIED.

Bazile, D., Dembélé, S., Soumaré, M., & Dembele, D. (2008). Utilisation de la diversité variétale du sorgho pour valoriser la diversité des sols au Mali. *Cahiers Agricultures, 17*(2), 86–94.

Beck, R. (2010). Farmers' rights and open source licensing. *Arizona Journal of Environmental Law and Policy, 1*(2). Marquette Law School Legal Studies Paper no. 10–28, SSRN. Retrieved May 13, 2013, from http://ssrn.com/abstract=1601574

Brahy, N., & Louafi, S. (2004). La convention sur la diversité biologique à la croisée de quatre discours. *Les rapports de l'Iddri*, 4.

Brockway, L. (1988). Plant science and colonial expansion: The botanical chess game. In J. R. Kloppenburg (Ed.), *Seeds and sovereignty: The use and control of plant genetic resources.* Durham: Duke University Press.

Brush, S. B. (1989). Rethinking crop genetic resource conservation. *Conservation Biology, 3*(1), 19–29.

Burel, F., & Baudry, J. (1999). *Écologie du paysage. Concepts, méthodes et applications* (p. 359). Paris: Tech et Doc.

Cassier, M. (2002). Bien privé, bien collectif et bien public à l'age de la génomique. *Revue internationale des sciences sociales, 1*(171), 95–110.

Chaïr, H., Cornet, D., Deu, M., Baco, M. N., Agbangla, A., Duval, M. F., et al. (2010). Impact of farmer selection on yam genetic diversity. *Conservation Genetics, 11*(6), 2255–2265.

Chevassus-au-Louis, B., & Bazile, D. (2008). Cultiver la diversité. *Cahiers Agricultures, 17*(2), 77–78.

Chiarolla, C., Louafi, S., & Schloen, M. (2012). An analysis of the relationship between the Nagoya protocol and instruments related to genetic resources for food and agriculture and farmers' rights. In M. Buck, E. Morgera, & E. Tsoumani (Eds.), *The 2010 Nagoya protocol on access and benefit-sharing: Implications for international law and implementation challenges*. Leiden, Boston: Brill Academic Publisher.

Coase, R. (1974). The lighthouse in economics. *Journal of Law and Economics, 17*(2), 357–376.

Collins, W. W., & Qualset, C. O. (Eds.). (1999). *Biodiversity in agroecosystems.* Boca Raton: CRC Press LLC.

Dale, V. H., & Polasky, S. (2007). Measures of the effects of agricultural practices on ecosystem services. *Ecological Economics, 64*, 286–296.

Dedeurwaerdere, T. (2004). Bioprospection, gouvernance de la biodiversité et mondialisation. De l'économie des contrats à la gouvernance réflexive. *Carnet du CPDR*, 104.

Dedeurwaerdere, T., Broggiato, A., Louafi, S., Welch, E., & Batur, F. (2012). Governing global scientific research commons under the Nagoya protocol. In M. Buck, E. Morgera, & E. Tsoumani (Eds.), *The 2010 Nagoya protocol on access and benefit-sharing: Implications for international law and implementation challenges*. Leiden, Boston: Brill Academic Publisher.

Deu, M., Rattunde, H. F. W., & Chantereau, J. (2006). A global view of genetic diversity in cultivated sorghums using a core collection. *Genome, 49*(2), 168–180.

Dounias, E. (1996). Sauvage ou cultivé? La paraculture des ignames sauvages par les pygmées Baka du Cameroun. In: C. M. Hladik, A. Hladik, H. Pagezy, O. F. Linares, G. J. A. Koppert, & A. Froment, (Eds.), *L'alimentation en forêt tropicale : interactions bioculturelles et perspectives de développement. 2. Bases culturelles des choix alimentaires et stratégies de développement* (pp. 939–960). Paris: UNESCO (L'homme et la biosphère).

Dutfield, G. (2011). *Food, biological diversity and intellectual property: The role of the international union for the protection of new varieties of plants (UPOV). Intellectual Property Issue Paper no. 9* (p. 24). New York: Quaker United Nations Office, Global Economic Issue Publications.

Elfstrand, S. Malmer, P., & Skagerfält, B. (2011). *Strengthening agricultural biodiversity for smallholder livelihoods. What knowledge is needed to overcome constraints and release potentials? Report to Hivos and Oxfam Novib, background document for the development of a knowledge programme, The resilience and development programme (SwedBio)*. Stockholm: Stockholm Resilience Centre.

Francis, C. A. (Ed.). (1986). *Multiple cropping systems*. New York: Macmillan.

Frey, B., & Jegen, R. (2001). Motivation crowding theory. *Journal of Economic Surveys, 15*(5), 589–611.

Garrison, Wilkes H. (1988). Plant genetic resources over ten thousand years: From handful of seed to the crop-specific mega-gene banks. In J. R. Kloppenburg (Ed.), *Seeds and sovereignty: The use and control of plant genetic resources*. Durham: Duke University Press.

Goëschl, T., & Swanson, T. (2002). The social value of biodiversity for R&D. *Environmental and Resource Economics, 22*(4), 477–504.

Guillaumet, J. -L. (1996). Les plantes alimentaires des forêts humides intertropicales et leur domestication: Exemples africains et américains. In C. M. Hladik, A. Hladik, H. Pagezy, O. F. Linares, G. J. A. Koppert, & A. Froment (Eds.), *L'alimentation en forêt tropicale: Interactions bioculturelles et perspectives de développement. 1. Les ressources alimentaires: production et consommation* (pp. 121–130). Paris: UNESCO (L'Homme et la biosphère).

Halewood, M., Lopez Noriega, I., & Louafi, S. (Eds.). (2012). *Crop genetic resources as a global commons* (pp. 311–328). London: Earthscan.

Hamon, P., Zoundjihekpon, J., Dumont, R., & Tio-Touré, B. (1992). La domestication de l'igname (*Dioscorea* sp.): Conséquence pour la conservation des ressources génétiques. *In: Complexe d'espèces, flux de gènes et ressources génétiques des plantes, Colloque international en hommage à Jean Pernès*, January 8–10, Paris-XI: BRG.

Harlan, J. R. (1971). Agricultural origins: Centers and non-centers. *Science, 174*, 468–474.

Hawkes, J. G. (1985). *Plant genetic resources: The impact of the international agricultural research centres, CGIAR study paper no. 3*. Washington: World Bank.

Heller, M. A., & Eisenberg, R. (1998). Can patents deter innovation? The anticommons in biomedical research. *Science, 280*(5364), 698–701.

Hladik, A., Bahuchet, S., Ducatillion, C., & Hladik, C. M. (1984). Les plantes à tubercules de la forêt d'Afrique centrale. *Revue d'écologie la Terre et la vie, 39*, 249–290.

Jackson, L. E., Pascual, U., & Hodgkin, T. (2007). Biodiversity in agricultural landscapes: investing without losing interest. *Agriculture, Ecosystems and Environment, 121*(3), 196–210.

Jackson, L., von Noordwijk, M., Bengtsson, J., Foster, W., Lipper, L., Pulleman, M., et al. (2010). Biodiversity and agricultural sustainagility: From assessment to adaptive management. *Current Opinion in Environmental Sustainability, 2*, 80–87.

Jarvis, D. I., Padoch, C., & Cooper H. D. (2007). *Managing bodiversity in agricultural ecosystems*. New York: Columbia University Press Book.

Kloppenburg, J. (1988). *First the seed: The political economy of plant biotechnology*. Cambridge: Cambridge University Press.

Kouressy, M., Bazile, D., Vaksmann, M., Soumaré, M., Doucouré, C. O. T., & Sidibé, A. (2003). La dynamique des agroécosystèmes: un facteur explicatif de l'érosion variétale du sorgho: le cas de la zone Mali-sud. In P. Dugué & P. Jouve (Eds.), *Organisation spatiale et gestion des ressources et des territoires ruraux: Actes du colloque international* (pp. 42–50). February 25–27, 2003. Montpellier: Cnearc-Sagert.

Kouressy, M., Traoré, S. B., Vaksmann, M., Grum, M., Maikano, I., Soumaré, M., et al. (2008). Adaptation des sorghos du Mali à la variabilité climatique. *Cahiers Agricultures, 17*, 95–100.

Leclerc, C. (Ed.). (2009). *Reproduire des plantes, reproduire une société. Structuration sociale de la diversité. Rapport scientifique Atp 06/01*. Montpellier: Cirad.

Leclerc, C., & Coppens, d' Eeckenbrugge G. (2012). Social organization of crop genetic diversity. The G × E × S interaction model. *Diversity, 4*, 1–32.

Louafi, S. (2012). Collective action challenges in the implementation of the multilateral system of the international treaty: What roles for the CG centres? In M. Halewood, I. Lopez Noriega, & S. Louafi (Eds.), *Crop genetic resources as a global commons*. London: Earthscan.

Marshall, D. R. (1977). The advantages and hazards of genetic homogeneity. *Annual Review of Plant Pathology, 27*, 77–94.

Parry, B. (2004). *Trading the genome*. New York: Columbia University Press.

Piraux, M., Silveira, L., Diniz, P., & Duque, G. (2012). Transição agroecológica e inovação socioterritorial. *Estudos Sociedade e Agricultura, 20*(1), 5–29. (UFRRJ, Rio de Janeiro).

Pistorius, R. (1997). *Scientists, plants and politics: A history of plant genetic movement*. Rome: International Plant Genetic Research Institute.

Rattunde, F., Vom Brocke, K., Weltzien, E., & Haussmann, B. I. G. (2009). Selection methods 4. Developing open-pollinated varieties using recurrent selection methods. In S. Ceccarelli, E. P. Guimaraes, & E. Weltzien (Eds.), *Plant breeding and farmer participation* (pp. 259–273). Rome: FAO.

Robson, J. P. (2007). Local approaches to biodiversity conservation: Lessons from Oaxaca, southern Mexico. *International Journal of Sustainable Development, 10*, 267–286.

Sabourin, E., Silveira, L., & Sidersky, P. (2004). Production d'innovation en partenariat et agriculteurs expérimentateurs au Nordeste du Brésil. *Cahiers Agricultures, 13*, 203–210.

Sabourin, E., Duque, G., Diniz, P. C. O., Oliveira, M. S. L., & Florentino, G. L. (2005). Reconnaissance publique des acteurs collectifs de l'agriculture familiale au Nordeste. *Cahiers Agricultures, 14*(1), 111–116.

Sagnard, F., Barnaud, A., Deu, M., Barro, C., Luce, C., Billot, C., Rami, J -F., Bouchet, S., Dembélé, D., Pomiès, V., Calatayud, C., Rivallan, R., Joly, H., Vom Brocke, K., Touré, A., Chantereau, J., Bezançon, G., & Vaksmann, M. (2008). Analyse multiéchelle de la diversité génétique des sorghos: compréhension des processus évolutifs pour la conservation in situ. *Cahiers Agricultures, 17*(2), 114–121.

Santonieri, L., Madrid, D., Salazar, E., Martinez, E. A., Almeida, M., Bazile, D., & Emperaire, L. (2011). Analyser les réseaux de circulation des ressources phytogénétiques: une voie pour renforcer les liens entre la conservation *ex situ* et locale. *In: Les ressources génétiques face aux nouveaux enjeux environnementaux, économiques et sociétaux. Actes du colloque FRB* (pp. 76–78). September 20–22, 2011. Montpellier: FRB. Retrieved May 13, 2013, from http://www.fondationbiodiversite.fr/images/stories/telechargement/actes_colloque_rg_web.pdf

Scherr, S. J., & McNeely, J. A. (2008). Biodiversity conservation and agricultural sustainability: Towards a new paradigm of "eco agriculture" landscapes. *Philosophical Transactions of the Royal Society B, 363*, 477–494.

Schloen, M., Louafi, S., & Dedeurwaerdere, T. (2011). *Access and benefit-sharing for genetic resources for food and agriculture. Current use and exchange practices, commonalities, differences and user community needs. Report from a multi-stakeholder expert dialogue, background study paper no. 59* (p. 42). Rome: Food and Agriculture Organization.

Shapiro, C. (2000). Navigating the patent thicket: Cross licenses, patent pools, and standard-setting. In A. Jaffe, J. Lerner, & S. Stern (Eds.), *Innovation policy and the economy* (pp. 119–150). Cambridge: MIT Press.

Shumann, G. L. (1991). Plant diseases are shifting enemies. *American Scientist, 35*, 321–350.

Soumaré, M., Kouressy, M., Vaksmann, M., Maikano, I., Bazile, D., Traoré, P. S., Traoré, S. B., Dingkuhn, M., Touré, A., Vom Brocke, K., Some, L., & Barro-Kondombo, C. P. (2008). Prévision de l'aire de diffusion des sorghos photopériodiques en Afrique de l'Ouest. *Cahiers Agricultures, 17*(2), 160–164.

Swanson, T., & Goëschl, T. (2000). Property rights issues involving plant genetic resources: implications of ownership for economic efficiency. *Ecological Economics, 32*(2000), 75–92.

Swinton, S. M., Lupi, F., Robertson, G. P., & Hamilton, S. K. (2007). Ecosystem services and agriculture: cultivating agricultural ecosystems for diverse benefits. *Ecological Economics, 64*, 245–252.

Tostain, S., Chaïr, H., Scarcelli, N., Noyer, J. L., Agbangla, C., Marchand, J. L., & Pham, J. -L. (2005). Diversité, origine et dynamique évolutive des ignames cultivées *Dioscorea rotundata* Poir. au Bénin. *In: Les Actes du Colloque national du BRG : Un dialogue pour la diversité génétique, 5* (pp. 465–482). Paris: BRG.

Vallaud, M. (2011). *Impact du développement des marchés de consommation du mil et du sorgho sur la diversité intraspécifique de ces deux céréales: le cas de trois villages situés dans la région de Sikasso au Mali. Rapport de stage de seconde année* (p. 77). Rennes: Agrocampus Ouest et Cirad Moisa, IER, Amedd.

Vallaud, M., Dury, S., & Coulibaly, H. (2011). *Market access of small-scale farms and biodiversity management of food crops. The case of sorghum and pearl millet in Mali. 5e Journées de recherches en sciences sociales*, December 8–9, 2011. Dijon: Cirad, SFER Inra.

Vom, Brocke K., Trouche, G., Zongo, S., Abdramane, B., Barro-Kondombo, C. P., Weltzien, E., et al. (2008). Création et amélioration de populations de sorgho à base large avec les agriculteurs au Burkina Faso. *Cahiers Agricultures, 17*(2), 146–153.

Wale, E., Drucker, A. G., & Zander, K. K. (Eds.). (2011). *The economics of managing crop diversity on-farm: Case studies from the genetic resources policy initiative*. London: Earthscan.

Zhu, Y., Chen, H., Fan, J., Wang, Y., Li, Y., Chen, J., et al. (2000). Genetic diversity and disease control in rice. *Nature, 406*(6797), 718–722.

Chapter 7
Towards Biodiverse Agricultural Systems: Reflecting on the Technological, Social and Institutional Changes at Stake

Estelle Biénabe

Agricultural biodiversity provides resources to simultaneously address multiple global challenges such as the resilience of agroecosystems, improving productivity and nutritional quality, and poverty reduction. These resources can contribute towards making agricultural systems more sustainable. Maintaining biodiversity in fields is therefore considered a 'global life insurance policy' by the Convention on Biological Diversity (CBD 2001). Trends that entail an improved mobilization of biodiversity in farming systems help address a large number of global challenges: food security, adaptation to global changes (especially climatic ones), management of natural resources and spaces, production of ecosystem services, etc. These changes transform agriculture's place in society, how it is perceived, how it is managed and how it contributes to social dynamics. They involve a wide range of social, economic and institutional issues which form the subject of this chapter.

The objective of this chapter is to provide insights into qualifying social transformations—considered in their broadest sense, i.e., social, economic and institutional—which accompany, orient or are influenced by the technological changes underway. This is intended to support policy makers and other stakeholders in taking these transformations better into account when taking decisions or planning action. Understanding and recognizing the role of biodiversity in its various dimensions (genetic and ecosystem in particular) in agroecosystems helps design and implement productive agricultural systems for an improved reconciliation of different societal expectations. How then can the institutionalization of these more biodiverse agricultural systems take place? Rethinking biodiversity in agricultural systems forms part of a variety of innovation processes, depending on both technical trajectories and social dynamics. At the heart of these innovation processes is a renewed way of designing systems to support and orient these processes and, in particular, to redefine the production processes and the flow of knowledge. This raises questions in terms of contexts and techno-economic

E. Biénabe (✉)
UMR Innovation, CIRAD, F-34398 Montpellier, France
e-mail: estelle.bienabe@cirad.fr

É. Hainzelin (ed.), *Cultivating Biodiversity to Transform Agriculture*,
DOI: 10.1007/978-94-007-7984-6_7, © Éditions Quæ 2013

trajectories of different actors, forms of organization, values and representations, and institutional frameworks.

Various regulatory mechanisms are at play: government intervention in different domains (agricultural policy, research and innovation policy, environmental policy), market mechanisms and territorial governance instruments. We will analyze in particular the evolution of practices and instruments related to markets and their regulation and governance. An improved understanding of these mechanisms' possibilities and limitations appears to be an important factor in steering agricultural transformations. With the increasing importance being accorded to quality and to the restructuring of food systems in response to environmental and social concerns, the scope of market dynamics in the agri-food domain has indeed become wider.

In the first part of this chapter, we will specifically address the dynamics at work, starting with those upstream of agriculture which underlies the design and promotion of new agricultural models, especially as pertaining to the development of agricultural innovation systems. In a second part, we will examine the dynamics downstream of agricultural systems relating to the governance of sectors and markets. The development of these phenomena—for the most part of recent origin—is addressed in very different social and institutional conditions in the North and in the South.

1 Co-evolution Between Technical Dynamics and Social Dynamics: An Analysis Which Starts Upstream of Agriculture

1.1 Biodiversity at the Heart of a Diversity of 'Alternative' Models of Agriculture

The preoccupations and issues concerning human societies are becoming increasingly diverse and agricultural issues are more and more interconnected with the broader dynamics at work. All this contributes widely to the emergence of social dynamics that underpin and drive various models, re-establishing ecological processes and biodiversity at the heart of agricultural production systems. Different social currents and movements, considered in the broader sense, are helping us to view biodiversity and ecology as engines of agriculture transformation.

Social dynamics that accompany this renewal of agricultural models have emerged from a questioning of conventional agriculture and have developed as 'alternatives' to the productivist and intensive industrial model also referred to as the conventional model. In the South, they reflect and promote a revalorisation and preservation of traditional agricultural systems based on the recognition of the role of local resources and of indigenous knowledge in particular. These dynamics reveal a wide variety of forms of agriculture such as organic farming, integrated and low input farming, conservation agriculture, various agroforestry systems and crop-livestock combinations. In view of the limitations of the conventional model, these alternatives offer solutions to a number of issues, especially environmental ones, such as scarcity and degradation of water and land resources and the future production capacity of the agroecosystem. Some questions are also being asked about the negative effects of the conventional model from the social point of view (Horlings and Marsden 2011). Conway (1997) questions, in particular, the replacement of manual labour by mechanization and increasing poverty in some rural areas. De Schutter (2011) also highlights the questions of employment, poverty and access to food as issues at the heart of the renewal of agricultural models: 'Taking measures to facilitate the transition to low-carbon, resource-efficient types of agriculture which benefits the poorest farmers'. He links agriculture and food clearly through the issue of the right to food.

The transformation of agriculture is therefore part of a broader social restructuring and is driven by major political and economic stakes. Bellon and Ollivier (2012) undertake a thorough analysis of agroecology as a social movement, a practice and a scientific discipline at the same time. For them, agroecology pertains to three related dynamics: 'social movements, agricultural projects, and research policy'. These authors, like many others, highlight the political dimension of alternative models constructed in order to develop and promote a new 'societal project', given the acknowledged limitations of the conventional and dominant agricultural development model. The positions adopted by the various stakeholders in this context lead to clashes between models and worldviews. They engender controversies on how to respond to various issues, especially those concerning the environment and food security. How can systems and sustainable agricultural practices contribute to developing an efficient agricultural sector, one that will be productive and profitable and which will meet the demands of a growing world population and be adapted to climate change?

1.2 New Ways of Designing Innovation Systems in Agriculture

The controversies that have arisen in many contexts, both in the North as in the South, between conventional agricultural models and alternative models to which are attached different social dynamics are strongly linked to the clash between a linear approach to innovation processes and a more systemic or network view (Box 1). These new ways of viewing agriculture that re-establish biological processes at the heart of models are partly built as an opposition to the 'agriculture of artificialization' as defined by Bonneuil et al. (2006). The latter is based primarily on the use of industrially produced external inputs in pursuit of productivity, predictability, stability of harvests, and adapted and designed to support the agro-industrial model. It goes hand in hand with the linear model of innovation that consists of a logic of specialization between researchers, producing knowledge and designing innovations, and support services which disseminate them and ensure that they are transferred to the farmers expected to adopt these innovations. This logic largely prevailed during the implementation of the first Green Revolution in the 1960s (Hall et al. 2003). Its successes and limitations have led to numerous calls for renewed investment and new collaborations in agriculture.[1]

Box 1 Innovation systems and agricultural dynamics

Ludovic Temple, Jean-Marc Touzard, Bernard Triomphe and Guy Faure
Beginning in the late 1980s, the social sciences undertook research on innovation and have since developed the concept of the innovation system to a great extent. This work is now increasingly serving as a reference for public policies for innovation and international development agencies (Organisation for Economic Cooperation and Development, European Union, World Bank, etc.). Originally used to study technological innovations in industry, then the development of 'knowledge economies' (Foray 2009), the concept is today extended to the analysis of agricultural and agri-food activities (Hall et al. 2006).

In general, this concept of the innovation system aims to explain how a set of institutions, networks and organizations can interact to foster innovation in a given national, regional or sectoral space, or in a space built by companies, or in order to develop a particular technology (Carlsson et al. 2002). It forms a systemic framework for understanding patterns in the complex network of actors and institutions working in innovation processes, and to better link these processes to the impact of the innovations, thus helping to better steer them. This concept seems to have found fertile ground in the agricultural sector because of the existence of specialized R&D institutions (in particular national, regional and international agricultural research institutes) and a renewal of interest in issues of agricultural innovation in the context of sustainable development (IAASTD 2009). In some least developed countries (Nigeria, Ghana), African authors have recently started mobilizing the innovation system to understand the dynamics of innovation endogenous to local societies.

Most alternative agriculture models are thus built on a radical criticism of the agricultural development model associated with the first Green Revolution and the phase of modernization that accompanied it. This phase was driven by a strong development of science and techniques which were designed as part of this logic of specialization.

---

[1] Griffon and Weber (1996) have thus developed the concept of the Doubly Green Revolution, as has Conway (1997). See also Swaminathan (2000), who speaks of an Evergreen Revolution. Recently, other studies and reports have also expressed optimism on the ability to increase agricultural production (World Bank 2007; IAASTD 2009).

New ways of producing agricultural knowledge accompany the definition of alternative agricultural projects. As described in more detail in the next section, these ways of producing knowledge seek to stand apart from the concept and techno-scientific frameworks of the agricultural sector which marked the modernization phase. All of this therefore pertains to the development and renewal of configurations of innovation systems that accompany transformations of agriculture.

More generally, this diffusionist linear approach, which underlies the conventional innovation systems and which may have helped isolate research, is now being discredited by a large number of actors actively participating in the renewal of these systems. They are involved in various approaches of innovation in agriculture and are implementing a wide range of processes built around the diversity of sources of innovation and of the localized nature of the processes. Sources include actors researching new practices in association with non-research partners (Faure et al. 2010), development actors, NGOs—especially in the South—and producer associations and collectives, as well as large companies, upstream and downstream of the sectors and supply chains. These trends are also evident from the different directions the designs of agricultural innovation systems are taking (Hall 2005): 'The need to take account of how scientific resources integrate with the rest of the economy and respond to society as a whole is now a major concern in the science, technology and innovation debate'. Transformations that are taking place in agriculture towards greater ecologization of practices therefore subscribe to the doubts on this linear, sequential and descending logic of links between research, agriculture and food and of the transformation of innovation processes and systems, and influence them.

An important aspect of the confrontation between conventional and alternative agricultural models has to do with how knowledge is produced and how it is used in the design and development of more biodiverse production systems. Several studies agree that the transformation of technical systems that rely more on biological regulation in the management of agroecosystems profoundly alter the dynamics of knowledge production and learning. These studies concur in highlighting the important and pivotal role played by knowledge in these processes. The amount of knowledge content in alternative agriculture models built around an ecologization of practices is often a key feature of these models. And it is assessed along with the role played by farmers with other actors in the production of knowledge, as observe Demeulenaere and Goulet (2012). They refer to the pioneering work by Kloppenburg (1991) on these issues: 'Some studies on the emergence of alternative models have argued that ecological innovation cannot occur without an epistemic change that would put farmers back at the centre of knowledge production.' Adopting the same perspective, Parrott and Marsden (2002) link the knowledge-intensive character of these ecological approaches to agriculture and the rebalancing of power in favour of farmers and away from the seed companies.

1.3 Knowledge Production and Farmers

Transformations towards biodiverse agricultural systems present particular technical difficulties, as described elsewhere in this book and as also discussed below in the case of conservation agriculture. They require the production of knowledge as a basis for new technical standards for these practices and for building new skills 'to make better use of biological control mechanisms, to better understand their effects, or to facilitate their adaptation to different environments and socio-economic conditions' (Triomphe et al. 2007). These changes require a local adaptation of practices, and therefore production of knowledge in local conditions, factors that are significant when the ability of conventional innovation systems to accompany these changes is questioned. Not only is the nature of the knowledge produced changed, but also how it is produced, distributed and exchanged. The ecologization of agricultural practices and cultivated biodiversity reinserted at the core of practices contribute to the renewal of places and forms of knowledge production.

Morgan and Murdoch (2000), comparing organic farming to conventional farming, stress the importance of knowledge-related dynamics. According to them, the transition from conventional farming to organic farming requires a radical change in the way knowledge and learning are distributed. And this is true for all knowledge, from knowledge produced upstream of agriculture by specialized actors to knowledge produced locally. This allows farmers to reclaim the production of knowledge which is adapted to their local production conditions: 'In the organic chain, we argue, farmers can once again become "knowing agents".'

In the same vein, Triomphe et al. (2007) document the role of precursor networks in which farmers play a vital role in the development of the techniques of conservation agriculture in the United States, Brazil (see also Coughenour 2003; Ekboir 2003), Argentina (see also Goulet and Hernandez 2011) and France. In response to the lack of support of research and extension services, these networks have embraced peer exchanges, sometimes over very long distances (e.g., exchange visits between Brazil and the United States, or between France and the United States or Brazil). They have also helped set up new collective forms of collaboration on technical matters (biophysicochemical functioning of soil, crop cover, plant services) and been instrumental in renewing socio-technical networks in agriculture. It has been possible to create these networks only through the advanced use of new information technologies (Internet discussion forums, photo and video sharing). These technologies have significantly increased the possibilities of exchanging information over distances and to overcome the need for geographical proximity for embarking on cooperative endeavours. They have allowed farmers who have become isolated locally because of their refusal to follow local technical norms to create social links with other such farmers elsewhere.

Collaborations thus developed play a structuring role in the formation of new agricultural dynamics and how these latter are socially driven and supported. They have allowed farmers and other actors involved in these transformations of agriculture to build new relationships with other farmers in similar situations. Farmers brought together on the basis of non-adherence to local conventional agricultural norms acquire a new identity and a sense of professional belonging (Goulet 2010). These groups are built around interactions and sharing of knowledge between peers—thus providing a space and opportunity to learn—and the consequent upgrading of the farmers' skills. This knowledge is key to designing new agricultural practices. It accords a place of honour to the farmer and other practitioners, their expertise, experience and capacities of observation. It helps upgrade the various skills necessary for agricultural work, and to convey and promote a vision of reinvention of technical references and bases by the farmer and practitioners. Indeed, these dynamics restore meaning to the farming profession. The farmer once again plays a primary role in knowledge production and in the innovation process.

Demeulenaere and Goulet (2012) analyze and question more thoroughly these processes of knowledge production, and the social dynamics of which they are part: 'When are the artificialization of environments and standardization of practices challenged? What happens when the complexity, diversity and unpredictability of nature are established as new pillars of efficiency?' Their work, based on the observation of alternative-agriculture networks in France organized around the topics of farmers' seeds and conservation agriculture, leads to a better understanding of the various social processes at work. They show that farmers engaged in these networks highlight their special relationship with biodiversity in their practices, for example, by enhancing seed diversity or by practising non-tillage, thus giving more importance to biology and the specificity of soils in their cropping systems. Adapting to the diversity and specificity of resources (seeds) and the environment (soil) is a key element of their approach to work. Indeed, this adaptability is what distinguishes these farmers from others. This relationship with biodiversity therefore contributes to constructing not only their individual identity but also a collective one. Both aspects, uniqueness and diversity of natural factors, are involved. The concept of biodiversity as an asset to be used to transform production systems echoes the valorisation of the social diversity of farmers (and of the knowledge they can mobilize within the socio-technical networks of which they are part), seen as enriching from the collective point of view. In some ways, biodiversity serves as a differentiator to demand a form of pluralism in the construction of their professional identity and their role in agricultural systems.[2] By incorporating the specific concept that these farmers convey through their unique relationship with nature, these authors offer a more comprehensive analysis of the social repositioning of the farmers participating in these dynamics and these movements: 'Rediscovery of the meaning of their profession, a combined effect of a sense of empowerment and freedom from adviser-prescribers, their increased

---

[2] '[…] permanent reference to biodiversity understood as a metaphor of an assumed pluralism as applicable to seeds as to farmers' (Demeulenaere and Goulet 2012).

understanding of the world around them, and an effective engagement with nature that they endeavour for' (Demeulenaere and Goulet 2012). A collective and individual identity for these farmers is built through a renewed relationship with nature.

This analysis is useful in understanding the dynamics at work, at least in countries in which these movements have social and institutional recognition and backgrounds—such as in France, with the existence of associations to promote such agricultural systems—and which manifest through networks such as the ones these authors have analyzed.[3]

Farmer collectives have created new social network configurations[4] which are driven by values of reciprocity and solidarity, and a refusal to specialize in different tasks. They therefore refuse to conform to the requirements of the hierarchical and Fordist conventional industrial model. These collective dynamics thus play an empowerment role as compared to other actors in agricultural systems. In particular, they eschew 'hierarchical' models of knowledge flow. They call into question the division of work between research, technical assistance, agricultural production, supply of inputs and marketing associated with the specialization-based conventional model. The social dynamics that accompany and steer these technical changes thus help us revisit and redefine the relationships between researchers and other actors in the system, especially the farmers. These latter 'regain' an active role in the production and dissemination of knowledge and technical innovations. The localized character of the practices related to a reduced artificialization of production systems is called upon to challenge modes of centralized production of knowledge based on experiments under controlled conditions. Triomphe et al. (2007) speak of a 'segment of this emerging profession, distinguishing itself through "ecological" practical techniques, organizations differentiating themselves from conventional R&D organizations, and cognitive practices challenging the Fordist division of knowledge production between scientists and laymen, between designers and users.'

These collectives serve as a space for organizing new systems of knowledge production, based on methods of validating and circulating knowledge primarily through field visits (on-location observation). In France, participation in these socio-technical networks is not limited to farmers. Scientists, public agencies (water agency, Ademe[5]), development institutions (cooperatives, chambers of agriculture) and private companies which sell inputs (agrochemicals, tools, seeds, etc.) can all become members. This results in the emergence of various schemes which combine contextualized data, i.e., observations on a network of plots made

---

[3] These networks are also referred to as 'the grouping of agricultural professionals around specialized arenas of technical options or of objectives, in which the questions relating to nature and knowledge form structuring elements' (Demeulenaere and Goulet 2012).

[4] The network is designed by farmers who are engaged in these approaches as a collective configuration in which 'unlike in a local professional group, individual behaviour is not governed by any rule or system of standards' (Demeulenaere and Goulet 2012).

[5] French Environment and Energy Management Agency.

by the farmers themselves, according to protocols which they design to a greater or lesser extent, and in-station experiments by research organizations (Demeulenaere and Goulet 2012; Brives and Tourdonnet 2010). These experimentation protocols are distributed throughout France with some standardization of observation methods built around shared conventional parameters. However, they also have provision for any observations deemed relevant by the observer in his specific context.

In the case of conservation agriculture, participation in these networks by actors such as manufacturers of seeders (a key tool for no-till agriculture) and agro-chemical companies forms an important part of the dynamics at work. Indeed, because of the special position they occupy in constructing specific support relationships with no-till agricultural producers, these actors can consolidate the observations and experiences of these producers to draw generic lessons useful to their activities. These actors sometimes associate themselves with schemes established to form links between associations promoting these types of agriculture and research institutions.

These new forms of designing practices involve the development of knowledge production schemes that command rather unusual testing. In fact, these schemes combine forms of knowledge codification—which allow comparability and which can be genericized—to local diagnoses which cannot be reduced to these codifications given the specificities linked to farmers' tricks-of-the-trade and environmental conditions. Behind the renewal of these knowledge-construction schemes is the tension between the robustness and scope of the knowledge produced and its relevance to local conditions, the latter being an engine of these renewals. This tension exists not only between scientists and farmers, but also within the scientific community with regards to these schemes which accord a more prominent place to farmers. Referring to agroecology, Bellon and Ollivier (2012) discuss in this regard the 'double rooting in science and in agriculture'. These changes contribute to the renewal of innovation systems, their design and the role of research and farmers in these systems.

1.4 Linkages Between Technical, Economic and Social Dimensions

In order to understand more comprehensively the trajectories and co-evolutions between different technical and social dynamics in the context of more biodiverse agriculture models, we have to analyze in greater detail how technical configurations differ and their possible economic and social implications. In this regard, the work of Triomphe et al. (2007) on no-till systems and conservation agriculture is especially illuminating. We follow FAO's definition of conservation agriculture here, i.e., an agricultural system based on three principles: minimal soil disturbance, permanent soil cover and crop rotations. This type of agriculture mobilizes cultivated biodiversity as an important tool in the technical system, in particular through enrichment of rotations and crop combinations, i.e., by introducing cover crops that can perform different functions (Scopel et al. 2012). Our interest in this broad interpretation lies in the fact that it can help transcend purely technical dimensions. Currently, these new ways of thinking about agriculture constitute 'both a proven innovation and a major issue in many agricultural systems around the world, in the South as in the North, in varied contexts and conditions' (Triomphe et al. 2007).

The driving forces behind conservation agriculture fall into one of three categories: limitations resulting from the degradation of natural resources (especially through erosion), the search for reduced production costs and a desire to reduce labour drudgery (Triomphe et al. 2007). As defined by FAO, conservation agriculture in most cases is a radical innovation for farmers who attempt to implement it for the first time. The elimination of tillage is indeed a significant technical change since tillage performs essential agroecosystem management functions: 'By relying on biological regulation, the farmer eliminates many technical issues he was faced with' (Triomphe et al. 2007). The abandonment of tillage requires a thorough review of the technical system as a whole and a rethink of the cropping system. Table 1 shows the diversity of technical paths that could be followed by farmers who decide to abandon tillage, depending on whether they rely on management methods based primarily on biological, chemical or mechanical means.

This allows us to characterize the combinations of production factors and technical choices that underlie different production systems. As Triomphe et al. (2007) point out, there can be multiple determinants of these choices, expressed through the diversity of technical systems observed. Decisions depend primarily on the access to production factors—labour, land, capital and inputs—on interactions with other activities, such as livestock agriculture or forestry, as well as on agri-environmental conditions. They also depend, as the previous section made clear, on the conditions of access to knowledge and on socio-economic factors such as educational level, risk perception and aversion, the heritage aspect with respect to land and nature, and membership of socio-technical networks in which these alternatives are developed and discussed.

The question of risk is critical in the transformations that this book envisages. The risks arising from the artificialization of agriculture can be very different in nature from the ones related to agroecological management. Risk management, especially during transitions from one production system to another, is a major aspect of the dynamics of transformation towards a more biodiverse agriculture. Variabilities pertaining to the different production factors may follow very distinct patterns. They relate in varying degrees to the actual commercial and physical environment (price volatility, climatic vulnerability, perishability of products, market unsuitability for some production factors). The choices concerning the evolution of technical systems are very clearly linked to the farmers' strategies and capacities to manage risk, and to the natural and institutional environments in which these farmers operate.

Table 1 Range of cropping systems and management methods used by the farmer

Management methods	Cropping systems				
	Tillage based	Simplified cultivation techniques, diverse weeding and scraping	Direct seeding	Direct seeding under permanent vegetation cover	Direct seeding under organic plant cover
Mechanical	+ + +	+ +	0 to −	0 to +	+ +
Chemical	+ +	+ + to + + +	+ to + + +	+	0
Biological			+	+ +	+ + +

Source Adapted from Triomphe et al. (2007)

By proposing keys to explaining the diversity of technical systems, the table above can help us understand better the evolution of different agricultural models. These models range from conventional industrialized farming systems, highly technicized and primarily reliant on chemical and mechanical means, to other agricultural systems, often still called alternative, which can be very sophisticated, but rely primarily on biological methods of management. This helps us analyze better the transitions involved in going from one technical system to another, and therefore to better understand and consider the social, economic and institutional implications of technological change in a context where most of the practices in question are, as of now, not fully stabilized.

It is important to note, however, that while a diversity of technical models does exist, the largely dominant conservation agriculture system is the one mainly developed during the last few decades in Brazil, the United States, Argentina and Australia—and more recently seen in Europe—in a framework of an agro-industrial and highly mechanized agriculture based on intensive use of synthetic inputs. It is practiced by large farms—from several hundred to several 1,000 ha in size—highly integrated with both upstream and downstream markets and agricultural support services. These agricultural systems, which are hardly conservation agriculture as per FAO's definition, are mainly practiced in monoculture in Brazil and Argentina and cannot be called complex systems with rotations and live cover. These are simplified cereal-growing systems based on maize and soybeans with a strong focus on reducing working time per hectare, on investment in agricultural equipment and, in countries where permitted, on the use of genetically modified organisms (GMOs). These systems have received substantial investments and benefitted from many innovations.

However, conservation agriculture is also practiced by small family farmers in a wide variety of conditions. These farmers are often poorly connected to markets and extension structures. They have to contend with particular soil and climatic conditions (high rainfall and rugged land in the case of slash and mulch[6]) and a high opportunity cost of labour. Their farming systems are based on the introduction of various cover crops into rotations, on strong agriculture-livestock

---

[6] This practice consists of clearing, without burning, the spontaneous or cultivated shrub or herbaceous cover to use as soil cover (Thurston 1996).

interactions (and tensions) through the production of forage biomass and on the availability of seedlings to overcome limitations of inadequate equipment and low investment capacity. This agricultural model which makes heavy use of cultivated biodiversity of both plant and livestock species, even though far from the limelight, is critical from a development perspective.

1.5 Social and Political Dynamics: Competing Agricultural Transformations

In the previous sections, the diversity of socio-economic factors involved in the choice of technical trajectories was identified. Developed too was the role played by the dynamics related to knowledge in co-evolutions between technical dynamics and social dynamics. The dynamics of change do not play out in an isolated and disembodied way but instead are strongly linked to political and economic issues. In addition to cognitive and collective-action dimensions and technical and economic constraints, the engines of—and impediments to—change also have a social and political aspect.

The work of Villemaine et al. (Box 2) first provides empirical evidence on the possibility of adapting conservation agriculture models in different contexts in the South. In addition, based on fieldwork at the Amazonian frontier, their study not only analyzes the generated social and technical dynamics but also illustrates the importance of socio-political and symbolic dimensions in laying out the processes of innovation and roadblocks on some innovation trajectories. This study shows how interpretations originating from a limited review of technical and economic efficiency and access to knowledge are inadequate.

Box 2 The limits of the adoption of direct seeding under mulch cover by family farmers in the Brazilian Amazon

Robin Villemaine, Éric Sabourin and Frédéric Goulet

Between 2006 and 2010, a Franco-Brazilian development project, coordinated by CIRAD and EMBRAPA (Empresa Brasileira de Pesquisa e Agropecuária), sought to introduce and develop techniques for direct seeding under mulch cover (DMC) adapted to the conditions of poor family farmers in Uruará municipality. A preliminary diagnosis conducted by the project team indicated that this Amazonian frontier would be a priori conducive to this innovation.

Farmers were seeking alternatives to help cope with the increasing criminalization of practices of shifting slash-and-burn agriculture. The proposed DMC model could be implemented manually and was relatively inexpensive. It offered them the possibility, a priori, to continue their self-consumption strategy and to market whatever surpluses they may produce.

It permitted the rejuvenation and enhancement of degraded pastures, while avoiding the soil-degradation problems associated with tillage. In addition, there already existed in this area an organized agri-supplies market, opportunities for family farmers to obtain credit and a public technical training scheme provided by EMATER (Empresa Brasileira de Asistência Técnica e Extensão Rural), all factors favourable to the sustainability of the process.

On-site experiments demonstrated the technical feasibility of these systems. Calculations of economic indicators with data from these experiment showed them to be very favourable, with a return on investment of up to 250 % in 6 months. The testimonies of farmers further strengthened the argument in favour of these techniques (improved work comfort, reduced risk of crop predation by wild fauna). Promotional activities therefore aroused great interest among the farmers, who sought, together with project stakeholders, to encourage technological institutions (EMATER), financial institutions (banks) and political institutions (local councils and town halls) to support this technique's development. Some ten farmers then adopted and adapted the DMC system to their goals. However, while some elements of the model were adopted (direct seeding, fertilizers, herbicides and improved seeds), the aspects pertaining to greater biodiversity in the plot—the intermediate legume crops—were not retained, with maize monoculture remaining largely dominant.

Moreover, local institutions have been reluctant to support the introduction of the proposed model. The analysis showed that the practice of direct seeding, as proposed by the project, conflicted with some of their interests and their ideas of agricultural progress. Pushed to invest in methods of crop intensification, they came out in favour of a competing model based on motorized tillage, according to them less dependent on herbicides and more in tune with the farmers' modernization aspirations—even though this model remains inaccessible to most of them.

Under these conditions, the difficulties in adopting the DMC model as expressed by farmers were the usual ones: cost of implementation and access to credit, access to knowledge and technical support, and competition with other activities. Nevertheless, this situation cannot be associated with the inadequacies of a fixed context or of the technique considered.* It stems instead from the confrontation of rationales of action of researchers, farmers and local elites, with each of these entities expressing—from its position in society, in terms of its identity, its interests and capacities for action—support for an intensification option whose terms and legitimacy are negotiable. It is thus the result of actors working to enforce or maintain their own interests and functioning in the guise of the common good (environment, food safety, progress, etc.). This study then shows the interlinkages between the technical and socio-political spheres, and inadequacies of analyses that over-emphasize economic efficiency and knowledge-access criteria, since these alone cannot explain the socio-technical dynamics.

> [*] 'The "context" is defined during negotiations, and reflects the technology-enabled social repositioning, within the limits of the possibilities of the biophysical, economic and socio-political environment' (Villemaine et al. 2012).
>
> **For further information:** Villemaine et al. (2012).

While the technical, economic and cognitive constraints are key factors in the contexts in the South in particular, the ability to respond to them depends primarily on the farmers themselves who, through their choices, favour certain technical options over others. They often 'hijack' the proposed technical itineraries—a phenomenon widely recognized in the social science literature on innovations and observed in the case studied—resulting in practices very different from those proposed. They also depend on the existence of support structures, which are often weak or inadequate in the South, but additionally, as is clearly illustrated by the study undertaken by Villemaine et al. on the orientations and choices that these structures recommend. Indeed, in the case presented here, these locally active structures advocated an agricultural model that least impacted their way of working and was more in tune with their stated positions on economic, political and identity issues. The issue of support and accompaniment in the case studied is significant since the family farmers perceive the technical and socio-economic changes involved in adopting direct seeding under mulch cover to be radical. They clearly ask questions relating to access to knowledge and changes in the socio-economic functioning of their farms (initial transition cost, labour constraints in interaction with other crops). Nevertheless, the lack of adaptability of these structures to the needs of farmers is not in itself the major obstacle in this case. It is rather socio-political and symbolic factors that are responsible, with the core issue being that of competition between different agricultural models. This competition is expressed socially by various actors with conflicting visions.

This raises the more general question of how the various technical models are promoted socially and favoured one over the other. To a large extent, this depends on the perception of each model's capacity to respond to challenges and also on the instrumentalization of these issues by actors and actor groups according to their own interests and positions. Alliances that can develop around new agronomic models can be very different, as shown by the study conducted by Andersson and Giller (2012) in southern Africa. In this case, the centres of the Consultative Group on International Agricultural Research (CGIAR) and FAO associated themselves with Christian NGOs to promote conservation agriculture, coming together in an alliance which was very different from those usually associated with alternative models of agriculture. These socio-technical dynamics were also analyzed with regard to the role played by agro-industrial firms in the conservation agriculture movements. In France and the Americas, the farmer innovator is strongly promoted by the seed and agrochemical companies to promote certain conservation agriculture models that rely heavily on herbicides (glyphosate-based) and on seeds which are genetically modified to resist these herbicides, as shown in particular by Goulet (2011). These companies which are acquiring an increasingly negative image, at least in the North, for environmental, health and ethical reasons are seeking an ecological legitimacy in this manner (Goulet and Vinck 2012). These same authors have also shown that, while some farmers have actually played a key role in the emergence of these new technical and social agricultural dynamics, as discussed in the previous section, these companies have also largely been present and active in constructing these schemes. They have therefore had a hand in adapting new production models to local requirements, based not only on a greater mobilization of biological regulation but also on synthetic inputs and specialized tools. And they have encouraged farmers to adopt these technical systems.

Environmental claims made by these companies—a return to an agriculture which helps protect soils, better use of natural processes and the role of the farmer—rely on, and instrumentalize in some way, the positions of some in the agronomical scientific community to promote the development of an ecologically intensive agriculture, one that includes a rational use of herbicides and promotes non-tillage. They choose this sort of agriculture over organic farming for purely performance reasons (Goulet 2012). The arguments advanced by these scientists when participating in the public debate form part of socially dominant and highly publicized issues on the need to increase production while reducing environmental impacts. Indeed, Goulet (2012) shows that this position is being increasingly echoed in France by agricultural stakeholders considered conventional (major farmer unions, etc.). These actors are increasingly forming part of discussions and debates on the issue. By adopting such a position, they can defend their profession while, at the same time, incorporating environmental concerns in their strategies.

While there do exist differences in the visions of the modernity-seeking actors, differences which are at the heart of the divisions and divergent positions, the technical concepts on which these visions are based can clearly evolve. The classic productivist paradigm has been largely vanquished, at least in the North, and the conventional agriculture sector, including innovative systems that are part of it, is increasingly taking environmental issues into account. As these concerns come to the fore, this sector has embraced and coopted them to increase or regain its social legitimacy and counter the attacks it is being subjected to. These phenomena are quite recent and much more socially and politically developed in industrialized and emerging countries than in developing countries. They have been studied and explored mainly through the viewpoint of the social sciences.[7] These changes are contributing to a renewal of the agro-industrial model. A specific vision of nature

---

[7] 'This is even true of actors involved in these movements in France, at least those who mobilize social science research on these movements in the construction and evolution of their positions, thus displaying considerable reflective activity.' (Demeulenaere and Goulet 2012).

and ecological modernization forms the heart of this renewal. In this context, Horlings and Marsden (2011) speak of a reductive vision of ecological modernization: '[…] a narrow interpretation of ecological modernization has become aligned to and adopted by the current dominant food paradigm'.

The questions being raised about progress and the doubts over a single pattern of modernization go hand in hand with new ideas on the role of agriculture in society and development. As noted by Horlings and Marsden (2011), the rationale behind conventional agricultural models is based on productivity resulting from economies of scale and from specialization, and on generic technological advances to reduce per-hectare costs and increase the price competitiveness of these models. This raises several questions about the social consequences of the evolution of the agro-industrial model—in particular on the loss of agricultural jobs and the threat to the autonomy of farmers—and its overall sustainability. These concerns also lead to criticisms of these models and to the development of other forms of agriculture (in particular, see section 'Biodiversity at the heart of a diversity of "alternative" models of agriculture' at the beginning of this chapter). Faced with the renewal of the agro-industrial model and with the global forces endeavouring to 'corporatize' food and agricultural production, various social movements have expressed support for another form of ecological modernization as a counter-balance, a modernization which is rooted locally and in the present, and which envisages increased autonomy of producers and processors (Horlings and Marsden 2011). According to these authors, this signifies a clash between weak and strong visions of ecological modernization. The former stands for a 'bio-economic' model that places improved agricultural production techniques, the use of new information technologies, genetic modification, etc. at the centre of agricultural development. The latter, conversely, favours an economic model built around a variety of local practices, NGO projects and farmer initiatives which are more locally rooted and more in tune with natural resources, thus allowing the expression of a wide range of agroecological approaches. Transformations which entail increased biodiversity in agroecosystems are at the heart of these different visions of progress and of ecological modernization as well as of the confrontations between models designed and promoted in response to major issues.

2 Recent Changes in Agriculture and Food Systems: Market Dynamics and New Directions

As discussed in the previous section, the changes observed in the agricultural world proceed from a co-evolution between technical and social dynamics. They are the result of multidimensional technological changes in society (Klerkx et al. 2012). Innovations in agricultural systems are determined not only by changes in techniques and technology, but also by the actors' needs, visions, capabilities and ambitions which find embodiment in their practices and positions. It is therefore a process of co-evolution between technological, social, economic and institutional changes (Klerkx et al. 2012). Changes in the way knowledge is produced and exchanged play an important role, but they are not sufficient conditions for transformation.

New forms of intervention in the agricultural sector are related to broader social reorganizations which involve a multiplicity of stakeholders (organizations, individuals). These entities' capacities for action depend on their legal and political environment, on infrastructure, on financing and organization of the markets and, most importantly, on the prevailing labour and land rights conditions (Leeuwis 2004; Röling 2009; Klerkx et al. 2010). In order to understand these agricultural dynamics and to steer them, we have to grasp this institutional and political environment as well as the many interactions between different segments of the food systems and the environment in which these systems insert themselves.

Farmers choose models which are based on the conservation and use of agricultural biodiversity as part of an economic and institutional framework. Their decisions pertaining to agricultural biodiversity are influenced by policies, markets and various institutions (Pascual and Perrings 2007). For these authors, as for many others who are interested in the links between biodiversity and agricultural dynamics, the solution to the decline of agricultural biodiversity lies not in technical solutions but in the creation of institutions which are able to take the role of this agricultural biodiversity in production and land use into account (Pascual and Perrings 2007).

The instruments resulting from public and private intervention can help guide actors towards more biodiverse agricultural transformations and can support them during the transitional phase. Actors can recognize agricultural biodiversity through various mechanisms which orient their actions: new legislation and standards, training and advisory services, agricultural and environmental subsidies (e.g., payments for ecosystem services), taxation, certifications, discursive resources that could influence representations and positioning of actors, development of specific markets, etc. In the first part of this chapter, the dynamics upstream of agriculture and those focused on the production and support of innovations were explored. As has been widely discussed in the literature on innovation systems, these dynamics of innovations are guided mainly by various research and innovation policies originating not only from governments but also increasingly from private entities. In this second part, we look at the changes primarily downstream of agriculture and, with them, at the market regulation mechanisms that can orient these processes. Indeed, the dynamics related to markets and private operators in agricultural sectors have developed considerably over the past two decades. Tied to the rise of environmental and social concerns and demands of sustainability, these dynamics encompass increasingly varied mechanisms that have accompanied the proliferation of quality-oriented initiatives and the restructuring of food systems. Market mechanisms therefore now influence the food industry much more than before. A better understanding of the potential and limitations associated with these processes appears to be essential in order to stimulate and support agricultural transformations towards increased biodiversity.

2.1 The Main Features of the New Market Dynamics: An Economy Focused on Quality

The new market dynamics are the manifestation of new linkages between, on the one hand, the position and structuring of agricultural sectors and of agricultural products on offer, in connection with the qualification of products—marks of quality and associated schemes (labelling, certification, etc.)—and, on the other hand, the diversification of demand, particularly in connection with changes in how consumption is perceived ('citizen-consumers').

Different ideas of quality form the basis of the qualification of products and their certifications, which are primarily meant for consumers but also serve intermediate operators within chains. They are represented by standards and labels that reflect the variety of actors and their strategies. A trend widely identified in the literature shows that these labels and standards increasingly refer not to attributes of products but to their production methods. These schemes bring together stakeholders from many different categories. Traditional actors of the chains are, of course, involved but increasingly so are non-governmental organizations (NGOs), especially environmental ones. Indeed, chain operators are using these schemes increasingly to differentiate themselves in the market, get out of price competition and increase their control over markets. Actors external to the chains, such as large NGOs, perceive these dynamics of quality as instruments to fulfil social expectations and steer agriculture towards greater sustainability. The proliferation of standards and labels related to sustainable agriculture (organic farming, UTZ Certified, Bird Friendly labels, etc.) thus highlights various attributes of modes of production, processing and trade. Other methods of addressing the links between production and food supply are also developing. They are often closely related to new patterns of consumption and purchases, and are often referred to as 'alternative' and help design new methods of organizing these links. These initiatives build new production and consumption chains and networks. They contribute therefore to the development of new economic activities that can help leverage and enhance varied and differentiated ecological resources more sustainably and in a more ecologically efficient manner. These different labelling and network dynamics therefore help redefine the relationships between production systems, organization of trade and consumption patterns, and orient changes in agricultural systems.

2.2 Proliferation of Standards and Transformation of Agriculture

All these labels and certifications (organic farming, geographical indications, fair trade, environmental labels such as Rainforest Alliance) pertain to different systems of production and trade designed and organized to guarantee the qualities claimed.

They are based on standards which are playing a growing role in the governance of sectors and, in conjunction with agricultural transformations, in the development of more biodiverse production practices. The proliferation of standards and the variety of schemes on which they are based have led to the development of a wide range of market dynamics around the world. According to Bonneuil et al. (2006), as far as developments in France are concerned, the dynamics of agricultural transformation related to marks of quality are no longer limited to niches: 'The productivist model combined mass production with mass consumption, standardization of environments, economies of scale and agreements to guarantee minimum quality standards. But initiatives once seen as niches—"AOC", PGI, organic, "Label Rouge"—now form an integral part of developments in agriculture and involve one farmer in five.' These standards and labels are predominantly present in the North and in international trade, but are also slowly being considered or implemented in the South, in particular in emerging countries (e.g., Brazil, India, South Africa).

2.2.1 How to Leverage and Enhance Various Labels? The Importance of the Specifications

Actors are increasingly perceiving these instruments as tools to recognize and create value for productions with differentiated practices which use specific local resources (varieties, landscapes, know-how, etc.) and thus to promote biodiverse production practices. It is a matter therefore of exploring the potential of these valorisation schemes and their linkages with local production practices and management of agricultural biodiversity. It is important to understand the conditions under which these qualification approaches can accompany more biodiverse agricultural transformations.

A study by Marie-Vivien et al. (Box 3) helps highlight the potential of various quality dynamics and labelling strategies to enhance biodiverse production systems. A wide diversity of agroforestry systems is present in India, in particular agroforestry systems under tree cover. This diversity is the result of a long history of culturally rooted agroforestry practices (Guillerme 2012). The coffee-cultivation systems in the Coorg region are part of these very biodiverse systems—which differentiates this region from many other coffee producing regions around the world. However, as observed by Marie-Vivien et al. several agricultural and ecological transformations are taking place in these systems in line with the recent moves towards intensification and simplification. Several quality-oriented strategies have been explored in the context of projects coordinated by CIRAD to support the preservation of these biodiverse agroforestry systems at both the technical level and by working on economic conditions and linkages with the producer's economic environment. This has mainly taken the form of comparing, in an exploratory manner, the advantages of environmental labelling (such as organic and Rainforest Alliance) over various modalities to protect and promote the name of the Coorg region through trademarks or even geographical indications (GI),[8] given that India has adopted a specific law on GIs.

Box 3 Maintaining biodiverse agroforestry systems in Coorg, the coffee producing region in south-western India: an exploration of the potential of different labelling strategies

Delphine Marie-Vivien, Claude Garcia, Béatrice Moppert, Cheppudira Kushalappa and Philippe Vaast

Coffee production is the main economic engine of the Coorg region and represents one-third of India's domestic production. Plantations in this region are among the most biodiverse in the world. Coffee is grown in this region most often in conjunction with pepper, an endemic variety of mandarin, cardamom and other citrus crops (Garcia et al. 2010). These agroforestry systems characterize this region and the cultural identity of the local coffee growers, the Kodavas, and contribute greatly to Coorg's image. However, the recent intensification of coffee production—access to chemical inputs, the introduction of new varieties and sprinkler irrigation systems*— is increasingly simplifying these systems. The native tree species are being replaced by a species of fast-growing native Australian oak. To counter these trends, the CAFNET and Biodivalloc projects have explored the potential associated with different labelling strategies: on the one hand, environmental certifications for coffee (organic, UTZ Certified, Rainforest Alliance, etc.) and, on the other, registration either of trademarks that incorporate the geographical name of Coorg to designate coffee or of a *sui generis* GI for different products originating from these agroecosystems whose reputation is well known, at least locally. Unlike trademarks, *sui generis* GIs, protected by a law passed in India in 1999, are associated with specifications that codify agricultural practices and define a geographic area.

Every quality-oriented strategy does not necessarily promote more biodiverse agroforestry or its sustainability. Only explicitly environmental labelling strategies include specifications which impose measures to manage biodiversity. However, two elements play a role in choosing a *sui generis* GI strategy based on the protection of the Coorg name. First, the reputation of Coorg coffee is rooted in these biodiverse production systems and the legal framework requires a substantial link to be demonstrated between origin and quality and/or reputation of the product to register the GI and protect it from misuse. Therefore, GI specifications will a priori have to incorporate practices favourable to agricultural biodiversity. This is all the more likely given that there are efforts being made to alter the region's current production systems in favour of a monoculture canopy consisting of exotic species. This is something that a specification formalizing the reputation of Coorg coffee

---

[8] Geographical indications are defined in the WTO as geographical names reserved to designate products whose origin and quality, reputation or other characteristics of practices are linked to the geographical area corresponding to this name.

should logically restrict or even ban. Second, the ability to implement labelling strategies is drastically constrained by the producers' limited negotiating power. Very few belong to processing or marketing collectives. And, as so often in such situations, exports—which currently form the main market—are controlled by a small number of traders who have no interest in developing dynamics of quality, especially those related to sustainability and biodiversity, even if these aspects could represent undeniable assets for the region. An important factor in this regard is that the coffee variety mainly grown in the region is Robusta, whose qualities are generally much less known on export markets than those of Arabica.

* In fact, this irrigation system produces an 'artificial' humid climate that maintains the buds during the dry season. It thus replaces this function of the canopy in traditional systems.

This study highlights the importance of including practices related to biodiversity in the technical specifications, i.e., in the specifications required to be fulfilled in order to obtain a certification or other form of control. While these aspects are an integral part of strategies for environmental labelling and associated certification, the study shows that in the case of Coorg coffee, establishing GIs would very likely help recognize and incorporate biodiverse practices in the specifications. Indeed, the region's image is closely linked to biodiverse agro-forestry systems and practices that underpin these systems are indissociable from the origin of this coffee. According to Indian law, it is on the basis of these practices that any GIs must be defined.

Moreover, these authors show that these strategies cannot be disassociated from the broader dynamics of marketing of which they are part. They highlight in particular the difficulties in this case of developing differentiation strategies for a coffee whose taste is a priori not well known (Robusta vs. Arabica). The potential associated with the dynamics of quality depends on opportunities and structuring of sectors downstream of production. The ecological value of these systems, i.e., pertaining to the provision of ecosystem services, does not constitute, for Indian market operators, a sufficient condition to justify market reorganization and valorisation. The difficulties encountered arise not only from the expectations of consumers but also, and more significantly, from the choices made by the intermediate operators in the chain. However, these issues are very dynamic and situations can change rapidly. In the case of the Indian coffee sector, quality-oriented dynamics are indeed rapidly developing and while, for the moment, they involve very few actors who operate in highly differentiated segments—mainly for export—they are slowly affecting the domestic sector too. The difficulties encountered could thus be reduced.

2.2.2 Framework for Constructing Standards and Actor Participation: Different Configurations

It is also important to note that even though technical specifications of environmental labelling incorporate criteria for valorising more biodiverse systems, these specifications are not all equal. Many of these labels are defined at a global level and standards that accompany them are usually constructed separately from the contexts of production. They are therefore not always suitable for local conditions, either at the environmental and/or social levels. They may even constitute barriers to entry in certain markets instead of helping to promote local production systems. The impact of these sustainability standards and the ability of different types of stakeholders to actually benefit from the various differentiation opportunities that these standards provide are a major issue which has been extensively studied.

Studies in agricultural economics are increasingly focusing on evaluating opportunities and constraints that different standards represent for different types of producers. It is a matter of assessing the short-term potential, on the one hand, of an increase in prices (premium and revenue) and incomes by comparing the increases in costs due to changed practices and, on the other, of securing markets or being excluded from them. These studies take special note of the difficulties faced by small producers—generally less technologically endowed and less capitalized—in adopting, conforming to and benefitting from these standards. Even though the adoption of these standards may be supposedly voluntary, the producers often do not have a choice because of the dominance of market operators who require their suppliers to adopt them. The standards are, in fact, being increasingly determined by requirements downstream of sectors or in arenas in which downstream actors are influential (see 'Production dynamics and market dynamics: what future?').

Even standards meant to promote market participation by small producers—fair trade standards in particular—can become barriers to entry for these producers when designed without reference to local conditions. Thus, for example, the requirement that coffee growers have to be family farmers and be part of producer organizations in order to participate in fair trade, while relevant in Latin America, is a major constraint for small Indian producers. In fact, from the fair trade viewpoint, the latter are considered plantations, with the vast majority of them relying on casual and/or permanent labour to farm on very small areas. And outside of a particular production area in the southern Indian state of Kerala, these producers are rarely organized collectively.

In this sense and depending on contexts, GIs whose specifications are constructed locally may be more suitable tools than some environmental labelling. Accompanying a process of action research in the development of a GI for rooibos tea produced in the Cape region of South Africa helps understand better the dynamics at work in the construction of GIs and gauge their implications (Box 4). As just mentioned, GIs are based on a process of codification of practices carried out by the proponents of the process of recognition of their product as a GI.

This has to be done on the basis of the institutional framework existing in the country in which protection is sought.

Box 4 The potential of GIs as a local construction of rules and valorisation of biodiversity in specific production systems: the case of rooibos tea

Estelle Biénabe, Maya Leclercq, Martine Antona, Patrick Caron and Pascale Moity-Maïzi

At the moment, there exists no specific institutional framework in South Africa for the protection of GIs or of local products registered as GIs. However, action-research initiatives undertaken jointly by research (CIRAD and University of Pretoria) and the Department of Agriculture of the Western Cape Province have, since 2006, supported local actors belonging to different sectors in developing GIs, thus helping to gauge their potential as a tool for sustainable development. It was a matter of understanding better the conditions under which GIs can be instruments for promoting practices conducive to the maintenance of biodiversity.* Among the products selected was rooibos. It is a tea produced in the fynbos, a highly biodiverse biome present only in South Africa. Rooibos is a plant endemic to this region and produced only in a very localized area around Cedarberg**, north of the Cape. Rooibos cultivation is a prominent feature of the landscape of this region and the local economy.

Historically harvested from the wild, rooibos has been cultivated since the 1930s. During the second half of the twentieth century, its consumption spread to all of South Africa and it has also become internationally popular over the past 20 years. The combination of Khoi Khoi and Afrikaner practices*** and the threat that the appropriation of the 'rooibos' name by an American company**** represented to the sector—a development that caused strong feelings—have contributed to anchor rooibos in South African heritage. They have been the driving forces in the process of developing a GI not only as an initiative of local stakeholders but also as a pilot approach at the national level. A recent export boom has led to a substantial expansion of the areas under cultivation and to the artificialization of practices, whereas this crop has traditionally relied little on external inputs (fertilizers and pesticides). Irrigation, again not used traditionally since rooibos is adapted to conditions in semi-arid zones, has also developed in areas of recent cultivation.

These dynamics threaten, on the one hand, the biodiversity of the fynbos and, on the other, the rooibos's quality and reputation. Rooibos producers and connoisseurs consider the taste differs significantly by region and also depends on production practices. According to some of them, the rooibos has been 'taken out of its *terroir*'. Actors of the rooibos sector involved in the process of building the GI increasingly felt concerned individually and collectively by such threats and their limited ability to cope with them. This has strengthened the process and led to its expansion in conjunction with another initiative of the sector's organizations for the conservation of bio-diversity (Biénabe et al. 2009a). Initial discussions on establishing specifi-cations relating to biodiversity have led to the recognition of the many varied forms of cultivation that mobilize biodiversity and which have a different impact on it. In this way, it was realized that the soft consensus on minimum rules—the result of a desire to include all stakeholders and to maintain the innovative capacities of actors in a sector in development—had to be exceeded. This has led to production systems being better qualified and codified (establishment of corridors within farms for farmers cultivating more than 50 % of their land, divider vegetation strips in cultivated fields) and to clarify the role biodiversity plays in the specificity of rooibos.

As highlighted by Biénabe et al. (2009b), the establishment of the GI provides opportunities for opening local spaces for discussions on managing collective resources which are directly or indirectly related to the product, and thus promotes the production of shared standards. In the case of rooibos, this local-negotiation approach has helped stakeholders of the sector and the territory to explain and integrate links between quality and biodiversity in production systems within the GI mechanism. An analysis of this case also shows that the ability to debate and arbitrate is essential in turning GIs into tools for the shared management and valorization of biodiverse and localized production systems. They are linked to the local conditions in which negotiations take place, and in particular to the stakeholders represented (different types of sector actors but also those not from the sector but still involved in related issues: biodiversity, local development, etc.).

[*] This study was conducted under the framework of the Biodivalloc project, 'Productions localized to geographical indications: what instruments to promote biodiversity in countries of the South?', coordinated by IRD and funded by ANR.

[**] Mountain massif located about 200 km north of the Cape.

[***] Although historical records are limited, rooibos combines a priori traditional uses by the local Khoi Khoi people with agricultural, trade and consumption practices closely associated with Afrikaner populations.

[****] A trademark of the rooibos name was registered in the United States by a South African company which then sold it to an American company. This company then tried to assert its rights by demanding royalties for the use of the name for commercial purposes by other market operators. While the sector managed to have this trademark cancelled, the process to do so was very time consuming and expensive.

As shown by the cases discussed above (Boxes 3 and 4), GIs can sometimes be useful instruments to impart value to biodiverse production systems but only when certain conditions are met. To begin with, the quality and/or reputation of the product protected by the GI must be recognized as being linked to biodiverse production practices—which is the case for the two products discussed in Boxes 3 and 4. Indeed, it is this condition that may justify a link between GI and biodiversity. That said, the type of the link itself could be very different, and that is the whole point of the local construction of the reference base associated with the GI for a particular product. Additionally, the nature of the legal framework that applies at the national level is important. In fact, as mentioned, the potential of the GI depends on the existence of specifications in which biodiverse production practices are codified. However, not all countries have adopted the same legal framework to recognize and protect GIs in spite of being required to do so to conform to the agreements on Trade Related Aspects of Intellectual Property Rights (TRIPS) under the framework of the World Trade Organization (WTO). The attachment of specifications to the GI is therefore not always a requirement (case of GI protection through trademarks). And even when it is, the inclusion of specifications linking quality and/or reputation to local production conditions is not always required to register the GI (especially true of systems for GI protection by certification marks, in which no substantive examination of the specifications is required). Finally, the local definition of specifications related to the GI can be a positive factor depending on the actors who will actually participate in the construction of the GI and the relationships of power between them. This issue assumes particular importance in developing countries where the drafting of rules for GIs not only depends on the choice of the legal framework, but also on the ability of the State to enforce them (Sautier et al. 2011).

In general, heterogeneity currently prevails in the regulatory frameworks for GIs, whether at the level of national legislation or of the conditions under which a GI is constructed for a specific product. This disparity is both this instrument's strength and its weakness for valorising biodiverse practices and production systems. It imparts flexibility to the instrument and provides the opportunity to adapt the codification of practices and quality attributes thus highlighted to local environmental and human conditions. As discussed in the first part of this chapter, these are important factors for the recognition of biodiverse agricultural systems. Conversely, it also constitutes a constraint because of the difficulties in actually establishing these instruments legally and enforcing them in different markets.

While the effects of labelling when considered separately seem interesting and may be significant in themselves, there is an increasing number of quality-assurance/certification schemes appearing in the same chains to differentiate products, with their overall implications varying depending on the actors who implement and govern them. When managed by different chain operators, these labels can even compete with each other. The modalities of this competition vary and are highly dependent on the alliances entered into with stakeholders who are external to the sector and who desire to promote specific models and practices through these labels (e.g., the NGO WWF which has partnered with Unilever to develop the Rainforest Alliance label). Looking at the interactions between the processes of developing a fair trade label on the one hand and a GI, on the other, Biénabe and Sautier (2008) show that they can also be complementary over time—i.e., be used sequentially—especially from the perspective of farmers and of their ability to valorise their production systems on the markets. In fact, the development of these schemes provides different opportunities (capacity building, quick return on investment, potential of differentiation) and presents different risks (in particular, types of exclusion) in the construction of markets. Thus, in the case of rooibos, schemes pertaining to fair trade in conjunction with the support of local NGOs have allowed small producers to build up their capacities for collective organization and marketing (developing networks with dedicated buyers and differentiation in end markets). These schemes have proven to be critical in building a highly differentiated chain benefitting from remunerative prices. After the establishment of this small-farmer-based fair trade chain, these producers were able to participate in the definition of GIs and bring their own specificities to the table in negotiations on the construction of the GI rules. Changes in the rules for participation in fair trade, which initially excluded large growers and now include them, posed a threat to small farmers who then found the need for the differentiation provided by GI even more compelling. This example illustrates the multiple interactions between different dynamics of quality. The analysis of the case of rooibos, in which a variety of recent labelling dynamics have manifested, helps us understand better the role of different schemes in promoting specific agricultural systems.

2.2.3 Labelling of Landscapes: A New Avenue to Explore

Box 4, which analyzes ways of establishing a GI, illustrates the difficulty of including and codifying practices considered relevant mainly from a biodiversity conservation perspective in the specification of a scheme whose purpose is primarily to establish the link between a product's quality and its origin. This is especially true of landscape-wide practices aimed at reducing the negative effects of a monoculture of rooibos on the surrounding areas and resources but for which the links between the land and the impact on quality are not always obvious.

This kind of observation is key to the ongoing debate on the potential for a labelling scheme which is based not on enhancing the value of a particular product, but instead that of an entire landscape. The main elements of what might constitute this approach and its rationale are presented in Box 5.

Box 5 Labelling of the landscape

Emmanuel Torquebiau
The multifaceted rural landscapes which combine agricultural production activities with environmental characteristics are composite landscapes that are capable of meeting two current challenges: food security and conservation of biodiversity. Assigning them a label could help in recognizing their value and in creating value-addition for products or services originating from such landscapes. These composite landscapes can be excluded from

traditional labelling strategies because they are sometimes seen as 'ordinary' landscapes as opposed to landscapes having an acknowledged heritage value. (The latter can sometimes benefit—indirectly—from traditional labelling strategies.)

The process of labelling a landscape does not pertain to a particular product, but to the processes that lead to the very existence of the landscape in question: for example, forests managed for the protection of an endangered species, or for the collection of natural products, in a landscape mosaic where there are also fields and human settlements. The original idea was advanced by Ghazoul et al. (2009). While the labelling of a product or a production process does not necessarily include an explicit goal of conserving biodiversity, landscape labelling must rely on specifications where this goal is clearly spelt out. The additional costs of labelling a landscape may be paid by a consumer of the landscape's products or a user of its services, for example, farm products or ecotourism activities. Compared to a 'conventional' payment for environmental management (e.g., payments for ecosystem services), the labelling of the landscape has the advantage of not relying on an institutional payment, often perceived as a disguised subsidy. Compared to a certification scheme such as a geographical indication, the landscape label necessarily includes an aspect of spatial heterogeneity having environmental attributes.

In order to label a landscape, one must first draft specifications to describe the criteria that will characterize the multifaceted nature of the landscape, for example, a landscape mosaic combining specific proportions of agriculture or of forest and interstitial areas, or a network of hedges separating plots. A reference framework to define the specifications criteria is provided by the concept of eco-agriculture (Scherr and McNeely 2008). An index can be used to characterize the multifacetedness of a landscape (Torquebiau et al. 2012). Dedicated institutions must then implement an evaluatory scale and a label allocation procedure.

The labelling approach presented in Box 5 forms part of a perspective of getting consumers to recognize directly the ecological value of an 'ordinary' landscape's biodiversity through market-based mechanisms. Developers of these approaches have to find alternatives to instruments such as payments for ecosystem services whose institutionalization can trigger problems of acceptability. However, these approaches require other forms of institutionalization. It is not just a matter of changing the payment process by tying it to different markets. Even though these approaches foresee remuneration mechanisms which may be more acceptable, the issue of their implementation remains. They must be linked effectively to market practices in order to fulfil their function and remunerate those who help maintain the landscape. In order to do this, the value imparted by the landscape must be recognized by the consumers. This means that chain operators should be capable of differentiating these approaches in the markets so that different consumers are able to recognize the utility of this 'ordinary' biodiversity in contexts where the qualities and requirements are usually based on iconic characteristics, or at least on attributes associated with the quality of the product in its broader sense. This applies even for local markets as envisaged principally by Torquebiau. Additionally, the construction of these approaches raises in particular the issue of the skills required. These appear to be crucial in the drafting of specifications reflecting the attributes of these multifaceted landscapes and they depend a lot on external knowledge. This in turn raises the question of who can implement these approaches and how. In general, this raises questions with regard to the form of mediation that, in this case, the market exchanges embody.

2.3 The Development of Conventional Agri-Food Systems: Historical Perspective and Current Stakes

For a more comprehensive understanding of the nature of transformations of food systems that are currently underway, and thus of the production-trade-food relationships, it may be necessary to provide a long-term historical context to the evolution of these relationships. To do so, we begin by analyzing the long history of the development of the agri-food industry and its links with agricultural production. We then cover the links between development of international trade, establishment of standards and production. The process discussed in the following two sections characterize the major features of the development of the agro-industrial system and clearly show their effects on agricultural production, whose purpose in the system has now become to produce widely standardized commodity products using substitutable ingredients. These two sections also discuss the implications of recent developments in relation to the requirements of sustainability.

2.3.1 Homogenization of Agricultural Commodity Products and Delayed Diversification in Agri-Food Systems: A Reappraisal in the Name of Sustainability?

Soler et al. (Box 6) explore the technological choices and innovations that are determining the direction that the agri-food industry is taking and which have led to a high level of homogenization of agricultural products on offer and of production systems. As explained in Box 6, the development of processes by the agri-food industry, which has relied on a logic of assembly and quality control and of the diversity of products offered at the processing stage, has largely contributed to the substitutionability of agricultural commodity products, and thus to their homogenization.

However, food, environmental and energy challenges facing humanity have led the agro-industrial system to rethink, both in the North as in the South, the relationship between agriculture and the agri-food industry. This relationship has been recast to promote foods that meet new consumer expectations and needs while ensuring increased sustainability concerning mainly the environment, production systems and the processing from which they are derived. As indicated by Soler et al. these challenges—of energy, in particular—have led to a rethink of some of the fundamentals of the agri-food industry, at least in France, and in particular of the processes allowing delayed diversification. Ways that could help reverse the trend of standardization to reintroduce diversity upstream of agro-industrial processing are being envisioned.

Box 6 Developments in the agri-food industry: challenging the structural link between homogenization of agricultural commodity products and diversification of the final products on offer?

Louis-Georges Soler, Vincent Réquillart and Gilles Trystram*

To ensure consistent and controlled product quality, the agri-food industry, initially built on conservation and stabilization techniques, has historically relied on a logic of assembly based on the deconstruction/reformulation pairing which remains at the heart of current industrial processes. This logic is based, on the one hand, on the production of intermediate products (ingredients, additives and processing aids) through a fractionation that aims to break down the agricultural commodity product and control the desired properties despite the commodity product's variability and, on the other hand, on the diversification and expansion of the offer of final products through formulation and reconstitution.

Fractionating the commodity product has therefore made agricultural commodity products more substitutable or fungible, helping to connect markets of these commodity products between themselves. These developments have been accompanied by the globalization of the origins of commodity products. This has helped industry in securing supplies and reducing costs. Developments in the agri-food sector have gone hand in hand with the standardization of agricultural commodity products in international trade and, more recently, with the efforts in genetics and agricultural practices to induce a shift towards commodity products that are reduced in number, less diverse and less variable over time. Delayed diversification, i.e., diversification taking place at the processing stage and not during production, is a major feature of current industrial agri-food systems. It has contributed greatly to the homogenization and the considerable reduction of the variety of agricultural products on offer. At the same time, it has helped increase sharply the diversification of finished products.

The concept of sustainability and the energy crisis have led to a relook at energy efficiency. Historically, the processes have changed through the addition of constraints and restrictions and therefore of new features (control of biological health safety, organoleptic attributes, search for nutritional attributes, or even health effects, and now sustainability). The compromises thus made have significantly reduced the industry's room for manoeuvre to a point that, given the present state of knowledge, no new requirements can be added without impacting earlier constraints. On the other hand, changing consumer behaviour, and more generally social expectations regarding the environment, health and nutrition, are forcing the industrial food system to restore the link between food and the consumer, between the upstream agricultural product and the final food product. The rationale of the economic trade-offs behind the choice of major upstream homogenization and delayed differentiation needs re-evaluating. On the technological front, the concept of minimal processing, initially designed to minimize adverse effects on sensory properties or nutritional value of foods during heat treatments, is now expanded to encompass a reduction in the amount of energy used to develop a food item and limit the cost of its processing. One way to do so is to limit the practice of fractionation and use the commodity product as is. Soler et al. (2011) ask the question of whether agriculture upstream of processing can reacquire a role in the production of product varieties: 'What "differentiating" characteristics can play a role in deciding the agricultural products on offer and what downstream contributions can be made through additional functionality provided by processes?' The challenge is to establish a different relationship between agriculture and industry.

[*] Adapted from Soler et al. (2011).

To rethink agriculture-processing relationships and change the agricultural products on offer, and consequently the properties of commodity products, we have to rethink production systems and processing methods in an integrated approach. This approach has to be based on an understanding of the 'genotype × environment × crop management × transformation process × quality' interactions. This involves a review of production practices and, on the upstream side, of the varietal supply and diversity, in a context of changing them through new processing strategies. It is a matter of mastering all the systems involved, ranging from the plant breeding process designed in conjunction with the technical itinerary to the development of suitable transformation and conservation processes. Bonneuil et al. (2006), who explore changes in the domain of seeds and varieties, show how these latter are increasingly becoming related to changes taking place downstream of food systems in an economy these authors call 'demand economy' as opposed to the historically dominant 'supply-side economy' discussed above. They confirm that the integration of these steps, which include 'research into plant breeding, definition of the desired product quality, and the construction of one's market [which] are no longer three successive stages', is an ongoing process in food systems, and suggest a 'co-construction of innovation and of the market in a single interactive process' (Bonneuil et al. 2006).

Given the challenges of ensuring greater sustainability of food systems in countries of the South, the valorisation of specificities of local products, especially for the local market, also orients innovations in processing procedures. Biodiversity in production systems is, in this way, identified as a potential asset for a diversified diet adapted to local resources and uses. These ways of thinking about the relationships between primary production and processing go hand in hand with the renewed importance being accorded to food diversity in a context of a wider approach to food security.

2.3.2   The Development of International Trade and the Role of Standards: An Ever-Increasing Rationalization and Control by Downstream Actors

Daviron and Vagneron (Box 7) discuss how the experience gained over a long period has helped us understand better the way standards are established. The major processes involved over the long term in the agro-industry in the North and in the development of international trade—given the interdependencies between these two segments—have converged to standardize, normalize and homogenize agricultural commodity products and pushed diversification to the downstream stage. This has allowed downstream actors to acquire a greater importance in taking decisions on matching the increasingly diversified products on offer with a demand whose expectations have changed. In this way, they have extended their control over the entire value chain. Over the last five decades, the retail sector has seen its importance increase, in the developed world, as widely recognized, but now also in countries of the South. Large retail chains have established themselves as the dominant actors in most of these value chains. Supply and demand equations are therefore resolved increasingly downstream of the chains.

A major long-term feature of these trajectories which highlight key current issues is the manner in which the process of standardization takes place. On this process depend upstream–downstream relationships. As shown by Daviron and Vagneron, standardization has historically been at the heart of commercialization and of the establishment of homogeneity which forms part of it. It plays a key role in how downstream actors control the disconnect between production and consumption in order to govern agricultural chains.

Standards developed recently in the context of international trade, and more generally the control of value chains by major retailers, help consumers differentiate between products on the basis of increasingly specific and technically precise production attributes. In particular, they have led consumers not only to consider and assess attributes associated with the product, but also the attributes associated with the production process and even how the product is traded (e.g., fair trade). However, these standards result in new disconnects and opaqueness between producers and consumers.

Box 7. Looking at standardization for the purposes of sustainable agriculture: a return to a historical process of commercialization and the decommodification of agricultural production

Benoît Daviron and Isabelle Vagneron
Placing the phase of standardization that is currently going on under the name of sustainability in the historical process of the development of international trade in agricultural commodity products changes our perception of the role played by these 'proliferating' standards. Indeed, this leads to bringing closer together the standardization and commercialization of agricultural products that accompanied the expansion of international trade. The standards, originally designed in the middle of nineteenth century for the grain futures markets in the United States and later for the futures market for other products such as cotton, coffee, cocoa, made it possible to rationalize these long-distance trades. They make them more efficient for processors and traders who organized these long-distance transactions. These standards are constructed by selecting attributes considered relevant during trades. They therefore determine the information transmitted along the supply chains. And they allow products to be homogenized according to grades, and thus create a basis for the substitutability between batches. During the mass production phase, the standards cover basic product information (colour, size, etc.).* Since the origin of suppliers and the production conditions are not retained as attributes of transactions, these standards also allow substitutability between suppliers and between modes of production. These standards thus contribute indirectly to making these long-distance transactions accessible to a greater diversity of producers, at least in this phase of creation. Conversely, they produce an opaqueness in the commercial relationships between producers and consumers. Originally created by market operators, these standards were then made public and mandatory in various countries in the early twentieth century through inclusion in product-defining national laws.**

The organic agriculture and fair trade sectors were initially developed based on brands belonging to small and specialized distributors, with an emphasis on quality and on specific modes of marketing. These chains endeavoured to re-establish links between producers and consumers and to bring about transparency in the production processes. In this way, they helped limit the substitutability of products and commodities. However, given the strong demand growth seen in the early twenty-first century and with massive investments in conventional marketing channels, marketing has now evolved towards the use of harmonized standards and labels. This has led to a redefinition, on the one hand, of organic farming as one not using synthetic inputs and, on the other, of associated oversight and control procedures. These latter have taken the form in particular of the third-party certifications, whose increasingly important role now even extends to the definition of the content of the specifications and to the evaluation of the producer's performance (Power 1997). This evolution in the direction of new expert systems—harmonized technical specifications and third-party certification—has contributed greatly to once again pushing producers and consumers further apart.

These labelling systems balance product homogeneity with requirements concerning transparency of production processes. By doing so, they have attracted many operators of the agri-food sector, thus contributing greatly to the current proliferation of standards and codes of good practice oriented towards sustainable production of many agricultural commodity products. In this historical perspective, these sustainability standards, based on a mechanism of producer substitutability and fungibility meet the required technical specifications. They thus constitute a new step—a radicalization even—in the process of rationalization of trade in agricultural products.

[*] The criteria for varieties and processability are not considered at this stage.

[**] This public intervention was initially strongly justified by the United States government and also by the colonial administration in Ghana, on behalf of producer interests. It also fulfils a widely recognized role for the benefit of consumers: quality and risk management, in particular through the development of health standards.

There seems to be a contradiction of sorts between, on the one hand, a requirement for greater transparency—which has also been promised to the consumer—of the more direct relationships with producers (in relation to the restructuring of chains and given the growing importance of the large retail sector) and, on the other, of an opaqueness at the producer level, with the construction and organization of the market taking place increasingly outside his scope of action in these chains. This transparency goes hand in hand with the implementation of procedures of traceability which can help raise barriers to market entry for some producers (e.g., barcodes). Major retailers and large agri-food companies gain domination over these chains through their control of the governance of these standards, either directly or through participation in standardization processes organized in other arenas (e.g., Roundtable on Sustainable Palm Oil). The governance of these standards is a major issue in agricultural transformations.

As was discussed in the section titled 'Proliferation of standards and transformation of agriculture', these quality-oriented strategies and the certification schemes associated with them present both opportunities and limitations for taking ecologization practices and biodiversity better into account. This depends on the characteristics of these standards and their formulation. The proliferation of standards controlled by downstream actors and by actors who are not even part of the chains, exogenous to the production areas, tends to make it more difficult for producers to access markets, especially in developing countries. In fact, compliance with these standards is becoming almost mandatory to access export markets and, increasingly, even domestic ones, especially when they are controlled by large retailers.

2.4 Production Dynamics and Market Dynamics: What Future?

In their analysis of the path organic farming has taken, Daviron and Vagneron (Box 7) show that this sector, constituted on the basis of a specific configuration of trades which allowed links between producers and consumers to be recreated, was gradually made compliant with and integrated into the prevailing trade structures through the evolution of standards. Conventionalization of organic agriculture has therefore become part of the renewal of the agro-industrial model. It has resulted in its processes becoming aligned with the conventional system at both the marketing level, mainly via mass distribution, as well as at the level of (organic) input supply and with modes of production in highly specialized farms whose operations are dominated by economies of scale.

Conversely, as mentioned in the introduction to this section and increasingly discussed in the literature on food systems, a diversification is taking place in how links between production and consumption are organized. The organic agriculture movement which continues to exist outside conventional marketing channels is a participant in this. This diversification is associated with that of production practices, particularly their ecologization, and that of patterns of consumption and purchase. A strong underlying basis for this diversification is provided by social dynamics which are building up in opposition to the conventional agro-industrial model, which has greatly stretched these links. Challenging mass consumerism—industrialized, productivist and globalized—and the desire to be proponents of a different model of development, linked to a certain form of political activism, are important drivers of these 'citizen' consumer movements. This is part of the same questioning logic adopted by some farmer movements endeavouring for a social reconfiguration in order to escape the agro-industrial logic of specialization and segmentation and to recreate links and alliances between producers and consumers. This manifests through specifically constituted production and consumption networks which allow participating actors to once again give meaning to their practices of production, trade and consumption. Examples include direct farm sales, associations for maintenance of peasant farming (AMAP), Food Councils, Ecovida network, farmers markets, etc. Products that circulate through these networks are no longer just 'commodified, but embody values which exceed their intrinsic qualities, i.e., give meaning to their consumption' (Verhaegen 2012). Within these networks, new codes of conduct, identities, rules and knowledge systems have been created. These new forms of marketing go hand in hand with specific quality marks and construction of specific quality-assurance mechanisms (for example, participatory certification). 'Participatory certification can be interpreted [in the Brazilian context] as the outcome of the criticism of socio-environmental consequences of Brazilian agricultural modernization' (Isaguirre and Stassart 2012). Through these initiatives, economic activities can therefore develop that help enhance varied and differentiated ecological resources in a more sustainable and environmentally efficient manner by proposing forms of co-evolution of new production methods and new consumption patterns.

The evolution of these movements and initiatives, and their associated potential to transform agriculture into more sustainable systems and more ecological practices, is the subject of much debate and divergent views. Even though these new models and social movements have developed mainly in the North, these dynamics of renewed ways of production, trade and consumption are also found in the South. The precedent created by the organic sector and its current prominence in various forms of ecologization of agriculture can make the analysis of this sector interesting to gauge the potential of different movements for helping the transition towards more biodiverse agriculture models. It is indeed a more developed model and one better established socially, economically, politically and institutionally. Various authors have mentioned the ongoing tensions between several competing trends within the organic farming movement: 'The relationships between protestive and economic aspects of organic farming are always unstable, in flux and in movement depending on periods and contexts' (Streith et al. 2012).

These tensions are closely linked to the issues that arise with the institutionalization of organic farming. Daviron and Vagneron (see Box 7) have analyzed an important stage in this institutionalization. Public authorities have been late in adopting the idea of the standardization of organic farming, much after the standardization phase pushed by private actors. In some countries, public authorities have not yet come around to this way of thinking (see Biénabe et al. (2011), for the case of South Africa), even though the harmonization pressures in this sector are significant. On the one hand, there is a desire to maintain local alternative practices and to exist in a space different from that of the conventional system by adopting more equitable social practices. Organic agriculture is perceived as a movement of protest against the relationships of power that exist in conventional agri-food systems and as an alternative to the conventional model. On the other hand there are pressures to give into institutionalization. Of the tension in between these two opposing sides, Van Dam and Nizet (2012) write, 'Subjected to certain social demands, sometimes these associations [of farmers, consumers and environmentalists] contribute to the institutionalization by seeking to influence public policy, other times they stand out by claiming an area of freedom and by reframing the "ecological" in the sphere of the social movement. The current momentum in favour of short marketing chains is exemplary in this regard.'

Behind these tensions is the issue of the transformative role of these networks and practices, both environmentally and socially. This goes hand in hand with the question of their ability to propose a path to a more sustainable food system in the fullest possible sense, i.e., environmentally more sustainable and socially more just. This means a system that is more encompassing than fragmented initiatives which form part of a niche economy and which have 'led to a diversification of modes of production and trade, with a two-tier system and separated production zones and social restratification of food consumption, with little prospect of fundamentally challenging the dominant system' (Verhaegen 2012). Can the heterogeneous diversity of sustainable practices, often existing at small agroecological scales, provide a viable alternative to the market and scientific dominance of the agro-industrial paradigm? (Horlings and Marsden 2011).

3 Conclusion

In this chapter, we have analyzed the co-evolutions between technical innovations towards more biodiverse systems and social dynamics that are present upstream or downstream of agriculture. Technical developments towards more biodiverse agricultural systems represent a priori trends that oppose the artificialization and standardization of agriculture, irrespective of whether they develop, at least initially, from upstream innovation systems or from downstream market dynamics. As is clear from this chapter, these artificialization and standardization processes characteristic of the agro-industrial model are strongly linked to the specialization of agriculture in the production of agricultural commodity products that are processed and valorised in increasingly sophisticated downstream processes. These processes have been promoted by the conventional agricultural innovation system in conjunction with an agricultural model which is highly integrated with—and dependent upon—the sector upstream of it. Many social movements which have helped to build and disseminate more ecological practices have developed in opposition to this model.

However, developments, both upstream and downstream of agriculture, show that changes related to ecological practices are taking place in the divisions between this productivist model—intensive, agro-industrial and widely criticized for its environmental and social impacts—and models often grouped under the term 'alternative' because of their opposition to the conventional productivist model. These alternative models employ a wide variety of locally rooted practices. While these divisions still structure the social dynamics at work, the transformations that have been made, whether in production systems—and in the innovation systems that accompany them—or in markets and food systems, have led, on the one hand, to a revival of the industrial model and, on the other, to an increased diversification of alternative systems and practices. More and more actors in the agro-industrial sector are mobilizing and enhancing certain agrobiodiverse production systems and practices, such as conservation agriculture for cereal monoculture in Brazil or 'conventionalized' organic farming. In this way, they address at least some of the environmental criticism they are subjected to.

At the global level, environmental and food issues are widely put forward. In contrast, social issues, equally important and forming part of debates and discussions on the impact of the agro-industrial model, are taken less into account and remain far from being resolved for now. In addition, food issues in particular can be approached in very different ways (for example, food sovereignty *vs* an increase in the global availability of food). This has led Horlings and Marsden (2011) to differentiate between weak and strong visions of ecological modernization. The former, as discussed at the end of the first part of this chapter, is part of a model called 'bio-economic' that places improved agricultural production techniques, the use of new information technologies, genetic modification, etc. at the centre of agricultural development. The latter, the strong vision, favours an economic model built around a variety of local practices, NGO projects and farmer initiatives which are more locally rooted, more in harmony with natural resources and based on a wide range of agroecological approaches. The social dynamics behind these two divergent models remain largely marked by different visions of progress and ecological modernization, but the confrontation between them is constantly evolving. Behind the judgement and evaluation of all these agricultural and food models, and the transformations that they propose, is the question of their performance in the broadest possible sense.

The ecologization of practices and social, economic and political realignments go hand in hand and influence each other. This confirms the importance of being able to understand the co-evolutions, to identify their driving forces and the manner in which they form part of various social and environmental issues (climate change, biodiversity, food security, but also increased disparities). Given the richness of the social and institutional processes linked to the ecologization of practices, we have been unable to cover here all the dynamics at work.

In this chapter, we have focused, in terms of governance and orientation of changes affecting agriculture, on market dynamics whose scope has expanded considerably with the development of a quality-based economy and the transition from a supply-side economy to a demand economy. These dynamics play out either at global scales, such as in the case of sustainable certifications like Roundtable on Sustainable Palm Oil (RSPO) or Rainforest Alliance, or at local ones, such as short circuit chains. With a view to respond to new social expectations and consumption patterns—citizen consumer movements in particular— these dynamics reflect new ways to organize and coordinate the interaction between production and consumption. These recent market dynamics therefore increasingly incorporate technical issues of agricultural production systems in the concept of quality, leading Bonneuil et al. (2006) to refer to a 'co-construction of the innovation and the market in a single interactive process.'

Technical innovations meant for increasing or using agricultural biodiversity in agricultural systems modify—or offer new opportunities to do so—the balance between, on the one hand, the production of an agricultural product for food purposes in the traditional sense, i.e., traded traditionally on markets, and on the other, the supply of goods and services (often grouped today under the term ecosystem services) that meet diverse expectations and social demands which are increasingly being clearly expressed.

Traditionally, the conventional offer of agricultural products covers needs met via supply chains, and the organization of the supply–demand equation by private operators has historically led to the structuring of these chains. This supply of agricultural 'goods' or commodities is now widely considered, politically and socially, as being subject a priori to regulation by the market, and more generally by the private sector, even though sometimes public interventions have been necessary or are considered in response to shortcomings. Conversely, the provision of ecosystem goods and services and the issues associated with them are generally considered, especially since the last decade, as pertaining a priori to the public sphere. These services are often designed as externalities, that is to say, outside the ambit of market dynamics. The desire to maintain these services, often viewed as public goods and in some cases as global public goods, has justified the development of specific policy instruments such as payments for ecosystem services (PES) to bridge the perceived gap between the supply of these services and demands for them.

However, it is interesting to note that recent developments in market dynamics have led to the inclusion in market exchanges of an increasing number of attributes, especially environmental ones, related to production systems. These developments raise questions as to the distinction between, on the one hand, market-regulated agricultural production and, on the other, the production of goods and services through specific public policy instruments. While not questioning the importance of the political dimension, the range of instruments which could change existing balances of power and steer transformations in agriculture towards more biodiverse systems largely surpasses the conventional technocratic vision. Indeed, this vision—especially prominent in regards to environmental issues—rarely exceeds incentive instruments related to taxation or subsidies (PES is an iconic example). More broadly, the nature of public action and the form it takes as well as the processes to take these issues into account can be increasingly diverse. This does not necessarily mean that reliance on market mechanisms is sufficient to ensure sustainable changes in production systems, as discussed in particular in the section on labelling the landscape. As far as conventional supply chains are concerned, this raises the question of the increasing dominance of actors downstream of these chains who can ally with NGOs and of the role that these actors can therefore play in the production and control of standards intented to help production systems become more sustainable. But this leads us to re-examine the place and role of policies and politics in these dynamics. In addition to the market and chain perspective taken in this chapter, the territorial dimension and the overlapping of scales, ranging from the very local to global, is an important issue, especially in terms of trade-offs in the use of space and landscape ecology between different production systems, dimensions which are as yet poorly integrated into market processes.

References

Andersson, J., & Giller, K. E. (2012). On heretics and God's blanket salesmen: Contested claims for conservation agriculture and the politics of its promotion in African smallholder farming. In J. Sumberg & J. Thompson (Eds.), *Contested agronomy: Agricultural research in a changing world*. London: Routledge.

Bellon, S., & Ollivier, G. (2012). L'agroécologie en France: L'institutionnalisation d'utopies. In F. Goulet, D. Magda, N. Girard, & V. Hernandez (Eds.), *L'agroécologie en Argentine et en France: Regards croisés* (pp. 55–90). L'Harmattan: Paris.

Biénabe, E., & Sautier, D. (2008). Commerce équitable et indications géographiques: Relations, tensions, complémentarités. Réflexions à partir du cas du rooibos en Afrique du Sud. *3e Colloque international sur le commerce équitable,* Montpellier, 14–16 May 2008.

Biénabe, E., Bramley, C., & Kirsten, J. F. (2009a). An economic analysis of the evolution in intellectual property strategies in the South African agricultural sector: The rooibos industry. In D. Kaplan (Ed.) *The economics of intellectual property in South Africa* (pp. 56–83). WIPO: Geneva.

Biénabe, E., Leclercq, M., & Moity-Maizi, P. (2009b). Le rooibos d'Afrique du Sud: Comment la biodiversité s'invite dans la construction d'une indication géographique. *Autrepart, 50,* 117–134.

Biénabe, E., Vermeulen, H., & Bramley, C. (2011). The food "quality turn" in South Africa: An initial exploration of its implications for small-scale farmers' market access. *Agrekon, 50*(1), 36–52.

Bonneuil, C., Demeulenaere, E., Thomas, F., Joly, P. -B., Allaire, G., & Goldringer, I. (2006). Innover autrement ? La recherche face à l'avènement d'un nouveau régime de production et de régulation des savoirs en génétique végétale. In P. Gasselin, O. Clément (Eds.), *Quelles variétés et semences pour des agricultures paysannes durables? Les Dossiers de l'environnement de l'Inra* (30, p. 186).

Brives, H., & Tourdonnet, S. (de). (2010, June 28–July 1). Comment exporter des connaissances locales? Une expérience de recherche intervention auprès d'un club engagé dans les techniques sans labour. Communication présentée lors du symposium international Innovation and Sustainable Development in Agriculture and Food (ISDA), Montpellier. Retrieved May 20, 2013 from http://www.projet-pepites.org/index.php/projets/content/download/10041/60666/file/Brives%20&%20De%20Tourdonnet%202010%20ISDA.pdf .

Carlsson, B., Jacobsson, S., Holmén, M., & Rickne, A. (2002). Innovation systems: Analytical and methodological issues. *Research Policy, 31,* 233–245.

CBD (Convention on Biological Diversity). (2001). Global biodiversity outlook. Retrieved May 20, 2013, from http://www.cbd.int/doc/publications/gbo/gbo-ch-06-en.pdf.

Conway, G. (1997). *The doubly green revolution: Food for all in the twenty-first century.* New York: Comstock Publishing Associates.

Coughenour, C. M. (2003). Innovating conservation agriculture: The case of no-till cropping. *Rural Sociology, 68*(2), 278–304.

De Schutter O. (2011, March 8). Agroecology and the right to food. Report submitted by the Special Rapporteur on the right to food to the 16th session of the UN Human Rights Council [A/HRC/16/49]. Retrieved May 20, 2013, from http://www.srfood.org/images/stories/pdf/officialreports/20110308_a-hrc-16-49_agroecology_en.pdf.

Demeulenaere, E., & Goulet, F. (2012). Du singulier au collectif. Agriculteurs et objets de la nature dans les réseaux d'agricultures alternatives. *Terrains et travaux, 20,* 121–138.

Ekboir, J. M. (2003). Research and technology policies in innovation systems: Zero tillage in Brazil. *Research Policy, 32*(4), 573–586.

Faure, G., Gasselin, P., Triomphe, B., Temple, L., & Hocdé, H. (2010). Innover avec les acteurs du monde rural: La recherche-action en partenariat, coll. Agricultures tropicales en poche, éditions Quae (p. 224).

Foray, D. (2009). *L'économie de la connaissance.* Paris: La Découverte.

Garcia, C. A., Bhagwat, S. A., Ghazoul, J., Nath, C. D., Nanaya, K. M., Kushalappa, C. G., et al. (2010). Biodiversity conservation in agricultural landscapes: Challenges and opportunities of coffee agroforests in the Western Ghats. India. *Conservation Biology, 24*(2), 479–488.

Ghazoul, J., Garcia, C., & Kushalappa, C. G. (2009). Landscape labelling: A concept for next generation payment for ecosystem service schemes. *Forest Ecology and Management, 258,* 1889–1895.

Goulet, F. (2010). Nature et ré-enchantement du monde. In B. Hervieu, N. Mayer, P. Muller, F. Purseigle, & J. Rémy (Eds.), *Les mondes agricoles en politique* (pp. 51–71). Paris: Presses de Sciences Po.

Goulet, F. (2011). Firmes de l'agrofourniture et innovations en grandes cultures: Pluralité des registres d'action. *POUR, 212*, 101–106.

Goulet, F. (2012). La notion d'intensification écologique et son succès auprès d'un certain monde agricole français. Une radiographie critique. *Le Courrier de l'environnement de l'Inra, 62*, 19–30.

Goulet, F., & Hernandez, V. (2011). Vers un modèle de développement et d'identités professionnelles agricoles globalisés? Dynamiques autour du semis direct en Argentine et en France. *Revue Tiers-Monde, 207*, 115–132.

Goulet, F., & Vinck, D. (2012). Innovation through withdrawal. Contribution to a sociology of detachment. *Revue française de sociologie, 532*, 195–224.

Griffon, M., & Weber, J. (1996). La révolution doublement verte: économie et institutions. *Cahiers Agricultures, 5*(4), 239–242.

Guillerme, S. (2012). L'agroforesterie en Inde: le défi de la diversité. In D. Van Dam, M. Streigh, J. Nizet, & P. M. Stassart (Eds.), *Agroécologie: Entre pratiques et sciences sociales* (pp. 179–200). Dijon: Educagri éditions.

Hall, A. (2005). Capacity development for agricultural biotechnology in developing countries: An innovation systems view of what it is and how to develop it. *Journal of International Development, 17*, 611–630. doi:10.1002/jid.1227.

Hall, A., Janssen, W., Pehu, E., & Rajalahti, R. (2006). *Enhancing agricultural innovation: How to go beyond the strengthening of research systems*. Washington DC: World Bank.

Hall, A., Sulaiman Rasheed, V., Clark, N., & Yoganand, B. (2003). From measuring impact to learning institutional lessons: An innovation systems perspective on improving the management of international agricultural research. *Agricultural Systems, 78*, 213–241.

Horlings, L. G., & Marsden, T. K. (2011). Towards the real green revolution? Exploring the conceptual dimensions of a new ecological modernization of agriculture that could 'feed the world'. *Global Environmental Change, 21*, 441–452.

IAASTD. (2009). In B. D. MacIntyre, H. R. Herren, J. Wakhungu, R. T. Watson (Eds.), *Agriculture at a crossroads, global report* (p. 606). Washington DC: International Assessment of Agricultural Knowledge, Science and Technology for Development , Island Press.

Isaguirre, K. R., & Stassart, P. M. (2012). Certification participative pour une ruralité plus durable: Le réseau Ecovida au Brésil. In D. Van Dam, M. Streigh, J. Nizet, & P. M. Stassart (Eds.), *Agroécologie: Entre pratiques et sciences sociales* (pp. 75–95). Dijon: Educagri éditions.

Klerkx, L., Aarts, N., & Leeuwis, C. (2010). Adaptive management in agricultural innovation systems: The interaction between innovation networks and their environment. *Agricultural Systems, 103*, 390–400.

Klerkx, L., van Mierlo, B., & Leeuwis, C. (2012). Evolution of systems approaches to agricultural innovation: Concepts, analysis and interventions. In I. Darnhofer, D. Gibbon, & B. Dedieu (Eds.), *Farming systems research into the 21st century: The new dynamic*. Dordrecht: Springer.

Kloppenburg, J. (1991). Social theory and the de/reconstruction of agricultural science: Local knowledge for an alternative agriculture. *Rural Sociology, 56*(4), 519–548.

Leeuwis, C. (2004). *Communication for rural innovation: Rethinking agricultural extension*. Oxford: Blackwell Science.

Morgan, K., & Murdoch, J. (2000). Organic versus conventional agriculture: Knowledge, power and innovation in the food chain. *Geoforum, 31*(2), 159–173.

Parrott, N., & Marsden, T. (2002). *The real green revolution: Organic and agroecological farming in the south*. London: Greenpeace.

Pascual, U., & Perrings, C. (2007). Developing incentives and economic mechanisms for in situ biodiversity conservation in agricultural landscapes. *Agriculture, Ecosystems and Environment, 121*(3), 256–268.

Röling, N. (2009). Pathways for impact: Scientists' different perspectives on agricultural innovation. *International Journal of Agricultural Sustainability, 7*, 83–94.

Power, M. (1997). *The audit society: Rituals of verification.* Oxford: Oxford University Press.

Sautier, D., Biénabe, E., & Cerdan, C. (2011). Geographical indications in developing countries. In E. Barham & S. Bertil (Eds.), *Labels of origin for food: Local development, global recognition* (pp. 139–153). Wallingford: CABI.

Scherr, S. J., & McNeely, J. A. (2008). Biodiversity conservation and agricultural sustainability: Towards a new paradigm of "ecoagriculture" landscapes. *Philosophical Transactions of the Royal Society of London. Series B, 363*, 477–494.

Scopel, E., Triomphe, B., Affholder, F., Macena Da Silva, F. A., Corbeels, M., Valadares Xavier, J. H., et al. (2012). Conservation agriculture cropping systems in temperate and tropical conditions, performances and impacts. A review. *Agronomy for Sustainable Development.* doi: 10.1007/s13593-012-0106-9.

Soler, L. -G., Requillart, V., & Trystram G. (2011). Organisation industrielle et durabilité. In C. Esnouf, M. Russel, & N. Bricas (Eds.) *Pour une alimentation durable—Réflexion stratégique duALIne* (pp. 109–122). éditions Quae: Versailles.

Streigh, M., Van Dam, D., & Nizet, J. (2012). L'agriculture biologique: Un champ en tension. In D. Van Dam, M. Streigh, J. Nizet, & P. M. Stassart (Eds.), *Agroécologie: Entre pratiques et sciences sociales* (pp. 155–163). Dijon: Educagri éditions.

Swaminathan, M. S. (2000). An evergreen revolution. *Biologist, 47*(2), 85–89

Thurston, D. (1996). *Slash-mulch systems: Sustainable methods for tropical agriculture.* Boulder CO: Westview Press.

Torquebiau, E., Garcia, C., & Cholet, N. (2012). Labelliser les paysages ruraux. Note de perspective. *Politiques de l'environnement*, 16, Montpellier, Cirad, p. 4.

Triomphe, B., Goulet, F., Dreyfus, F., & Tourdonnet, S. (de). (2007). Du labour au non-labour: Pratiques, innovations et enjeux au Sud et au Nord. In R. Bourrigaud, & F. Sigaut (Eds.), *Nous labourons. Actes du colloque Techniques de travail de la terre, hier et aujourd'hui, ici et là-bas*, Nantes, Nozay, Châteaubriant, 25–28 October 2006, Éditions du Centre d'histoire du travail (pp. 371–384).

Van Dam, D., & Nizet, J. (2012). Les agriculteurs bio deviennent-ils moins verts ? In D. Van Dam, M. Streigh, J. Nizet, & P. M. Stassart (Eds.), *Agroécologie: Entre pratiques et sciences sociales* (pp. 249–264). Dijon: Educagri éditions.

Verhaegen, E. (2012). Les réseaux agroalimentaires alternatifs: Transformations globales ou nouvelle segmentation du marché? In D. Van Dam, M. Streigh, J. Nizet, & P. M. Stassart (Eds.), *Agroécologie: Entre pratiques et sciences sociales* (pp. 265–279). Dijon: Educagri éditions.

Villemaine, R., Sabourin, E., & Goulet, F. (2012). Limites à l'adoption du semis direct sous couverture végétale par les agriculteurs familiaux en Amazonie brésilienne. *Cahiers Agricultures, 21*(4), 242–247.

World Bank. (2007). *World development report 2008: Agriculture for development.* New York: Oxford University Press.

Printed by Printforce, the Netherlands